河南省科技著作出版资助项目

昆虫刺吸电位技术及其应用

闫凤鸣　王满囷　主编

河南科学技术出版社
·郑州·

内容提要

本书系统介绍了昆虫刺吸电位技术（EPG）的原理和应用，分为“原理篇”“实验篇”和“应用篇”三大部分。“原理篇”介绍刺吸电位技术的历史，概述刺吸式昆虫的取食行为和传毒机制，译述刺吸电位技术的原理和波形的生物学意义等；“实验篇”详细介绍实验的具体步骤，包括实验准备、实验过程、数据分析、软件使用、注意事项、常见问题处理等；“应用篇”是我国学者利用 EPG 在昆虫与植物的关系、植物抗性机制、传毒机制，与人工饲料技术结合等方面的研究工作总结。书末附有中、英文名词索引等。

本书可作为高等院校、科研院所的生物学、生态学、植物保护、昆虫学和农学高年级本科生和研究生，以及相关专业科研人员的参考资料和实验指导。

图书在版编目（CIP）数据

昆虫刺吸电位技术及其应用/闫凤鸣，王满困主编．—郑州：河南科学技术出版社，2017.10

ISBN 978-7-5349-8463-1

Ⅰ.①昆… Ⅱ.①闫… ②王… Ⅲ.①昆虫-研究 Ⅳ.①Q96

中国版本图书馆 CIP 数据核字（2017）第 018958 号

出版发行：河南科学技术出版社
地址：郑州市经五路 66 号　　邮编：450002
电话：（0371）65737028
网址：www.hnstp.cn
策划编辑：杨秀芳
责任编辑：杨秀芳
责任校对：丁秀荣
封面设计：张　伟
版式设计：栾亚平
责任印制：朱　飞
印　　刷：河南新华印刷集团有限公司
经　　销：全国新华书店
幅面尺寸：185 mm×260 mm　　印张：17　　字数：380 千字
版　　次：2017 年 10 月第 1 版　　2017 年 10 月第 1 次印刷
定　　价：198.00 元

《昆虫刺吸电位技术及其应用》编写人员名单

主　　编　闫凤鸣（河南农业大学）

　　　　　王满囷（华中农业大学）

副 主 编　罗　岚（中国科学院动物研究所）

　　　　　薛　堃（中央民族大学）

　　　　　黄　芳（浙江湖州检验检疫局）

　　　　　汤清波（河南农业大学）

编　　委（以姓氏拼音为序）

Elaine A. Backus　岑伊静　陈法军

陈巨莲　何文波　黄翠虹　黄　芳

江守林　李　杰　李静静　刘　芳

刘树生　刘银泉　苗　进　罗　晨

罗　岚　罗孝竹　卢少华　潘建斌

孙　凯　孙　泽　万贵钧　王满囷

王雪丽　王　炜　温　娟　吴莉莉

乌天宇　邢玉清　薛　堃　闫凤鸣

杨成良　张振飞　郅军锐

主要参编人员简介

（以姓氏拼音为序）

Elaine A. Backus，女，博士，昆虫学家，目前在美国农业部农业研究院圣金谷农业科学中心作物病虫害遗传实验室工作。1983 年获加州大学戴维斯分校昆虫学博士学位。长期利用 AC-EPG 进行叶蝉等头喙亚目昆虫的取食行为、传播病原机制的研究，近年来与其他合作者研发 AC-DC EPG 仪器。联系方式：USDA Agricultural Research Service，San Joaquin Valley Agricultural Sciences Center，9611 So. Riverbend Ave，Parlier，CA 93648，U. S. A.；e-mail：elaine. backus@ ars. usda. gov。

岑伊静（1966. 11—），女，博士，华南农业大学农学院昆虫学系副教授，硕士研究生导师。2002 年毕业于华南农业大学昆虫生态研究室，长期从事柑橘害虫研究，近 10 年来重点研究柑橘木虱传播黄龙病的机制。联系方式：广州市天河区五山路 483 号华南农业大学农学院，邮编：510642；e-mail：cenyj@ scau. edu. cn。

陈法军（1974. 05—），男，博士，南京农业大学植物保护学院昆虫学系副主任，教授、博士生导师。毕业于中国科学院动物研究所，现主要从事昆虫生态与预测预报，以及环境昆虫学和气候变化生物学等方面的研究。联系方式：江苏省南京市卫岗童卫路 6 号，南京农业大学植物保护学院，邮编：210951；e-mail：fajunchen @ njau. edu. cn。

陈巨莲（1965. 09—），女，中国农业科学院植物保护研究所研究员。2001 年毕业于中国农业科学院研究生院农业昆虫与害虫防治专业，获农学博士学位。长期从事小麦害虫生物学、小麦抗蚜及诱导抗蚜生化与分子机制、小麦—麦蚜—天敌互作及协同进化、小麦害虫生态调控技术等研究。联系方式：北京市海淀区圆明园西路 2 号，邮编：100193；e-mail：jlchen@ ippcaas. cn。

黄翠虹（1979. 01—），女，博士，顺德职业技术学院讲师。毕业于北京大学动物学专业，现主要从事农业害虫（玉米螟、烟粉虱等）综合防治及 L-阿拉伯糖抗虫机制等研究。联系方式：广东省佛山市大良德胜东路顺德职业技术学院医药卫生学院药学教研室，邮编：528333；e-mail：cuihongh@ 139. com。

黄　芳（1981. 10—），女，博士，湖州检验检疫局副研究员。毕业于浙江大学环境生物学专业，现主要从事经济作物害虫（小菜蛾、烟粉虱等）综合防治及检疫性昆虫（扶桑绵粉蚧等）的入侵机制等研究。联系方式：浙江省湖州检验检疫局，邮编：313000；e-mail：huangfang_ch@ hotmail. com。

刘　芳（1972. 04—），女，农学博士，扬州大学园艺与植物保护学院教授。毕业于浙江大学农业与生物技术学院，现主要从事昆虫生态和化学生态学等方面的研究。联系方式：扬州市文汇东路48号，扬州大学园艺与植物保护学院，邮编：225009；e-mail：liufang@ yzu. edu. cn。

刘树生（1955. 06—），男，博士，浙江大学教授，校学术委员会副主任，“植物保护”一级学科负责人。毕业于澳大利亚国立大学。近年来的研究工作重点为入侵生物学、媒介昆虫—病毒—植物互作、昆虫行为和化学生态学、蔬菜害虫生物防治和综合治理。联系方式：杭州市西湖区余杭塘路866号，浙江大学紫金港校区农生环大楼C座，邮编：310058；e-mail：shshliu@ zju. edu. cn。

刘银泉（1972. 02—），男，博士，浙江大学昆虫科学研究所副教授。2009年毕业于浙江大学农业昆虫与害虫防治专业。主要从事昆虫行为生态与化学生态，蔬菜害虫生物防治，害虫综合治理以及烟粉虱入侵生态学等理论与应用方面的研究工作。联系方式：杭州市西湖区余杭塘路866号，浙江大学紫金港校区农生环大楼C座，邮编：310058；e-mail：yqliu@ zju. edu. cn。

苗　进（1977. 07—），男，博士，河南省农业科学院植物保护研究所副研究员。2008年毕业于中国农业科学院研究生院农业昆虫与害虫防治专业，主要从事昆虫行为学、昆虫生态学方面的研究。联系方式：河南省郑州市花园路116号，邮编：450002；e-mail：miaojin1977@ 163. com。

罗　晨（1970. 10—），女，博士，北京市农林科学院研究员。2006年获中国农业大学博士学位。主要从事烟粉虱生物型、生物学特性及寄生性天敌等方面的研究。联系方式：北京市海淀区曙光花园中路9号植物保护研究所，邮编100097；e-mail：luochen1010@ 126. com。

罗　岚（1982. 01—）女，硕士，中国科学院动物研究所工程师。2004年获得华中科技大学生物医学工程学士学位。2007年获得华中科技大学生物物理学硕士学位。主要研究方向为昆虫行为学，昆虫与寄主植物互作，细胞电生理和昆虫电生理，原代细胞培养及细胞系的建立。联系方式：北京市朝阳区北辰西路1号院5号，邮编：100101；e-mail：luolan@ ioz. ac. cn。

潘建斌（1962. 04—），男，河南农业大学教授。1982年毕业于华中师范大学物理系，长期从事智能仪器及农产品加工方面的研究，近年来参与EPG仪器改进工作。联系方式：河南省郑州市文化路95号河南农业大学理学院，邮编：450002；e-mail：panjb01@ 163. com。

孙　凯（1988. 04—），男，硕士。2011年获得安徽农业大学植物保护学士学位，2014年获得湖南农业大学农业昆虫与害虫防治硕士学位。2014年至2016年，在中国农业科学院麻类研究所任实习研究员。现在湖南省森林病虫害防治检疫总站工作，主要从事有害生物综合治理。联系方式：湖南省长沙市湘府东路一段1001号，邮编：410004，e-mail：sunkai424@ foxmail. com。

汤清波（1975. 07—），男，河南农业大学教授。2006年获中科院动物所博士学位，主要从事昆虫与植物关系、昆虫味觉电生理方面的研究。联系方式：河南省郑州市文化路95号河南农业大学植物保护学院；邮编：450002；e-mail：qingbotang@ 126. com。

王满囷（1972.09—），男，博士，华中农业大学教授，博士生导师；教育部新世纪优秀人才；兼任中国昆虫学会常务理事，湖北省昆虫学会理事长。主要从事昆虫与植物互作、昆虫嗅觉感受机制等方面的研究。联系方式：湖北省武汉市华中农业大学植物科技学院，邮编：430070；e-mail：mqwang@ mail. hzau. edu. cn。

王　炜（1984—），男，博士。2009 年 7 月获得湖北民族学院生物工程专业学士学位。2014 年 7 月获得北京林业大学植物学专业博士学位。2014 年 9 月进入中国科学院动物研究所开展博士后研究工作。主要从事昆虫—病毒—植物相互作用的机理研究。e-mail：031150796@ 163. com。

吴莉莉（1977.11—），女，河南农业大学副教授。近年来一直从事纳米材料、气敏传感器、信号处理和模式识别等方面的研究，近年来参与 EPG 仪器改进工作。联系方式：河南省郑州市文化路 95 号，河南农业大学理学院，邮编：450002；e-mail：wullzju@ 126. com。

邢玉清（1981.09—），男，硕士，河南农业大学讲师。2005 年获河南大学数学与信息科学学院学士学位，2008 年获中山大学硕士学位，研究方向为网络安全与密码学。近年来参与 EPG 仪器改进工作。联系方式：河南省郑州市文化路 95 号，河南农业大学理学院，邮编：450002；e-mail：issxyq@ 163. com。

薛　堃（1980.04—），男，博士，中央民族大学生命与环境科学学院副教授。毕业于北京大学生命科学学院，现主要从事动物生态和转基因生物安全性评价等方面的研究。联系方式：北京市海淀区中关村南大街 27 号，中央民族大学生命与环境科学学院，邮编：100081；e-mail：xuekun@ muc. edu. cn。

闫凤鸣（1962.07—），男，博士，教授，博士生导师。1993 年获北京大学博士学位，1993~2007 年在北京大学生命科学学院任教，2007 年至今在河南农业大学工作。目前的主要研究方向为化学生态学、刺吸式昆虫取食行为、传毒机制及防控新技术，以及生物安全、种衣剂副作用防控技术等。联系方式：河南省郑州市文化路 95 号，河南农业大学植物保护学院，邮编：450002；e-mail：fmyan@ henau. edu. cn。

张振飞（1981.01—），男，博士，广东省农业科学院植物保护研究所副研究员。毕业于华南农业大学农业昆虫与害虫防治专业。现主要从事水稻抗虫品种与褐飞虱互作分子机制、水稻害虫生态调控等方面的研究。联系方式：广东省广州市天河区五山金颖路 7 号，邮编：510640；e-mail：zhangzf2013@ gmail. com 或 zzf@ gdppri. cn。

郅军锐（1965.06—），女，博士，贵州大学昆虫研究所教授。毕业于华南农业大学农业昆虫与害虫防治专业。现主要从事系西花蓟马与寄主植物互作机制、天敌小花蝽的利用等方面的研究。联系方式：贵州省贵阳市花溪贵州大学南校区（农学院）昆虫研究所，邮编：550025；e-mail：zhijunrui@ 126. com 或 agr. jrzhi@ gzu. edu. cn。

前　言

刺吸式口器昆虫是一类非常特殊的昆虫类群，其直接为害作物和作为植物病毒的介体，一直是农林生产上的重要害虫类别。研究刺吸式昆虫与植物之间的关系、植物抗性机制以及昆虫传毒机制，是治理这类昆虫及其所传播植物病毒病的基础。昆虫刺吸电位（EPG）技术是进行这些研究的有力工具。

与咀嚼式昆虫相比，刺吸式昆虫取食过程很难从外观上直接观察到，因而就不容易评价昆虫对植物的喜好程度和植物对昆虫的抗性水平。为了能够直观研究刺吸式昆虫在植物上的取食行为，科学家们一直在探索和改进研究方法。20 世纪 60 年代，EPG 技术由美国科学家 D. L. McLean 和 M. G. Kinsey（1964）发明，使用交流电，当时称为 electrical monitoring system（EMS）（电监测系统），以后不断进行改进；20 世纪 70 年代，荷兰瓦赫宁根农业大学（现名瓦赫宁根大学 Wageningen University, the Netherlands）的 W. F. Tjallingii（1978，1988）对该系统进行了改进，使用直流电，称为 electrical penetration graph（EPG）。后来，两套系统都称为 EPG，美国的交流电系统称为 AC-EPG，荷兰的直流电系统称为 DC-EPG。两套系统各有特点（Reese et al., 2000），DC-EPG 对于昆虫取食的细节识别更好一些，但对干扰敏感，需要使用屏蔽笼；而AC-EPG 使用更方便些，不需要屏蔽笼。目前美国已经成功研发可以使用两套系统的 AC-DC-EPG 系统（Backus and Bennett，2009）。

关于 EPG 名称中文翻译的讨论

Electrical penetration graph（或 electropenetrography）（EPG）或 electrical insect feeding monitor，我国学者曾经分别翻译为“昆虫刺探图谱技术”（雷宏和徐汝梅，1996）或“电子取食监测仪”（陈建明等，2002）。目前，EPG 这个英文名称被国内外广泛接受，笔者建议将 EPG 翻译为“刺吸电位仪”，是基于以下考虑：①利用该技术研究的昆虫类别是刺吸式口器昆虫，名称中的“刺吸”既是指刺吸式昆虫，又指昆虫的“刺入”植物组织内部的过程和“吸食”植物汁液的取食行为；②penetration（动词 penetrate）意思是“穿过”（through）、“进入”（into），没有“探”的意思；③graph 为“记录”（recording）、“书写”（writing），作为构词成分，-graph 表示“记录仪器”。同时，参照“触角电位”（electroantennograph，EAG）的译法，EPG 中的 graph 应翻译为“电位”。

EPG技术在研究植食性刺吸式昆虫取食行为、昆虫与植物的关系、昆虫传毒机制、作物抗虫机制等方面得到了越来越广泛的应用。利用EPG研究的昆虫种类不断增加，从最初的蚜虫、叶蝉、粉虱，逐步拓展到蓟马、螨类；美国开始利用EPG进行吸血昆虫的取食行为研究。近年来，随着烟粉虱和棉盲蝽的大发生及其带来的植物病毒病的暴发，刺吸式昆虫传毒机制研究成为热点，这些研究多用到EPG技术。

我国从20世纪90年代开始引进和应用DC-EPG，目前我国已经有许多单位购买和正在使用EPG进行各类研究工作。1992年的第19届国际昆虫学大会在北京召开期间，世界各地的科学家就两套EPG系统的比较、名词和概念的统一等进行了研讨，并在北京举办了EPG培训班。随后，EPG技术在我国得到了广泛应用，但单位之间一直缺乏沟通，有些单位由于人员变动，EPG应用也时断时续，但需要使用EPG解决的课题日益增多，很多专家和研究生要求学习EPG技术。2007年和2010年，笔者和W. F. Tjallingii教授分别在郑州（河南农业大学）和杭州（浙江省农科院植保及微生物研究所）举行了第一届和第二届全国EPG培训班和研讨会，由W. F. Tjallingii主讲，笔者负责组织和翻译。2012年在河南农业大学举办了第三届全国EPG培训班和研讨会，我借准备培训材料的机会，对EPG的技术原理、试验设计、应用技巧、数据分析等方面，进行了一次系统的梳理，以提纲的形式作为培训教材，同时，来自全国各地80多名学员在培训过程中提出了许多应用中遇到的问题。这些年经常收到不同单位的研究生提出的EPG使用中的疑惑、仪器操作方面的细节或者波形辨识技巧等。笔者在帮大家解决问题的同时，就在思考在培训教材的基础上出版一本EPG技术实验指导书。同时笔者在为许多杂志审稿时，发现不少稿件涉及EPG试验及结果分析的内容，在波形识别、名词理解、指标选择、结果阐释中存在一些误区。因此，出版一本EPG理论和实验方面的中文参考书就显得十分必要。

在此背景下，笔者邀请了国内使用EPG技术较多的专家和研究生参与编写了《昆虫刺吸电位技术及其应用》一书，参与编写本书的均是对EPG技术应用有丰富经验的专家、中青年学术骨干，或者有一定使用经验的研究生。这应该是我国第一本EPG理论和实验方面的中文参考书，是作者多年应用体会的总结和参编者多方面应用的成果。可以说，本书既是一本实验指导书，也是我国EPG技术应用的全面展示，期望能够对EPG技术在我国的进一步应用起到有效的推动作用。希望使用本仪器的同行加强合作，使EPG技术在科研中真正发挥应有的作用。

本书在筹划和撰写过程中，得到了本书编者的积极响应。薛堃博士对“原理篇”、罗岚女士和黄芳博士对“实验篇”、王满困教授和汤清波博士对“应用篇”的各章节进行内容和格式方面的编辑；王满困教授对全书编辑加工后，最后由我对全书进行统稿。美国农业部的Elaine A. Backus教授专门为本书撰写“History, Principles, and Summary of Applications of Electropenetrography (EPG)”一章，其中介绍了我国学者不太熟悉的AC-EPG和AC-DC EPG的原理和应用。据Backus教授介绍，台湾农业研究院Tsien tzung Shih和他的研究生Po An Lin（林普安）将该章翻译为中文，美籍台湾学者C. J. Chen进行了校对。我通读整个章节后发现，台湾学者的翻译有些地方不太符合大陆读者的阅读习惯，同时存在一些翻译错误。因此，我委托薛堃博士对本章进行了重译，我对译文进行了校对。

河南省科技厅特设立河南省科技著作出版基金，用于资助优秀科技著作出版，并组织专家对申报的书稿进行评审遴选，本书有幸获得资助出版。河南科学技术出版社杨秀芳编辑为本书出版倾注了大量心血。

本书撰写、编辑的过程其实也是一个学习的过程，本书涉及昆虫形态学、昆虫行为学、昆虫生理学、植物解剖学、植物病毒学、电生理学、计算机等相关知识，笔者和其他参编者在撰写和编辑书稿的过程中，阅读了大量文献和书籍，自感专业知识水平都得到了一定程度的提升。然而，我们尽管付出了极大努力，但因限于专业水平和能力，还有很多理解不透的地方；而且本书各章节由多位学者或研究生撰写，表述方面很难完全统一。因此，希望读者在使用本书过程中，如发现问题，请及时沟通，以便在重印或再版时更正。

闫凤鸣

2015 年 7 月 23 日于河南郑州

目　录

第一部分

原理篇

第一章　昆虫刺吸电位技术的历史、原理及应用概述

Elaine A. Backus[a]

[a] 文中提及的商业名称或商业化产品仅为提供相应的信息，不代表美国农业部推荐或支持。美国农业部在供需双方间保持中立。

提要：研究半翅目昆虫的取食、对植物的损伤过程和植物病原的传播机制等存在极大挑战，这主要是因为这类昆虫特化的刺吸式口器对植物组织进行刺探时不能被直接观察到。直至 50 多年前昆虫刺吸电位技术（electropenetrography 或 electrical penetration graph，EPG），一种能有效地识别半翅目害虫取食行为技术的发明才解决了上述问题。目前，EPG 主要从三个方面推进半翅目害虫综合治理（integrated pest management，IPM）。第一，在取食损伤或植物病原传播机制不明时，EPG 可以作为鉴别手段；第二，已经了解了取食相关的损伤或病原传播，EPG 可以用来阐明杀虫剂、拒斥剂或其他化学成分对损伤或传毒相关特定取食行为的影响；第三，EPG 可以用来鉴定作物的抗性或感性品系，包括表达生物杀虫剂的转基因植物。本章的目的是：①综述 EPG 的电路原理和历史，特别是最新的第三代 AC-DC EPG 的发展；②介绍各种波形的生物学意义；③以蚜虫为例，简述 EPG 在基础昆虫学研究和害虫治理应用中的方法。

由于刺吸式口器昆虫刺探不透明的植物组织时，人们很难直接观察到口针，这就导致半翅目害虫的取食、造成的植物损伤和植物病原传播（特别是获毒、持毒和传毒）等研究面临巨大挑战。早期关于半翅目昆虫取食的研究，由于技术的限制，只能事后研究取食分泌物。例如，冰冻被取食植物，可以发现取食活动发生之后残留的唾液鞘（Leopold et al.，2003），或收集分泌的液滴（Dugravot et al.，2008），记录害虫长时间的刺探从而确定植物病原的传播（Purcell et al.，1979，Almeida and Purcell，2006）。直至 50 多年前一个革命性的技术发明之前，研究者是没有办法在取食行为发生的时候

注：本章为 History，Principles，and Summary of Applications of Electropenetrography（EPG）的中文翻译，为适应图书体例编排，略有改动。

直接实时观察半翅目昆虫的取食的。这个技术就是研究昆虫取食的刺吸电位技术（Walker，2000，Almeida and Backus 2004，Backus et al.，2005）。大多数关于半翅目昆虫取食行为的研究进展都是在最近 50 多年中通过使用 EPG 技术取得的。

如今，EPG 在新的半翅目害虫综合治理，特别是植物病原介体防控中的应用主要集中在三个方面：第一，在取食损伤或植物病原传播机制不明时，EPG 可以作为鉴别手段；第二，已经了解了取食相关的损伤或病原传播，EPG 可以用来阐明杀虫剂、拒斥剂或其他化学成分对损伤或传毒相关特定取食行为的影响；第三，EPG 可以用来鉴定作物的抗性或感性品系，包括表达生物杀虫剂的转基因植物。快速对 EPG 数据进行计算分析，可以量化比较抗性和感性植物对介体昆虫的反应。据此，研究者可以预测病原在抗性植株间的传播，进而发现新的寄主植物抗性机制。

本章的目的是：①综述 EPG 的原理和历史，特别是第三代 AC-DC EPG 系统的发展；②以蚜虫作为模式，EPG 可以获得什么波形和信息；③EPG 如何与农业的特殊需求相契合，特别是具有明显经济价值的 7 种蚜虫。本综述首次描述并讨论了用于蚜虫研究的三种类型的 EPG 系统。未来也将对使用 EPG 研究与农业相关的叶蝉、飞虱、木虱和粉虱的取食进行综述。本章关于 EPG 原理的部分内容取自 Backus 的报告（Backus in press），EPG 应用于蚜虫研究的部分选取自 Backus et al.（2015），读者可以从上述报告中获取更多的信息。

一、昆虫刺吸电位技术的历史和原理

（一）基本原理

如图 1.1a 所示，使用导电胶将细金丝（10~60μm）粘在昆虫背板上，并将昆虫接入与监测仪相连的放大器的输入端，昆虫就成为电路的一部分了。监测仪还可监测植物电势，图 1.1a 中 V_s 即为植物—土壤间的电势差。当昆虫口针刺入植物组织，口针的食物和唾液通道中的流动液体产生的电信号通过虫体进入监测仪，信号被放大，如图 1.1b 中的信号生成回路，并以数字信号的形式输出至计算机。在昆虫—植物界面产生的可变生物电势（biopotential，biological voltage，或 emf）和（或）电阻将恒定信号转化成了可变的输出信号，持续一段时间的信号就以波形的形式呈现出来了。

（二）历史和仪器设计

EPG 最初出现于 20 世纪 50 年代末至 60 年代初，50 多年间，随着电子工业的革命性发展而进步。最早的第一代 EPG 监测仪，由 McLean 和 Kinsey 发明，使用了 20 世纪 50 年代典型的玻璃管放大器（glass-tube amplifiers），到 20 世纪 60 年代则使用早期的固体晶体管（solid-state transistor）（McLean and Kinsey，1964）。仪器设计思路来自于 McLean 熟悉的无线电技术，使用交流电（alternating current，AC）提供信号，放大器敏感度较低，输入阻抗为 $10^6\Omega$（McLean and Weigt，1968）。这类早期的具有低输入阻

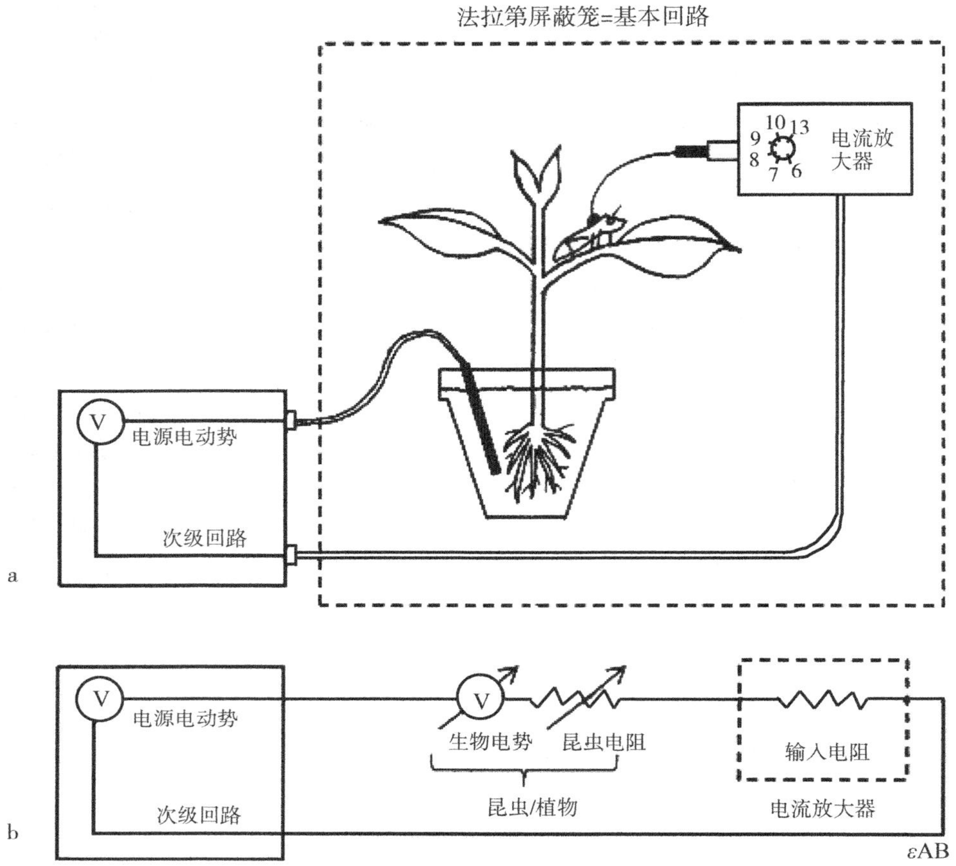

图 1.1　EPG 的基本回路示意图

a. 植物和昆虫的模型；b. 基本电路图，包括可变的生物电势（*emf*）和可变电阻（R_a）。（本图源自 G. P. Walker，有修改）（引自 Walker，2000，本图经美国植物病理学会惠允使用）

抗的 AC 监测仪输出的信号主要是由电阻或通过口针液体（如唾液、植物汁液）中的离子电荷产生的（类似于可变电阻 R_a，图 1.1b）（Walker，2000）。因此，早期的 AC 监测仪第一次实时监测到了重要信息，如口针刺探的开始与终止、唾液分泌和唾液鞘的形成（目前被认为是刺探路径活动）、口针伸缩、口针与维管组织的接触等（Backus et al.，2005）。理论上，一些 AC 监测仪已经可以监测回路中的生物电位了（Backus et al.，2000），但实际上，AC 监测仪受低输入阻抗水平和低反应速率的条带记录仪，以及某些设计中使用的滤波器的限制，输出信号中不含电动势信息（Backus et al.，2000，Tjallingii，2000）。

到 20 世纪 70 年代末，电子工业由于固态晶体管技术（以前称为运算放大器，op

amps）而产生巨大变革，先进的放大器和记录设备应用于 EPG 监测仪。Schaefers（1966）设计了第二代 EPG（使用直流电作为电源，所以称为 DC 系统），Tjallingii 对其进行了改良（Tjallingii，1978），包括使用运算放大器、直流电源、法拉第笼控制噪声，以 FM 记录或高反应速率的条带记录仪作为输出设备，以及更高的放大器灵敏性（输入阻抗为 $10^9 \sim 10^{13}\Omega$）（Tjallingii，1978；1985）。此外，Tjallingii 也通过引入电阻成分和电动势成分的概念，以及两者在输出信号中的混合，建立了现代 EPG 科学的理论（Walker，2000）。电阻成分和电动势成分决定了波形的组成来源，也是现代 EPG 科学的基础。

植物中或植物—昆虫界面上的电动势成分有两种已知机制：其一是植物细胞膜内外存在电位差异，口针尖端刺破植物细胞膜会导致输出波形表现出两种电学效应，即当口针刺破细胞膜时产生突然的电势下降，电压水平的正负分别表示口针的尖端位于胞外和胞内；第二种生物电势机制是液流电势，即含电解质液流快速通过细管，如口针的食物和唾液通道，而产生电荷分离而形成的微小电压变化（Walker，2000）。液流电位产生规律的波形，是肌肉节律性收缩的结果（Walker，2000），例如食窦扩张肌（the cibarial dilator muscles）控制液体进入前肠（Dugravot et al.，2008）。相比 AC 系统，DC 系统提高了输入阻抗的水平，使通过 EPG 输出波形监测电阻组分和电动势组分成为可能。区分波形的电学来源对理解其生物学意义有很大的帮助。

Tjallingii 的另一个有价值的理论是电位变化曲线（Tjallingii，1978；1985），后命名为电阻/电动势变化曲线（Backus and Bennett，2009；Backus，in press）。这个曲线表示输入阻抗 R_i 的水平（$10^6 \sim 10^{13}\Omega$）变化引起的昆虫 EPG 输出波形中电动势成分的比例（Walker，2000；Backus and Bennett，2009）。R_i 较低时，总信号中电动势成分较小（图 1.2）。由于电阻/电动势的比率在总 EPG 信号中具有相关性，电动势成分越小，则电阻成分越大（Backus et al.，2000；Walker，2000；Backus and Bennett，2009）。此外，电阻/电动势变化曲线的位置会随试虫大小而变化（图 1.2；比较了蚜虫、小型叶蝉与大型叶蝉的差异）。一般情况下，试虫体形大（食物唾液通道的直径大，则具有较强的导电性），曲线会向左方偏移（图 1.2）。因此，相对于小型昆虫，低 R_i 水平使得大型半翅目昆虫的记录信号中的电动势成分较高。当 R_i 水平在 1∶1 平衡点时，可以记录到两个电学来源的波形信息，因此，所有可能的刺探行为都会反映出来。据此，Tjallingii 为其 DC 系统选择中间的 $10^9\Omega$ 的输入电阻，以平衡小型蚜虫的电阻成分和电动势成分。这样，电阻/电动势变化曲线也解释了为什么输入阻抗为 $10^6\Omega$ 的 AC 系统几乎不可能在蚜虫刺探中监测到电动势成分（Backus and Bennett，2009），而在其他大型昆虫的波形中可以反映出电动势成分（Backus et al.，2005）。

到 20 世纪 90 年代中期，所有的 EPG 研究者都使用计算机的模拟—数字转化系统来记录波形，这极大地提升了 AC 和 DC 系统波形结构中的细节特征的呈现（Reese et al.，1994）。也是在那时，EPG 开始出现了分化，即 AC 系统主要用来记录叶蝉、飞虱和其他头喙亚目的昆虫，而 DC 系统则用于记录小型昆虫，特别是蚜虫、粉虱、蓟马、木虱和其他胸喙亚目昆虫（Backus，1994）。这种分化的情况直至 AC 系统停产为止（在 2000 年左右），之后，多数研究都在使用 DC 系统。回顾过去，电阻/电动势变化曲线解释了 AC 系统和 DC 系统在记录不同体形昆虫时的分化，因为胸喙亚目的小型昆虫的信号在输入阻抗为 $10^9\Omega$ 时包含了电阻成分和电动势成分，而如头喙亚目这类的大型昆虫的记录只有在低输入阻抗（R_i 为 $10^6\Omega$）的情况下才能包含两种成分。然而，分化的另一个原因可能是对不同类型的电源耐受程度的差异。这类耐受性的差异是研究的活跃领域，目前还知之甚少，但是有观察表明在低输入阻抗时，大型叶蝉似乎对 AC 系统比对 DC 系统的耐受程度更高（Backus，未发表数据）。许多以 R 成分为主的波形需要在较高电压（几百毫伏）和较低输入阻抗（R_i 为 $10^6 \sim 10^7\Omega$）的情况下才能被检测到。当使用高电压时，昆虫必须能够耐受较大的电流。因为某些原因（可能与细胞内外的取食刺探有关，见后文），不同昆虫在高电位、低输入阻抗时对 DC 或 AC 表现出不同程度的耐受。

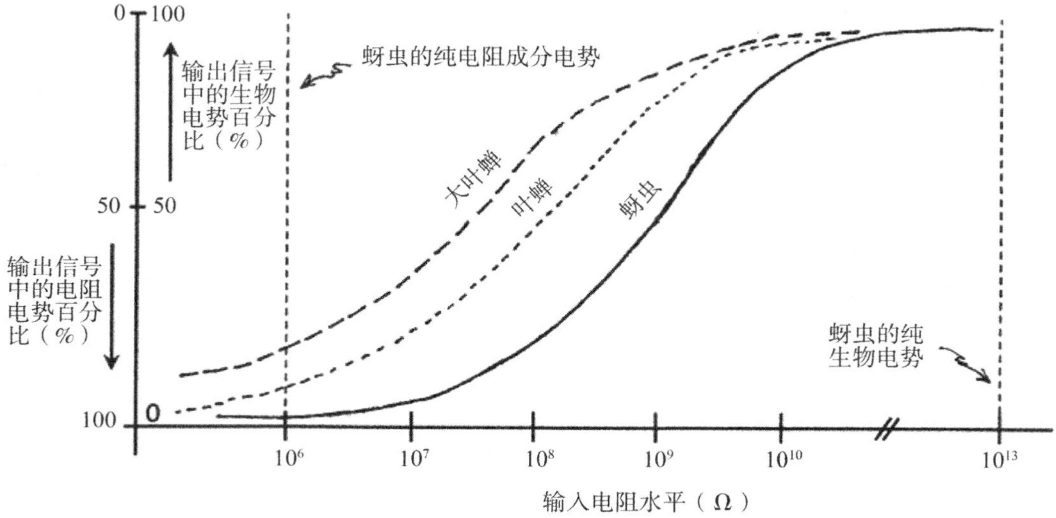

图 1.2　几种代表性昆虫的理论电阻/电动势变化曲线

［文中有简要介绍，Backus 和 Bennett（2009）对其理论细节进行了解释。本图经美国植物病理学会惠允使用］

为了适应更多类型和体型的昆虫，Backus 和 Bennett 研发了第三代通用型 EPG 监测仪，即 AC-DC 系统（Backus and Bennett，2009）。其输入阻抗可选择 $10^6 \sim 10^{10}\Omega$ 和 $10^{13}\Omega$，可选择 AC 或 DC 作为电源，使用现代最新的电子元件，如仪表运算放大器、标准化的印刷电路板等。该仪器消除了人为影响和其他老式仪器的问题（Backus et al.，2000；Backus and Bennett，2009），具有更高的敏感性和可扩展性。尽管被命名为“AC-DC”，但这个通用的仪器与原来的 AC 系统具有明显的不同。AC-DC 系统包含了第二代 DC 系统的所有功能，包括使用交流电作为电源时电动势成分的检出。例如，AC 系统信号处理回路不再使用高通/带通滤波器、耦合电容或其他目前已知会消除输出信号中电动势成分的因素（Backus et al.，2000；Tjallingii，2000；Backus and Bennett，2009）。虽然 AC 信号处理过程中仍然包括整流，但电压补偿模块可以让使用者去除由负极波形整合而导致的叠相（fold-over）失真的情况，恢复 DC 系统输出波形的保真度。此外，第三代 EPG 在技术上的扩展包括，可调输入阻抗、高放大倍数，以高分辨率和电压控制来实现设置的精确输出，使实验具有很高的可重复性（Backus and Bennett，2009）。

基于上述设计上的创新，AC-DC 系统综合了前两代仪器的优点，排除了两者的缺陷。研究者可以根据每种昆虫记录的特殊需要来调整仪器设置，建立不同输入阻抗水平的输出信号"波形库"，更精确地分辨电动势成分和电阻成分（Backus et al.，2013；Pearson et al.，2014）。

使用 AC-DC 系统时，不同输入电源（AC 或 DC）产生的蚜虫波形没有差异。但是，正如电阻/电动势变化曲线预测的那样，输入阻抗的改变会导致差异的产生（Backus and Bennett，2009）。这些结果与这类仪器的设计目的相符。与第二代仪器类似，电阻/电动势变化曲线的效应体现在波形强度的变化上，通过改变输入阻抗可以分析波形中的电阻成分和电动势成分信息。

二、确定 EPG 波形生物学意义的基本原则

EPG 波形生物学意义的确定可以通过以下方式：①通过唾液鞘的组织学研究、切断刺探中的口针和（或）截断口针的溢出液体分析等手段确定口针尖端在植物组织中的确切位置（例如，维管组织和非维管组织）；②通过透明人工饲料观察的方法和其他方法确定昆虫的口针活动（如口针移动、分泌唾液等）；③上文解释过的波形电学特征（Walker，2000）。

波形的电学特征包含口针深度信息。根据输入阻抗的水平，可以分为两种方式。低输入阻抗（根据昆虫种类的差异，一般为 $10^6 \sim 10^7\Omega$）情况下记录的波形通常为正，波形的位置在 0V 基线之上，与口针深度有关。口针在植物组织中回撤或深入分别会造成记录电压的升和降（Jiang and Walker，2001；Backus et al.，2005）。由此可见，口针在植物中的深度具有很强的电阻成分。在高输入阻抗（$10^9 \sim 10^{13}\Omega$）时，波形通常既有正值，也有负值，电压水平变化反映了口针尖端周围的电荷情况。细胞内外的植物电荷形成了生物电势，也就是所谓的电动势成分。负值波形（在 DC EPG 文献中称为"胞内"水平）（Walker，2000）代表共质体（symplastic）区域（活细胞内部，未被口针穿刺或被唾液鞘封闭的完整质膜保持了细胞内外环境中的电荷分离）。正值波形（"胞间"水平）代表植物中非原生质体区域（细胞间隙、细胞壁与细胞膜之间、成熟木质部或被口针活动破坏了电荷分布的细胞内）（Backus et al.，2009；Miranda et al.，2009；Pearson et al.，2014）。在输入阻抗值为中间的 $10^8\Omega$ 时，记录波形根据昆虫（可能是由于金丝连接质量差异）或刺探的不同，可能为正或兼有正负。

另外一个特征是标志性的 X 波的出现（Backus et al.，2009）。McLean 和 Kinsey（1967）最早将 X 波识别为一种复杂的、常见的、重复的、具有种特异性的波，只在能分泌唾液鞘的昆虫记录中出现（McLean and Kinsey，1967；Backus and Bennett，2009）。X 波表示口针接触并刺入可持续取食的细胞中。刺探路径阶段和持续取食阶段的发生

均有目标细胞参与，一个真正的 X 波代表路径阶段与持续取食阶段的转换。对于所有分泌唾液鞘的半翅目昆虫来说，X 波可以作为一个取食位点接受过程的判断依据。其代表的一系列行为包括：①尝试并判断韧皮部筛管或木质部导管（根据昆虫种类的取食偏好）是否可以接受；②建立与该细胞的稳固接触；③克服植物细胞的防御机制（Backus et al.，2009）。叶蝉 X 波的鉴定是首次对木质部吸食昆虫对该组织导管的特异性接触和接受行为的研究，其他的 X 波鉴定均是在韧皮部吸食昆虫上进行的。

三、蚜虫 EPG 应用于 IPM 研究

使用 EPG 研究半翅目害虫的行为特征可以广泛应用于各个领域，蚜虫的研究就是一个很好的例子。EPG 的一个主要应用就是评价寄主植物的抗性。蚜虫在敏感植株和抗性植株或寄主植物与非寄主植物上的行为特征为研究者提供了一个快速评价抗性等级（Nielson and Don，1974；Lohar and Kawada，1987；Peters et al.，1988；Montllor and Tjallingii，1989；Annan et al.，2000；Wang et al.，2004）或鉴别抗性位点的工具（van Helden and Tjallingii，1993；Caillaud et al.，1995；Kaloshian et al.，2000；Zehnder，2001；Sandanayaka et al.，2003；Le Roux et al.，2008；Lightle et al.，2012；Koch et al.，2015）。EPG 在农业领域的另一个应用是评价杀虫剂或拒食剂的使用对昆虫取食行为的影响（Holbrook，1977；Reuter et al.，1993；Harrewijn and Kayser，1997；Abraham and Epperlein，1999；Kang et al.，2012）。蚜虫是很多植物病原的主要携带者，会造成全球性的重大经济损失。蚜虫的传毒行为可以利用 EPG 进行深入研究。与病原的获毒或传毒相关的蚜虫行为被鉴定出来（Hodges and McLean，1969；Holbrook，1978；Powell，1991；Prado and Tjallingii，1994；Martin et al.，1997），对带毒和不带毒蚜虫的取食行为差异，或者其在染毒的和无毒的寄主植物上的行为差异都进行了研究（Ullman et al.，1988；Blua and Perring，1992；Quiroz et al.，1992；Boquel et al.，2011；Boquel et al.，2012；Hu et al.，2013）。

EPG 也可用于生态学和进化生物学研究。EPG 实验可以快速推断蚜虫和植物的互作结果。蚜虫的种内和种间相互作用也可以利用 EPG 记录其行为而进行研究（Chongrattanameteekul et al.，1991；Formusoh et al.，1992；Prado and Tjallingii，1997；Morris and Foster，2008；Brunissen et al.，2009）。此外，行为变化可以用来表示遗传变异，检验寄主专化性、多样化以及蚜虫的同域/异域成种等假说（Caillaud，1999；Powell and Hardie，2000；Funk and Bernays，2001；Tosh et al.，2001；Tosh et al.，2003；Halarewicz and Gabry，2012；Kordan et al.，2012）。

引用文献（略）

（薛堃译，闫凤鸣校）

History, Principles, and Summary of Applications of Electropenetrography (EPG)

Elaine A. Backus[a]

[a] Mention of trade names or commercial products in this article is solely for the purpose of providing specific information and does not imply recommendation or endorsement by the U. S. Department of Agriculture. USDA is an equal opportunity provider and employer.

ABSTRACT

Studying the feeding, plant damage, and transmission (i. e. acquisition, retention, and inoculation) of plant pathogens caused by Hemipteran insect pests is challenging. Hemipteran mouthparts, the stylets, are probed into opaque plant tissues precluding direct observation. This challenge was overcome by the invention of electropenetrography, or electrical penetration graph (EPG) monitoring, the most rigorous method to identify feeding behaviors of Hemipteran crop pests. Today, EPG is used in three main ways for the development of novel integrated pest management (IPM) tactics for hemipteran pests. First, in cases where the fundamental mechanisms of feeding damage or transmission of a plant pathogen are unknown, EPG is instrumental in identifying such mechanisms. Second, once the feeding - related causes of damage or pathogen transmission are understood, EPG can be used to demonstrate the effects of insecticides, antifeedants, or other chemical compounds on specific feeding behaviors responsible for the damage or transmission. Third, EPG can similarly identify the effects of resistant vs. susceptible varieties of crop plants, including transgenic plants genetically engineered to express biopesticides. The purpose of this paper is to: 1) review electronic principles and history of EPG, especially development of the new, third-generation AC-DC monitor, 2) review principles underlying biological meanings of waveforms, and 3) briefly summarize the ways that EPG can be used for both basic entomological research and applied improvements in insect pest management, using aphids (the best-studied system using EPG) as examples.

Key words: Hemiptera, Aphididae, electrical penetration graph, feeding, and transmission

INTRODUCTION

Studying the feeding, plant damage, and (especially) transmission (i. e. acquisition, retention and inoculation) of plant pathogens caused by Hemipteran insect pests has always been a challenge. This is primarily because their specialized, piercing-sucking mouthparts, the stylets, perform probing/penetration action into opaque plant tissues in which the stylets cannot be directly visualized. Early studies of Hemipteran feeding were forced (by the technical limitations of their time) to study snapshots of feeding, after the fact. For example, one could view deposits of hardened saliva *in planta* left behind after feeding activities had ceased (Leopold et al. 2003), in essence, frozen in time. Or, one could quantify collections of excretory droplets (Dugravot et al. 2008) or document transmission of plant pathogens from long-past stylet probes (Purcell et al. 1979, Almeida and Purcell 2006). However, a researcher could not directly study or quantify Hemipteran feeding in real time, as it was occurring, until the invention of a revolutionary technology over 50 years ago. That technology is electropenetrography, also known as electrical penetration graph monitoring of insect feeding (both names abbreviated as EPG) (Walker 2000, Almeida and Backus 2004, Backus et al. 2005). Most advances in understanding Hemipteran feeding behavior have been made possible by use of EPG over the last 50 years.

Today, EPG is used in three main areas for the development of novel integrated pest management (IPM) tactics for Hemipteran pests, especially vectors of plant pathogens. Firstly, in cases where the fundamental mechanisms of feeding damage or transmission of a plant pathogen are unknown, EPG is instrumental in elucidating such information. Secondly, once such mechanisms are understood, EPG can be used to demonstrate the effects of insecticides, antifeedants, or other chemical compounds on specific feeding behaviors responsible for damage or transmission. Thirdly, EPG can similarly identify the effects of resistant vs. susceptible varieties of crop plants, including transgenic plants genetically engineered to express biopesticides. Rapid computerized analysis of EPG data can provide quantitative comparisons of, for example, the responses of vectors to resistant and susceptible plants. A researcher can then predict whether a pathogen will be transmitted from/to a putatively resistant host plant, leading to a novel mechanism for host plant resistance.

The purpose of this paper is to review: 1) principles and history of EPG, especially the development of the new and improved, third-generation AC-DC monitor, 2) what waveforms and types of information can be gained via EPG, using aphid pests as a model system, and 3) how EPG can be applied to the special needs of agriculture, especially for seven species of aphids of economic significance. For the first time in print, our review describes and discusses in details the information of all three types of EPG monitors used for research on aphids. Future papers will review the use of EPG in studies of leafhopper, planthopper, psyllid, and whitefly feeding in relationto Taiwanese agriculture. Some of the narrative herein on EPG principles is excerpted and/or adapted from (Backus *in press*). The summary of applications from aphid EPG is excerpted from (Backus et al. 2015). The reader is referred to these publications for more information.

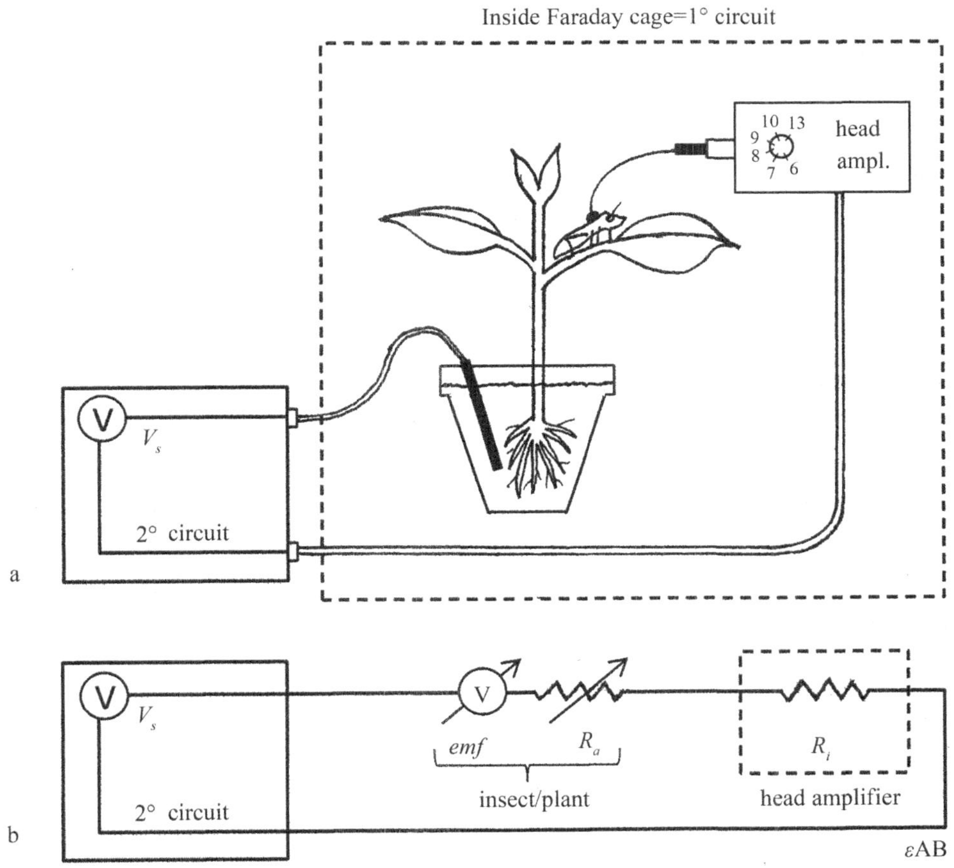

Fig. 1. 1　Diagrammatic representation of the primary (1°) circuit of an EPG monitor

a. Realistic model of the plant and insect. b. Electronic block diagram of primary circuit, including variable biopotentials (emf) and variable resistance (R_a). Modified from an original drawing by G. P. Walker.

2° = secondary circuit (i. e., signal processing circuitry); head ampl. = head stage amplifier; *emf*= electromotive force (biopotential); R_a= insect (e. g., aphid) resistance; R_i= input impedance of the head amplifier; V_s = source voltage. (Derived from an original drawing by G. Walker, in Walker 2000.) Reprinted with permission of the American Phytopathological Society.)

HISTORY AND PRINCIPLES OF ELECTROPENETROGRAPHY

Basic principles. As diagrammed in Fig. 1. 1a, the insect is made a part of an electrical circuit by attaching a thin (10–60μm) gold wire to its dorsal surface with conductive glue or paint, then connecting the insect to the input of a head stage amplifier attached to a monitor that also electrifies the plant (V_s to the electrode in the plant soil, Fig. 1. 1) . When the insect's stylets probe the plant tissues, fluids in the stylet food and salivary canals ionically conduct the electrical signal through the insect to the monitor, where it is amplified (secondary signal processing circuit, Fig. 1. 1) and outputted to a computerized digital display (not depicted in Fig. 1. 1) . Variable biopotentials (i. e. , biological voltages or emf [see below] in Fig. 1. 1b) and/or electrical resistances to fluid flow (Ra [see below] in Fig. 1. 1b) generated by the insect–plant interface instantaneously transform the constant applied signal into a fluctuating–voltage output signal that is graphed over time as a waveform.

History and monitor designs. First developed in the late 1950's to early 1960's, EPG has advanced over the last 50 years along with the revolution in electronics. The earliest, first–generation EPG monitors, developed by McLean and Kinsey, used technology typical of the time, i. e. , glass–tube amplifiers in the late 1950's, later evolving into early solid–state transistors by the 1960's (McLean and Kinsey 1964) . They also used AC (alternating current) applied signal, and a low amplifier sensitivity or input impedance (R_i) of 10^6 Ohms (McLean and Weigt 1968), in a design inspired by radio technology, with which McLean was familiar. These early "AC monitors" with low Ri outputted signals that were caused primarily by electrical resistance to/conductance of ionic charges carried in fluids (e. g. saliva, plant fluids) passing through the stylets (modeled as a variable resistor, Ra, in Fig. 1. 1b) (Walker 2000) . Thus, the earliest AC monitors detected for the first time, in real time, vital information such as beginning and termination of stylet probing, saliva secretion and salivary sheath formation (today termed pathway activities), stylet movements such as extension, retraction, and partial stylet withdrawal, and stylet contact with vascular tissues such as phloem and xylem (Backus et al. 2005) . Some AC monitor designs actually could detect a small fraction of the biopotentials (voltages) in the circuit (emf, defined below) (Backus et al. 2000); however, practically speaking, AC monitors outputted virtually no emf, due primarily to low Ri level and the then–prevalent use of slow–response strip chart recorders, but also secondarily to band pass filters found in certain AC monitor designs (Backus et al. 2000, Tjallingii 2000) .

By the late 1970's, electronics had been revolutionized with improved solid–state transistor technology (by then termed operational amplifiers, or op amps), so that more sophisticated amplifiers and recording devices were available and affordable. The design of the second–generation electropenetrograph (termed "DC monitor" or "DC system") was introduced by Schaefers (1966), but further developed by Tjallingii (Tjallingii 1978). Tjallingii used op amps in simple printed circuits, DC (direct current) applied signal, Faraday cages to control noise, FM tape recorders or rapid-response strip chart recorders as output devices, and (most importantly) higher amplifier sensitivity (Ri of either 10^9 or 10^{13} Ohms) (Tjallingii 1978, 1985). Tjallingii's elegant analyses (Tjallingii 1978, 1985) also established the modern theoretical understanding for EPG science by introducing the concepts of the R (or resistance) component (R_a, described above) and the electromotive force, emf (nearly synonymous with biological voltage or biopotential) component, blended together in the output signal (Walker 2000). The R and emf components are also termed electrical origins of a waveform, and are the foundation of modern EPG science.

There are two known mechanisms underlying the emf component in the plant or plant-insect interface. The first is disruption of the plant cell membrane's charge separation between external and internal cell environments, by stylet tips breaking living plant cell membranes. Such membrane breakages lead to two electrical effects on waveform output: 1) sudden voltage drops as the stylets puncture a membrane, and 2) positive or negative voltage levels that indicate extracellular (apoplastic) vs. intracellular (symplastic) stylet tip positions, respectively. The second biopotential mechanism is streaming potentials, i. e., tiny voltages developed by charge separation that occurs nearly instantaneously in ionic fluids rapidly moving through thin capillary tubes, such as the stylet food and salivary canals (Walker 2000). Streaming potentials cause regular-frequency waveforms that result from rhythmic pumping of muscles (Walker 2000), such as the cibarial dilator muscles controlling uptake of fluid into the functional foregut (Dugravot et al. 2008). Increased Ri level in the DC monitor (compared with AC monitor designs) made it possible to detect both R components (previously detected by the AC monitor) and additional emf components in the EPG output waveform. Identifying the electrical origin (s) of a waveform greatly aids in defining its biological meaning.

Another valuable theory of Tjallingii was the sigmoidal emf responsiveness curve (Tjallingii 1978, 1985), later renamed the R/emf responsiveness curve (Backus and Bennett 2009, Backus *in press*). The curve is produced when the proportion (0-100%) of emf in an insect's total EPG output signal (Walker 2000) is graphed in relation to Ri level (on a scale of 10^6 to 10^{13} Ohms) (Backus and Bennett 2009). The lower the Ri level, the smaller the proportion of the total signal that consists of emf (Fig. 1. 2). Because the ratio of R: emf is reciprocal within the total EPG signal, the smaller the emf proportion, the larger the R proportion (Backus et al. 2000, Walker 2000, Backus and Bennett 2009). In addition, the position of the R/emf responsiveness curve with respect to R_i can shift with the size of the insect being recorded (Fig. 1. 2; compare small aphids vs. small leafhoppers vs. large sharpshooter leafhoppers [Cicadellidae: Cicadellinae]). Generally, the larger an insect's body (i. e., with larger-diameter food and salivary canals, therefore greater ionic conductivity), the more the responsiveness curve shifts towards the left on the chart (Fig. 1.2). Thus, lower Ri levels allow detection of more emf in the signal of large hemipterans like sharpshooter leafhoppers, compared with smaller species like aphids. The R_i level that represents the 50: 50 R: emf balance point will detect the maximum number of waveforms from both electrical origins, and thus present the best representation of all possible probing behaviors. Tjallingii chose an intermediate R_i level of 10^9 Ohms for his (DC monitor) design, to balance R and emf components for small aphids. Thus, the R/emf responsiveness curve explains why an AC monitor (Ri 10^6 Ohms) detected almost no emf component in aphid probing (Backus and Bennett 2009), but slightly more in waveforms of sharpshooters and other large insects (Backus et al. 2005).

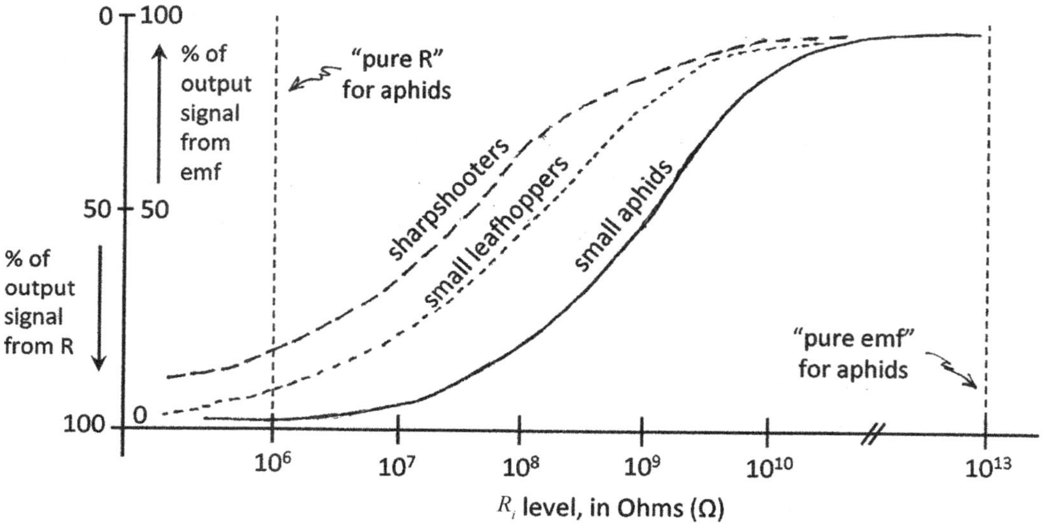

Fig. 1.2 Theoretical R/emf responsiveness curves for selected example insects

See text for summary, and (Backus and Bennett, 2009) for more detailed explanation of theory. (Reprinted with permission of the American Phytopathological Society.)

By the mid-1990's, all EPG researchers were using computerized analog-to-digital waveform display, greatly improving waveform fine-structure detail from both AC and DC recordings (Reese et al. 1994). Also by that time, EPG had gradually become specialized so that AC monitors were used primarily for medium-to-large insects such as leafhoppers, planthoppers, and other auchenorrhynchans, while DC monitors were used for smaller insects, especially aphids, whiteflies, thrips, psyllids, and other sternorrynchans (Backus 1994). Such specialization lasted until AC monitors were no longer manufactured (about 2000); since then, most research has used DC monitors until recently. In retrospect, the R/emf responsiveness curve explains part of the specialization of AC versus DC recordings for differently-sized insects, because signals from small insects like sternorrhynchans would contain both R and emf at R_i 10^9 Ohms, while recordings of larger insects like auchenorrhynchans would contain both components at lower input impedance (R_i 10^6 Ohms). However, another reason for past specialization may be due to a difference in tolerance to type of applied signal. Although such differential tolerance is poorly known and an active area of research, anecdotal observations suggest that large leafhoppers like sharpshooters seem to tolerate AC better than DC at low R_i levels, while aphids may tolerate DC better (Backus *unpub. data*) . This is important because many R-dominant waveforms can only be detected with rather high applied voltages (hundreds of mV) combined with low R_i (R_i 10^6or 10^7Ohms). When high voltages are used, the insect's physiology must tolerate such high current densities. For some reason (perhaps related to intercellular versus intracellular style of feeding, see below), different insects seem to tolerate DC or AC better at low Ri combined with high voltages.

To increase flexibility for all types and sizes of insects, a third-generation, "universal" EPG monitor (termed AC-DC) was developed by Backus and Bennett (Backus and Bennett 2009). It provides selectable Ri levels of 10^6-10^{10} plus 10^{13} Ohms, choice of AC or DC applied signal, and modern, up-to-date electronics such as instrumentation-quality op amps on standardized, commercially-manufactured printed circuit boards. This instrument removes artifactual voltages and other problems of older electronics (Backus et al. 2000, Backus and Bennett 2009), and has controls that are highly sensitive and scalable. Although termed " AC - DC ", the "universal" instrument is not the same as the old AC monitor technology. The AC-DC monitor was designed to incorporate all functions of the second-generation DC monitor, including retaining emf despite making AC applied signal available (for use with insects that don't tolerate DC at high voltages). For example, the AC signal processing circuit no longer has high pass/band pass filters, coupling capacitors, log amplifiers, or other components now understood to remove emf from the output signal (Backus et al. 2000, Tjallingii 2000, Backus and Bennett 2009). Although rectification is still performed for AC signal processing, a voltage offset knob allows the user to remove the waveform fold-over caused by rectification of negative-going waveforms, restoring perfect waveform fidelity to output of the DC monitor. In addition, the third-generation monitor expands EPG technology to include the first-ever, user-defined choice of Ri levels, very high amplification capabilities, and precise reproducibility of settings with high-resolution dials for gain and applied voltage control, allowing settings to be exactly reproduced among recordings (Backus and Bennett 2009) .

As a result of the above design innovations, the AC - DC monitor combines all the advantages of both previous generations of EPG monitor, with virtually none of the disadvantages. Thus, a researcher can precisely tailor the monitor settings to the specific needs of each insect species recorded, allowing characterization of a "waveform library" of output signals at different R_i levels, more accurately and correctly defining emf- versus R-components (Backus et al. 2013, Pearson et al. 2014) .

Using the AC-DC monitor, it was found that there were no differences in appearance of aphid waveforms based on type of applied signal *per se*, i. e. , AC versus DC. However, large differences occurred based on R_i level, as predicted by the R/emf responsiveness curve (Backus and Bennett 2009). These results support the success of the design goals for this monitor. In a similar manner, differences in amplitude among waveforms can be used to demonstrate the effects of the R/emf responsiveness curve, to determine electrical origins of waveforms by changing the R_i level.

Principles underlying biological meanings of EPG waveforms. The biological meanings of EPG waveforms are defined by correlation with: 1) stylet tip locations in the plant (e. g., vascular versus non-vascular tissue termini), via histology of salivary sheaths, cut stylets in probed plant tissues, and/or fluid exudation of cut stylets, 2) intricate stylet activities (e. g. stylet movements, salivation, etc.) performed by the insect, via observation in transparent artificial diets, and other methods, and 3) electrical characteristics of the waveforms, some of which have been explained above (Walker 2000).

Additional electrical characteristics of waveforms include those that reveal stylet depth. This can occur in two ways, depending upon R_i level. Waveforms recorded at lower R_i levels (10^6 or 10^7 Ohms, depending on species) are usually positive-going, and the position of the waveform above 0 V baseline (termed voltage level) is correlated with stylet depth (see figures in Backus *in press*) In other words, the voltage level rises or falls as the stylets are partially withdrawn or pushed more deeply into the plant tissue, respectively (Jiang and Walker 2001, Backus et al. 2005). Thus, stylet depth in the plant is a strong R component. At higher R_i settings ($10^9 - 10^{13}$ Ohms), waveforms usually become both positive- and negative-going (Fig. 1. 1a, b), and the biological meaning of voltage level shifts so that it reflects the electrical charge immediately surrounding the stylets tips; thus plant electrical charges internal or external to cells create a biopotential, i. e., an emf component. Negative voltage level (termed "intracellular" level in the DC EPG literature) (Walker 2000) denotes symplastic space (i. e. inside living cells whose charge separation between external and internal environment is maintained by an intact plasmalemma that is either not disrupted by stylet penetration or sealed by sheath saliva). Positive voltage level (termed "extracellular") denotes apoplastic space in the plant (i. e. intercellular spaces, interior of cell walls between living cell membranes, mature xylem, or living cell interiors whose electrical charge separation has been destroyed by stylet action) (Backus et al. 2009, Miranda et al. 2009, Pearson et al. 2014). Interestingly, at the intermediate R_i level of 10^8 Ohms with the AC-DC monitor, waveform voltage level can be either positive-going only or both negative- and positive-going, depending upon the insect (i. e., probably quality of wiring job) or probe.

Another general characteristic is presence of a landmark waveform termed the X wave (Backus et al. 2009). Originally identified by McLean and Kinsey (1967), an X wave is a complex, highly visible, stereotypically repeating waveform that is species-specific and (to date) performed only by salivary sheath feeders (McLean and Kinsey 1967, Backus and Bennett 2009). An X wave marks contact with and penetration of the stylets into a preferred cell type for eventual, sustained ingestion. Accordingly, a true X wave marks the transition between pathway and sustained ingestion phases; both are performed in an acceptable "target" cell of the preferred ingestion cell type. For all sheath-feeding hemipterans, X waves provide a visual portrayal of the behavioral acceptance process for an ingestion site. Thus, they represent a series of behaviors that: 1) sensorially test and judge the acceptability of a phloem sieve element or xylem vessel element (depending on the ingestion preference of the species), 2) secure a firm attachment to that cell, and 3) begin overcoming any challenges or defenses presented in the cell by the plant (Backus et al. 2009). Identification of the sharpshooter X wave marked the first time that such specialized contact and acceptance behaviors have been demonstrated for xylem vessels by preferential xylem-ingesters; all other X waves have been defined for preferential phloem-ingesters (e. g. aphids, deltocephaline leafhoppers).

Applications of aphid EPG to IPM research

Behavioral characteristics of hemipteran pests uncovered by EPG can be applied across a wide discipline of practical studies. Some of the best examples are from the extensive studies of aphids. One of the major applications of EPG is in evaluation of host plant resistance. Aphid's behavioral characteristics on aphid-susceptible and -resistant cultivars or host and non-host plants provides the researcher a time-saving tool to evaluate the degree of resistance (Nielson and Don 1974, Lohar and Kawada 1987, Peters et al. 1988, Montllor and Tjallingii 1989, Annan et al. 2000, Wang et al. 2004), or to identify the location of resistance (van Helden and Tjallingii 1993, Caillaud et al. 1995, Kaloshian et al. 2000, Zehnder 2001, Sandanayaka et al. 2003, Le Roux et al. 2008, Lightle et al. 2012, Koch et al. 2015). Another agricultural application of EPG is evaluation of feeding behavioral changes by the application of pesticides or repellents (Holbrook 1977, Reuter et al. 1993, Harrewijn and Kayser 1997, Abraham and Epperlein 1999, Kang et al. 2012) . Aphids are major vectors of many plant pathogenic viruses that cause enormous loss worldwide. The viral transmission behaviors of aphids have been extensively studied via EPG. Aphid behaviors that are correlated with viral acquisition or inoculation can be identified (Hodges and McLean 1969, Holbrook 1978, Powell 1991, Prado and Tjallingii 1994, Martin et al. 1997), and differences in feeding behavior of viruliferous/non-viruliferous aphids or on virus-infected/virus-free host plants can be disclosed (Ullman et al. 1988, Blua and Perring 1992, Quiroz et al. 1992, Boquel et al. 2011, Boquel et al. 2012, Hu et al. 2013) .

EPG can also be applied to studies of ecology and evolutionary biology. The results of interactions between aphids and plants can often be quickly inferred by EPG, which has been mentioned in previous paragraph. In addition, intraspecific interactions or interspecific interactions of aphids were studied by comparing behaviors recorded using EPG monitors (Chongrattanameteekul et al. 1991, Formusoh et al. 1992, Prado and Tjallingii 1997, Morris and Foster 2008, Brunissen et al. 2009). Moreover, behavioral variations were used to reflect genetic variations and to test hypothesis of host specialization, diversification, and sympatric/allopatric speciation of aphids (Caillaud 1999, Powell and Hardie 2000, Funk and Bernays 2001, Tosh et al. 2001, Tosh et al. 2003, Halarewicz and Gabry's 2012, Kordan et al. 2012).

LITERATURE CITED

[1] ABRAHAM K, EPPERLEIN K. Influence of imidacloprid after seed-treatment of maize on the sucking behavior of the bird-cherry aphid (*Rhopalosiphum padi* L.) and on the transmission of BYDV using electrical penetration graph technique in laboratory investigations. Gesunde Pflanzen, 1999, 51: 90-94.

[2] ALMEIDA R P P, BACKUS E A. Stylet penetration behaviors of *Graphocephala atropunctata* (Signoret) (Hemiptera, Cicadellidae): EPG waveform characterization and quantification. Annals of the Entomological Society of America, 2004, 97: 838-851.

[3] ALMEIDA R P P, PURCELL A H. Patterns of *Xylella fastidiosa* colonization on the precibarium of sharpshooter vectors relative to transmission to plants. Annals of the Entomological Society of America, 2006, 99: 884-890.

[4] ANNAN I B, WARD M. Stylet penetration activities by *Aphis craccivora* (Homoptera: Aphididae) on plants and excised plant parts of resistant and susceptible cultivars of cowpea (Leguminosae). Annals of the Entomological Society of America, 2000: 93.

[5] BACKUS E A. History, development, and applications of the AC electronic monitoring system for insect feeding, pp. 1-51. In M. M. Ellsbury, E. A. Backus and D. E. Ullman (eds.), History, Development, and Application of AC Electronic Insect Feeding Monitors. Entomological Society of America, Lanham, MD. 1994.

[6] BACKUS, E A. Sharpshooter feeding behavior in relation to transmission of *Xylella fastidiosa*: A model for foregut-borne transmission mechanisms. In J. K. Brown (ed.), Vector-Mediated Transmission of Plant Pathogens. American Phytopathological Society (in press).

[7] BACKUS E A., and BENNETT W H. The AC-DC Correlation Monitor: New EPG design with flexible input resistors to detect both R and emf components for any piercing-sucking hemipteran. Journal of Insect Physiology, 2009, 55: 869-884.

[8] BACKUS E A., M. J. DEVANEY, BENNETT W H. Comparison of signal processing circuits among seven AC electronic monitoring systems for their effects on the emf and R components of aphid (Homoptera: Aphididae) waveforms,. 102-143. In G. P. Walker and E. A. Backus (eds.), Principles and Applications of Electronic Monitoring and Ohter Techniques in the Study of Homopteran Feeding Behavior. Entomological Society of America, Lanham, MD. 2000.

[9] BACKUS E A, HABIBI J, YAN F, et al. Stylet penetration by adult *Homalodisca coagulata* on grape: Electrical penetration graph waveform characterization, tissue correlation, and possible implications for transmission of *Xylella fastidiosa*. Annals of the Entomological Society of America, 2005, 98: 787-813.

[10] BACKUS E A, LIN P A, CHANG C J, et al. Electropenetrography (EPG): A new diagnostic technol-

ogy for study of feeding behavior of piercing-sucking insects. *In*, Symposium on Trans-Century Plant Protection and Quarantine Technologies, 2015, Taichung, Taiwan.

[11] BACKUS E A, HOLMES W J, SCHREIBER F, et al. Sharpshooter X wave: Correlation of an electrical penetration graph waveform with xylem penetration supports a hypothesized mechanism for *Xylella fastidiosa* inoculation. Annals of the Entomological Society of America, 2009, 102: 847-867.

[12] BACKUS E A, RANGASAMY M, STAMM M, et al. Waveform library for chinch bugs (Hemiptera: Heteroptera: Blissidae): Characterization of electrical penetration graph waveforms at multiple input impedances. Annals of the Entomological Society of America, 2013, 106: 524-539.

[13] BLUA M J, PERRING T M. Effects of zucchini yellow mosaic virus on colonization and feeding behavior of *Aphis gossypii* (Homoptera: Aphididae) alatae. Environmental Entomology, 1992, 21: 578-585.

[14] BOQUEL S, GIORDANENGO P, AMELINE A. Divergent effects of PVY-infected potato plant on aphids. European Journal of Plant Pathology, 2011, 129: 507-510.

[15] BOQUEL S, DELAYEN C, COUTY A, et al. Modulation of aphid vector activity by potato virus y on in vitro potato plants. Plant Disease, 2012, 96: 82-86.

[16] BRUNISSEN L, CHERQUI A, PELLETIER Y, et al. Host-plant mediated interactions between two aphid species. Entomologia Experimentalis et Applicata, 2009, 132: 30-38.

[17] CAILLAUD C M. Behavioural correlates of genetic divergence due to host specialization in the pea aphid, *Acrythosiphon pisum*. Entomolgia Experimentalis et Applicata, 1999, 91: 227-232.

[18] CAILLAUD C M, PIERRE J S, CHAUBET B, et al. Analysis of wheat resistance to the cereal aphid *Sitobion avenae* using electrical penetration graphs and flow charts combined with correspondence analysis. Entomologia Experimentalis et Applicata, 1995, 75: 9-18.

[19] CHONGRATTANAMETEEKUL W, FOSTER J E, SHUKLE R H, et al. Feeding behavior of *Rhopalosiphum padi* (L.) and *Sitobion avenae* (F.) (Hom., Aphididae) on wheat as affected by conspecific and interspecific interactions. Journal of Applied Entomology, 1991, 111: 361-364.

[20] DUGRAVOT S, BACKUS E A, REARDON B J, et al. Correlations of cibarial muscle activities of *Homalodisca* spp. sharpshooters (Hemiptera: Cicadellidae) with EPG ingestion waveform and excretion. Journal of Insect Physiology, 2008, 54: 1467-1478.

[21] FORMUSOH E S, WILDE G E, REESE J C. Reproduction and feeding behavior of greenbug biotype E (Homoptera: Aphididae) on wheat previously fed upon by aphids. Journal of Economic Entomology, 1992, 85: 789-793.

[22] FUNK D J, BERNAYS E A. Geographic variation in host specificity reveals host range evolution in *Uroleucon ambrosiae* aphids. Ecology, 2001, 82: 726-739.

[23] HALAREWICZ A, GABRY'S B. Did the evolutionary transition of aphids from angiosperm to non-spermatophyte vascular plants have any effect on probing behaviour? Bulletin of Insectology, 2012, 65: 77-80.

[24] HARREWIJN P, KAYSER H. Pymetrozine, a fast-acting and selective inhibitor of aphid feeding. In-situ studies with electronic monitoring of feeding behaviour. Pesticide Science, 1997, 49: 130-140.

[25] HODGES L R, MCLEAN D L. Correlation of transmission of bean yellow mosaic virus with salivation activity of *Acyrthosiphon pisum* (Homoptera: Aphididae). Annals of the Entomological Society of America, 1969, 62: 1398-1401.

[26] HOLBROOK F R. Aldicarb and thiofanox: effect on the feeding activity of green peach aphids. Journal of Economic Entomology, 1977, 70: 742-744.

[27] HOLBROOK F R. Transmission of potato leaf roll virus by the green peach aphid. Annals of the Entomological Society of America, 1978, 71: 830-831 (832).

[28] HU Z, ZHAO H, THIEME T. Modification of non-vector aphid feeding behavior on virus infected host plant. Journal of Insect Science, 2013: 13.

[29] JIANG Y X, WALKER G P. Pathway phase waveform characteristics correlated with length and rate of stylet advancement and partial stylet withdrawal in AC electrical penetration graphs of adult whiteflies. Entomologia Experimentalis et Applicata, 2001, 101: 233-246.

[30] KALOSHIAN I, KINSEY M G, WILLIAMSON V M, et al. Mi-mediated resistance against the potato aphid *Macrosiphum euphorbiae* (Hemiptera: Aphididae) limits sieve element ingestion. Environmental Entomology, 2000, 29: 690-695.

[31] KANG M A, SEO M J, HWANG I C, et al. Insecticidal activity and feeding behavior of the green peach aphid, *Myzus persicae*, after treatment with nano types of pyrifluquinazon. Journal of Asia-Pacific Entomology, 2012, 15: 533-541.

[32] KOCH K G, PALMER N, Stamm M, et al. Characterization of greenbug feeding behavior and aphid (Hemiptera: Aphididae) host preference in relation to resistant and susceptible tetraploid switchgrass populations. Bioenergy Research, 2015, 8: 165-174.

[33] KORDAN B, DANCEWICZ K, WRÓBLEWSKA A, et al. Intraspecific variation in alkaloid profile of four lupine species with implications for the pea aphid probing behaviour. Phytochemistry Letters, 2012, 5: 71-77.

[34] LE ROUX V, DUGRAVOT S, CAMPAN E, et al. Wild *Solanum* resistance to aphids: Antixenosis or antibiosis? Journal of Economic Entomology, 2008, 101: 584-591.

[35] LEOPOLD R A, FREEMAN T P, BUCKNER J S, et el. Mouthpart morphology and stylet penetration of host plants by the glassy-winged sharpshooter, *Homalodisca coagulata*, (Homoptera: Cicadellidae). Arthropod Structure and Function, 2003, 32: 189-199.

[36] LIGHTLE D M, DOSSETT M, BACKUS E A, et al. Location of the mechanism of resistance to *Amphorophora agathonica* (hemiptera: Aphididae) in red raspberry. Journal of Economic Entomology, 2012, 105: 1465-1470.

[37] LOHAR M K, KAWADA K. Probing behaviour of the aphid, *Schizaphis graminum* (Rondani), *Rhopalosiphum maidis* (Fitch) and *Longiunguis sacchari* (Zehntner) on resistant and susceptible sorghum plants. Ber. Ohara Inst. Iandw. Biol., Okayama Univ., 1987, 19: 137-144.

[38] MARTIN B, COLLAR J L, TJALLINGII W F, et al. Intracellular ingestion and salivation by aphids may cause the acquisition and inoculation of non-persistently transmitted plant viruses. Journal of General Virology, 1997, 78: 2701-2705.

[39] MCLEAN D L, KINSEY M G. A technique for electronically recording aphid feeding and salivation. Nature, 1964, 202: 1358-1359.

[40] MCLEAN D L, KINSEY M G. Probing behavior of the pea aphid, *Acyrthosiphon pisum*. I. Definitive correlation of electronically recorded waveforms with aphid probing activities. Annals of the Entomological Society of America, 1967, 60: 400-405.

[41] MCLEAB D L, WEIGT W A, Jr. An electronic measuring system to record aphid salivation and ingestion. Annals of the Entomological Society of America, 1968, 61: 180-185.

[42] MIRANDA M P, FERERES A, APPEZZATO-DA-GLORIA B, et al. Characterization of electrical penetration graphs of *Bucephalogonia xanthophis*, a vector of *Xylella fastidiosa* in citrus. Entomologia Experi-

mentalis et Applicata, 2009, 130: 35-46.

[43] MONTLLOR C B , TJALLINGII W F. Stylet penetration by two aphid species on susceptible and resistant lettuce. Entomologia Experimentalis et Applicata, 1989, 52: 103-111.

[44] MORRIS G, FOSTER W A. Duelling aphids: Electrical penetration graphs reveal the value of fighting for a feeding site. Journal of Experimental Biology, 2008, 211: 1490-1494.

[45] NIELSON M W, DON H. Probing behavior of biotypes of the spotted alfalfa aphid on resistant and susceptible alfalfa clones. Entomologia Experimentalis et Applicata, 1974, 17: 477-486.

[46] PEARSON C C, BACKUS E A, SHUGART H J, et al. Characterization and correlation of epg waveforms of *Bactericera cockerelli* (Hemiptera: Triozidae): Variability in waveform appearance in relation to applied signal. Annals of the Entomological Society of America, 2014, 107: 650-666.

[47] PETERS D C, KERNS D, PUTERKA G J, et al. Feeding behavior, development, and damage by biotypes B, C, and E of *Schizaphis graminum* (Homoptera: Aphididae) on 'Wintermalt' and 'Post' barley. Environmental Entomology, 1988, 17: 503-507.

[48] POWELL G. Cell membrane punctures during epidermal penetrations by aphids: consequences for the transmission of two potyviruses. Annals of Applied Biology, 1991, 119: 313-321.

[49] POWELL G, HARDIE J. Host-selection behaviour by genetically identical aphids with different plant preferences. Physiological Entomology, 2000, 25: 54-62.

[50] PRADO E, TJALLINGII W F. Aphid activities during sieve element punctures. Entomologia Experimentalis et Applicata, 1994, 72: 157-165.

[51] PRADO E, TJALLINGII W F. Effects of previous plant infestation on sieve element acceptance by two aphids. Entomologia Experimentalis et Applicata, 1997, 82: 189-200.

[52] PURCELL A H, FINLAY A H, MCLEAN D L. Pierce's disease bacterium: mechanism of transmission by leafhopper vectors. Science, 1979, 206: 839-841.

[53] QUIROZ C, LISTER R M, SHUKLE R H, et al. Selection of symptom variants from the NY-MAV strain of barley yellow dwarf virus and their effects on the feeding behavior of the vector *Sitobion avenae* (Homoptera: Aphididae). Environmental Entomology, 1992, 21: 376-381.

[54] REESE J C, MARGOLIES D C, BACKUS E A, et al. Characterization of aphid host plant resistance and feeding behavior through use of a computerized insect feeding monitor, 52-72. In M. M. Ellsbury, E. A. Backus and D. L. Ullman (eds.), History, Development, and Application of AC Electronic Insect Feeding Monitors. Entomological Society of America, Lanham, 1994.

[55] REUTER L L, TOSCANO N C, PERRING T M. Modification of feeding behavior of *Myzus persicae* (Homoptera, Aphididae) by selected compounds. Environmental Entomology, 1993, 22: 915-919.

[56] SANDANAYAKA W R M, BUS V G M, CONNOLLY P, et al. Characteristics associated with woolly apple aphid *Eriosoma lanigerum*, resistance of three apple rootstocks. Entomologia Experimentalis et Applicata, 2003, 109: 63-72.

[57] SCHAEFERS G A. The use of direct current for electronically recording aphid feeding and salivation. Annals of the Entomological Society of America, 1966, 59: 1022-1024.

[58] TJALLINGII W F. Electronic recording of penetration behaviour by aphids. Entomologia Experimentalis et Applicata, 1978, 24: 721-730.

[59] TJALLINGII W F. Electrical nature of recorded signals during stylet penetration by aphids. Entomologia Experimentalis et Applicata, 1985, 38: 177-186.

[60] TJALLINGII W F. Comparison of AC and DC systems for electronic monitoring of stylet penetration activi-

ties by homopterans, pp. 41 - 69. In G. P. Walker and E. A. Backus (eds.), Principles and Applications of Electronic Monitoring and Other Techniques in the Study of Homopteran Feeding Behavior. Entomological Society of America, 2000.

[61] TOSH C R, WALTERS K F A, DOUGLAS A E. On the mechanistic basis of plant affiliation in the black bean aphid (*Aphis fabae*) species complex. Entomolgia Experimentalis et Applicata, 2001, 99: 121-125.

[62] TOSH C R, POWELL G, HARDIE J. Decision making by generalist and specialist aphids with the same genotype. Journal of Insect Physiology, 2003, 49: 659-669.

[63] ULLMAN D E, QUALSET C O, MCLEAN D L. Feeding responses of *Rhopalosiphum padi* (Homoptera: Aphidae) to barley yellow dwarf virus resistant and susceptible barley varieties. Environmental Entomology, 1988, 17: 988-991.

[64] VAN HELDEN M, TJALLINGII W F. Tissue localisation of lettuce resistance to the aphid *Nasonovia ribisnigri* using electrical penetration graphs. Entomologia Experimentalis et Applicata, 1993, 68: 269-278.

[65] WALKER G P. A beginner's guide to electronic monitoring of homopteran probing behavior, pp. 14-40. In G. P. Walker and E. A. Backus (eds.), Principles and Applications of Electronic Monitoring and Other Techniques in the Study of Homopteran Feeding Behavior. Entomological Society of America, 2000.

[66] WANG Y M, ZHANG P F, CHEN J Q. Host-preference biotypes of the cotton aphid, *Aphis gossypii* Glover and the behavioral mechanism in their formation. Acta Entomologica Sinica, 2004, 47: 760-767.

[67] ZEHNDER G W, NICHOLS A J, EDWARDS O R. et al. Electronically monitored cowpea aphid feeding behavior on resistant and susceptible lupins. Entomologia Experimentalis et Applicata, 2001, 98: 259-269.

第二章　刺吸式昆虫取食行为及传毒机制

近年来，刺吸式昆虫（烟粉虱、棉盲蝽等）种群大暴发，成为目前世界范围内最重要且最难控制的害虫类别。而且，刺吸式昆虫传播大部分的植物病毒，造成植物病毒病的大流行，成为农业生产上的严重问题。通过控制介体从而控制病毒，是有效和经济的措施。EPG 技术通过研究刺吸式昆虫的取食行为和植物抗性机制，从而可以为利用生态途径和植物抗性控制害虫奠定基础；EPG 技术还可以用于研究刺吸式昆虫的传毒机制，有助于阐明植物—介体—病毒互作机制，从而制定刺吸式害虫和植物病毒的分子和生态控制技术。

一、刺吸式昆虫的主要类别及特点

（一）刺吸式昆虫类群

按照食性，昆虫可以分为植食类、肉食类、腐食类。按照目前的分类系统（袁锋等，2006），昆虫纲包括 29 个目，其中 8 个目是植食性的（直翅目 Orthoptera，革翅目 Dermaptera，鳞翅目 Lepidoptera，半翅目 Hemiptera，鞘翅目 Coleoptera，膜翅目 Hymenoptera，缨翅目 Thysanoptera 和双翅目 Diptera）。而按照昆虫口器的类型，昆虫可以分为咀嚼式、刺吸式、锉吸式、舐吸式、虹吸式等。昆虫口器类型与取食行为密切相关，也是其食性分化的形态基础。植食类昆虫中，咀嚼式昆虫破坏植物器官组织，对植物损害大；而刺吸式昆虫对植物机械损伤小，只取食植物汁液，一般认为与植物之间协同进化程度较高。

表 2.1　吸食汁液昆虫的主要类别

不同大类	种类
典型的吸血类刺吸式昆虫	半翅目蝽类，双翅目蚊类、虻类
典型的植食类刺吸式昆虫	
半翅目	
半翅类	蝽类
同翅类	头喙类（叶蝉、飞虱、蝉等）
	胸喙类（蚜虫、粉虱、介壳虫、木虱等）
双翅目	瘿蚊科（吸浆虫）
植食类的锉吸式昆虫	缨翅目（蓟马）

具有刺吸式口器的昆虫，又可以分为吸血类和吸食植物汁液类；目前 EPG 技术主要应用于蚜虫、粉虱等植食性刺吸式昆虫，但有些研究开始利用 EPG 记录吸血类昆虫的取食行为。蓟马（属缨翅目 Thysanoptera）口器为锉吸式，与有些刺吸式昆虫的为害有相似之处，故利用 EPG 研究其取食行为的也比较多。刺吸式昆虫的主要类别如表 2.1 所示。这里，我们主要讨论植食性刺吸式昆虫的取食特点。

（二）口器形态特点

一般认为，刺吸式口器是从咀嚼式口器演化来的，上唇退化，上颚、下颚延长为口针，下唇延长为分节的喙，其前壁内陷为唇槽，口针不用时就缩入唇槽内（图 2.1）。蝉、蚜虫、粉虱等昆虫的上颚、下颚的内颚叶转化为口针。上颚口针较粗，端部有倒刺，主要起刺入寄主组织的作用；2 条下颚口针较细弱，内侧各有 2 条纵槽，合在一起时形成 2 条管道，分别为食物道和唾液道（彩万志等，2011；Tjllingii，1988），但在口针最前端两个管道合二为一。

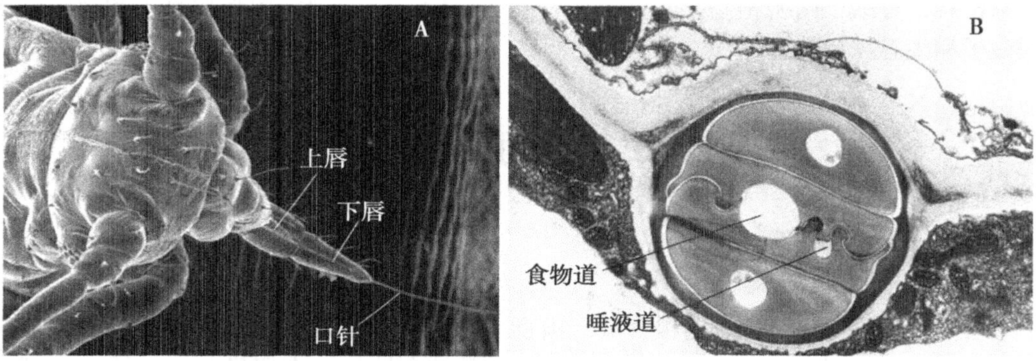

图 2.1　蚜虫口器外部形态（A）和口针横切图（B）
（引自 Tjallingii，1988）

（三）生物学特征及为害

半翅目刺吸式昆虫多数属于渐变态类，少数为过渐变态，经历卵、若虫和成虫三个阶段，但雄性介壳虫和粉虱有类似蛹的虫态。多数蚜虫类有周期性孤雌生殖现象，一些介壳虫经常进行孤雌生殖。若虫和成虫均可以吸食植物汁液，一般在嫩芽或嫩叶上为害，多数在叶片的背面（叶背面少蜡层、表皮细胞间隙大，利于昆虫口针刺入叶片组织）。也有不少刺吸式昆虫在茎秆上或叶片基部苞叶上吸食，如褐飞虱。由于从若虫到成虫要经历多次蜕皮现象，因此在利用 EPG 进行取食行为记录时，一般使用成虫，因为成虫不再蜕皮，连接的金丝不会因昆虫蜕皮而脱离。如果使用若虫，最好在刚蜕皮时进行记录。

刺吸式昆虫之所以成为世界农林业生产上的重要害虫类别，主要原因包括以下几个方面：第一，吸食植物汁液，影响植物的营养积累，造成植物生理状态不佳，如萎蔫、黄化、矮缩等；第二，分泌蜜露，招致霉菌寄生，影响植物光合作用和产品外观质量；第三，体形小，繁殖快，进化迅速，易对农药产生抗性，且多为孤雌生殖，种群建立快；第四，也是最重要的方面，刺吸式昆虫是植物病毒最重要的介体，50%以上的植物病毒以这类昆虫作为介体（Hull，2002），造成植物病毒病的大发生。这也是这类昆虫受到关注的最重要原因。

二、刺吸式昆虫的吸食部位和吸食方式

（一）吸食部位

刺吸式昆虫一般在植物的叶背面，因为叶背细胞疏松，蜡质层薄，有利于口针刺入。植物顶部的鲜嫩部位也是刺吸式昆虫喜欢取食的部位。但不同的昆虫种类在取食叶片部位上会有差异，有的种类在植物上部叶片取食，而有的喜欢在中下部叶片上取食。在进行 EPG 记录前，要观察昆虫在自然状态下的喜食部位，在进行实验的时候，把昆虫放在它最喜欢取食的植物部位，才能取得令人满意的 EPG 结果。

不同的刺吸式昆虫吸食植物的组织部位是不同的，这点在进行 EPG 记录的时候特别重要，初学 EPG 的研究者容易忽略这一点，以为刺吸式昆虫的取食部位均为韧皮部。可以用分析蜜露成分、口针截断、唾液鞘染色等方式确定昆虫在植物中的取食部位（具体见 EPG 新波形的研究章节）。不同类群的昆虫取食部位有一定规律，可以作为参考。

Tonkyn & Whitcomb（1987）根据刺吸式昆虫的唾液鞘或口针在植物组织里的部位、蜜露性质等特点，结合这些昆虫的传毒特性、形态特征及生态学信息，将刺吸式昆虫的取食部位分为三个类型，即韧皮部吸食类（phloem feeders）、木质部吸食类（xylem feeders）和叶肉吸食类（mesophyll feeders）。

头喙类 Auchenorrhyncha：主要包括蝉、蜡蝉、叶蝉、飞虱等类群。这类昆虫体形相对较大，口器相对强壮，取食部位为韧皮部、木质部、叶肉，或者混合方式（韧皮部和木质部均有）。头喙类昆虫吸食部位有很大差异，进行 EPG 记录时要先做波形与生物学意义关联的研究，不能简单类比其他昆虫的波形生物学意义。Wayadande（1994）则将叶蝉（leafhoppers）和飞虱（planthoppers）的取食策略（feeding strategies）划分为唾液鞘吸食（sheath feeding）和伤流吸食（lacerate and flush feeding）两类。唾液鞘吸食类主要包括叶蝉科和飞虱科昆虫，昆虫在刺吸过程中形成唾液鞘，从维管束（木质部或韧皮部）吸食汁液；伤流吸食类主要指叶蝉科的小叶蝉亚科（Typhlocybinae）种类，昆虫只在叶肉组织吸食，取食过程不形成唾液鞘，用口针刺吸叶肉细胞（薄壁组织）（mesophyll/parenchyma）汁液，由于这类取食方式破坏植物叶绿体或通过分泌的唾液间接阻止了韧皮部营养运输，因此此类昆虫对植物破坏较大，受害植物通常表现为叶表有大量斑点（stippling）。

胸喙类 Sternorrhyncha：这类昆虫主要包括各类蚜虫（蚜次目 Aphidmorph）、粉虱、介壳虫（蚧次目 Coccomorph）、木虱等小型刺吸式昆虫（袁锋等，2006）。这类昆虫主要吸食植物韧皮部，但有时也吸食木质部。胸喙类昆虫在 EPG 波形方面表现比较一致，因此波形的类别和生物学意义可以相互借鉴。

蜡类昆虫主要在植物韧皮部吸食，但也有在木质部和叶肉取食的种类。由于EPG在这类昆虫上应用相对较少，在对具体种类记录EPG波形时，要通过蜜露分析、唾液鞘染色等方式确定其取食部位（见“新波形的研究方法”一章），不能简单套用蚜虫、粉虱等的EPG波形的生物学意义。

（二）吸食过程

这里主要以蚜虫和粉虱为例，介绍刺吸式昆虫刺探吸食过程。

1. 评价叶表性质阶段 昆虫在将口针插入植物组织之前，会利用触角、口针前端及足前跗节上的化学（嗅觉、味觉）和触觉感器综合评价植物表面的物理、化学状况，如植物的挥发物、叶表化学物质、表面毛的性质和密度等。

2. 尝试取食阶段 将口针插入叶表皮内，抽吸少量胞间汁液，转换位置重复同样的行为，所以我们通常看到昆虫在叶片表面不断移动，而EPG波形则显示出C波和非刺探波（non-penetration）交替出现，反复几次后，口针就插入叶肉组织的细胞间隙，EPG就记录到比较长的C波。

3. 叶肉取食阶段 这个阶段和前一个阶段没有截然的界限，只不过在叶肉组织内的取食时间较长，在EPG记录的波形中称为路径波（pathway waveforms）。这个阶段有两个显著的特点：一是昆虫分泌鞘唾液（sheath saliva），在口针周围逐步固化形成唾液鞘（salivary sheath），其功能主要是形成有利于口针来回抽动的通道。鞘唾液的分泌和固化，使得EPG记录中的电阻由小到大，电压则由高到低，波形表现为起伏较大，可以细分为A、B、C三种波形，统称为C波。随着唾液鞘的不断延长，A、B、C波形不断出现，直至韧皮部。

这个阶段的第二个特点是昆虫的口针会对表皮或叶肉细胞进行短暂刺探，持续时间虽然很短（几秒到十几秒），但包括了水溶性唾液分泌和植物汁液吸食两个动作。由于植物细胞内部的电压为负，这种刺入和抽出很迅速，持续时间又短，在EPG记录中表现为电压突然下降，迅速回复到正常水平，这种波形被称为“电势降落”（potential drop，pd）。pd波与昆虫传播非持久性病毒密切相关。蚜虫类在EPG记录中会产生大量的pd波，因而蚜虫多传播非持久性病毒，而粉虱类则很少产生pd波，多传播半持久和持久性病毒。具体见本章“刺吸式昆虫传毒机制”部分。

路径波的持续时间与叶表皮性质（物理、化学特性）、叶肉组织营养和次生物质等相关，可以作为评价植物抗虫性的指标之一。

4. 韧皮部吸食阶段 昆虫口针顶端到达韧皮部后，首先分泌水溶性唾液（watery saliva），然后开始吸食（ingestion）。昆虫分泌的水溶性唾液含有多种酶，可以把植物汁液的营养成分进行一定的体外消化，还用于对付植物的诱导反应［具体过程可参见《化学生态学》（第2版）的相关章节］（闫凤鸣，2011）。由于韧皮部由筛管细胞组成，内部电压为负，因此EPG记录的韧皮部分泌唾液波（E1）和吸食波（E2）均为负电压，而汁液压力很大，因而昆虫的吸食则为被动吸食（passive ingestion）。有时也会出现蚜虫或粉虱在木质部吸食的情况，但昆虫取食木质部汁液需要主动吸食（positive ingestion），这种主动吸食需要食窦强有力的肌肉相配合，因此主要取食木质部汁液的昆虫一般体型相对较大，口器较为强壮（如蝉）。韧皮部取食过程可以获取和传播半持久或持久性植物病毒。

三、刺吸式昆虫传毒机制

由于植物病毒不能突破植物细胞壁，故大多数的植物病毒都是依赖于昆虫介体进行传播的（图2.2），而植食性刺吸式昆虫传播75%以上的虫传植物病毒（Nault，1997）。根据植物病毒在刺吸式昆虫内的滞留位点、滞留时间以及病毒粒子在体内的循回情况，介体昆虫传播植物病毒的方式分为非持久性（non-persistent）、半持久性（semi-persistent）和持久性（persistent）三类。介体的传毒方式与病毒的特性、流行、在植物上的为害等密切相关。因此，研究刺吸式昆虫的传毒机制、介体与病毒互作规律，从而阐明传毒的关键环节，将为植物病毒病的有效防控奠定理论基础（施艳等，2013）。

植物病毒的昆虫介体包括半翅目、鞘翅目、缨翅目、直翅目、革翅目、鳞翅目和双翅目昆虫，其中半翅目是迄今最重要的植物病毒介体类群。半翅目可以分为三个亚目：异翅亚目、胸喙亚目和头喙亚目。80%以上的植物病毒虫传介体都包含在后面的两个亚目中，统称为同翅类（Richards and Davis，1977；Nault，1997），蚜科和粉虱科是同翅类中传播病毒的主要介体（Nault，1997；Mitchell，2004）。

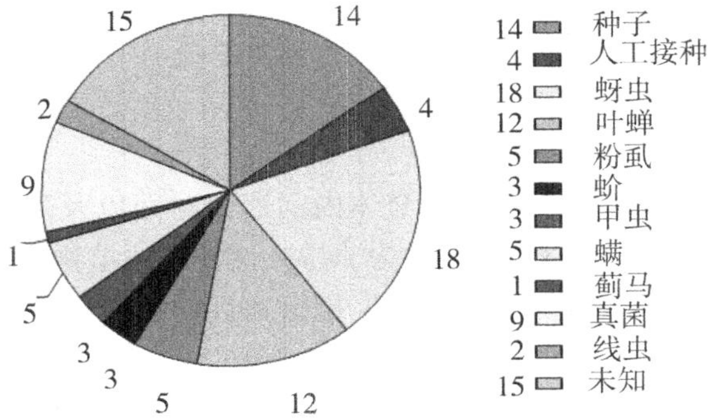

图2.2　病毒的介体组成（Hull，2002）

（一）昆虫介体传播病毒的方式

介体昆虫传播植物病毒存在一定的特异性。不同的病毒在介体内的滞留时间、滞留位点以及病毒粒子在体内的循回情况决定了介体的传毒模式。植物病毒的传播模式分为循回型（circulative）和非循回型（non-circulative）传播两类，其中非循回型分为非持久性（non-persistent）和半持久性（semi-persistent），循回型分为循回增殖型（circulative propagative）和循回非增殖型（circulative non-propagative）。非持久性病毒在介体中滞留时间很短，一般在几小时以内，而半持久性病毒或持久性在体内滞留几天、几周甚至一年；非持久性和半持久性病毒一般在介体获毒后几秒或几分钟内就可被传播到寄主植物上；此外，它们都没有潜育期，不在介体内复制，不能垂直传播

（传给子代），蜕皮后就失去传毒能力。循回型病毒一般可以垂直传播。

（二）介体获取和接种病毒

病毒的传播一般分为三个步骤，获毒（acquisition）、持毒（retention）和传毒（inoculation）。通过分析蚜虫的 EPG 波形，可以明显区分这三个不同阶段，口针传病毒通常是通过短暂的刺吸表皮细胞来获取，最初的刺吸不但可以获取非持久性病毒还可以获取半持久性病毒（Palacios，2002），病毒粒子都存在于口针的最端部或者前肠。当蚜虫的取食时间超过 10min，非持久性病毒的获毒率会明显降低，而对于半持久性病毒而言，当蚜虫口针穿刺到维管束吸食时，可以明显地增加获毒率。

蚜虫在经过一系列胞内穿刺后，口针开始通过胞间的空隙往深层穿刺，直到韧皮部，在确定取食位点前会先穿刺几个筛管分子，筛管穿刺后常伴随着长时间的唾液分泌，持续时间从 5s 到 30min，分泌的唾液能够平衡植物伤口的反应，从而避免韧皮部蛋白堵塞筛管分子（Tjallingii，2006）。维管束病毒只有通过介体的韧皮部取食来帮助传播。这类病毒包括蚜虫传播的黄症病毒科（Luteoviridae）病毒、长线形病毒属（*Closterovirus*）病毒和粉虱传播的菜豆金色花叶病毒属（*Begomovirus*）病毒和毛形病毒属（*Crinivirus*）病毒。它们的传播是在介体对筛管分子分泌唾液的过程中进行的。已经证明几种病毒的传播是在韧皮部进行的，譬如粉虱传播的莴苣褪绿病毒（lettuce chlorosis virus，LCV）（Johnson et al.，2002）和番茄黄曲叶病毒（tomato yellow leaf curl virus，TYLCV）（Jiang et al.，2000），以及蚜虫传播的大麦黄矮病毒（barley yellow dwarf virus，BYDV）（Prado et al.，1994）等。

（三）病毒在介体中的滞留

不同病毒的传播模式取决于病毒在介体内的滞留时间以及位点，同翅类传播的主要植物病毒的滞留位点见图 2.3。一般认为非持久性传播的植物病毒是口针传的，半持久性传播的病毒是前肠传播的。Ng（2006）提出口针的前端是非持久性病毒附着的位置，前肠是半持久性病毒附着的位置。而最近的研究发现，非循回型半持久性传播的病毒花椰菜花叶病毒（cauliflower mosaic virus，CaMV）存在于口针公共通道（common duct）中以前没有描述过的一个区域，在蚜虫口针的最顶端，而不是在前肠。通过荧光定位的方法研究莴苣侵染性黄化病毒（lettuce infectious yellows virus，LIYV）的病毒粒子在介体中的定位，发现病毒粒子存在于前肠的前端，通过用 4 种不同的外壳蛋白重组体饲喂烟粉虱发现大多数介体中都存在小外壳蛋白 CPm（minor capsid protein），CPm 的突变体直接影响病毒在介体中的滞留，从而影响病毒传播，表明了 LIYV 的传播是通过 CPm 介导的病毒粒子在前肠短暂的滞留机制发挥作用（Chen et al.，2011），CPm 提前终止的突变体不影响病毒在寄主中的系统侵染，但是失去了介体传播的能力（Stewart et al.，2010）。

持久性传播的病毒与介体关系非常紧密。病毒粒子需要到达中肠和辅助唾液腺才能进行传播，而穿越中肠壁和唾液腺壁的障碍是病毒所特异的。循环性病毒必须能穿过一系列障碍才能传播（Gray and Gildow，2003）（图 2.3）。

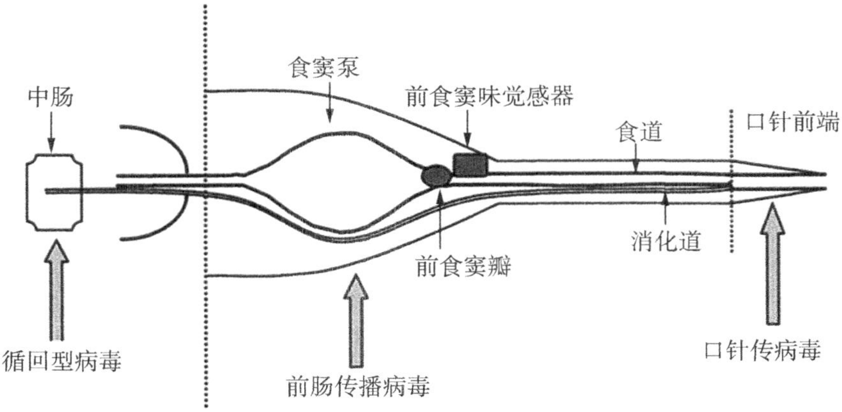

图 2.3　病毒在昆虫介体中的滞留位点（Stewart et al.，2010）

（四）病毒影响介体行为和表现

一方面，昆虫介体的行为及其与植物病毒的互作，和病毒的传播方式、传毒效率密切相关；另一方面，许多实验证明植物病毒通过侵染寄主植物和介体，从而影响植物的特性和介体的行为。一般而言，寄生物通过改变寄主的特性（营养组成）来影响介体的取食、带毒以及传毒，或者是通过改变吸引寄主定位的引诱化学物质组成（Eigenbrode et al.，2002；Jimenez-Martinez et al.，2004），从而影响介体对于病毒的传播。寄主植物释放的挥发性化学物质组成或浓度的变化最有可能影响介体的传播，因为昆虫主要是通过寄主植物的挥发性化学物质来定位寄主（Birkett et al.，2004）；此外，病原物的侵染能够改变寄主的气味谱，从而影响气味介导的介体与寄主的互作（Eigenbrode et al.，2002；Jimenez-Martinez et al.，2004；Kavaliers et al.，2005；McLean et al.，2005）。病毒侵染能够改变寄主挥发性有机物的浓度和组分，使得介体更趋向于到病毒侵染的寄主上取食。目前有关气味信号物质在昆虫的传毒途径中的作用的研究还相对较少。

从病原的角度来说，病毒侵染的植物对介体的吸引力越大，就越容易促使介体的取食，从而帮助病毒传播；但当介体获毒后，介体在植物上的滞留时间越短，就越有利于病毒的传播。

从介体的角度来看，延长在病毒侵染的植株上的滞留时间对介体是有利的，因此，病毒和介体之间在滞留时间上是相互矛盾的。对于多数“病毒—介体”的组合来说，介体从取食病毒侵染的植株中受益，通过增加有翅昆虫的比例从而促进它们的迁飞。此外，介体也可以通过取食病毒侵染的植物间接受益，带毒蚜虫能够影响天敌的正常发育。然而，也有报道关于病毒侵染的植物对介体表现的负面影响（Donaldson，2007），大多都是关于增殖循环型病毒的报道。增殖循环型病毒可能对它的介体有毒，从而影响蚜虫的寿命。简言之，同翅类昆虫取食非循环和循环非增殖型病毒侵染的植物对介体的生活史是有利的，而取食增殖循环型病毒侵染的植物会增加介体的致死率。增殖型病毒对同翅类昆虫的消极影响不适用于蓟马，有研究表明，蓟马取食增殖型病毒番茄斑萎病毒（tomato spot wilt virus，TSWV）对其是有利的（Belliure et al.，

2005）。大多数植物病毒并不是持久性传播的，而是非持久性传播的，由于介体的传播模式不同，病毒侵染引起的寄主表型的变化也是不一样的（Perring et al.，1999），非持久型病毒可能会引起寄主味觉信号的变化从而驱赶蚜虫而不是促使蚜虫定殖。这种病毒侵染后对植物和昆虫互作的模式与持久性病毒是不同的。

参考文献

[1] 彩万志，庞雄飞，花保祯，等．普通昆虫学．2 版．北京：中国农业大学出版社，2011.

[2] 袁锋，张雅林，冯纪年，等．昆虫分类学．2 版．北京：中国农业出版社，2006.

[3] 施艳，王英志，汤清波，等．昆虫介体行为与植物病毒的传播．应用昆虫学报，2012，50（6）：1 719-1 725.

[4] 闫凤鸣．化学生态学．2 版．北京：科学出版社，2011.

[5] BELLIURE B，JANSSEN A，MARIS P C，et al. Herbivore arthropods benefit from vectoring plant viruses. *Ecol. Lett.*，2005，8：70-79.

[5] DONALDSON J R，GRATTON C. Antagonistic effects of soybean viruses on soybean aphid performance. *Environ. Entomol.*，2007，36：918-925.

[6] EIGENBRODE S D，DING H，SHIEL P，et al. Volatiles from potato plants infected with potato leafroll virusattract and arrest the virus vector，*Myzus persicae*（Homoptera：Aphididae）. *Proc. Biol. Sci.*，2002，269：455-460.

[6] GRAY S，GILDOW F E. Luteovirus-aphid interactions. *Ann. Rev. Phytopathol.*，2003，41：539-566.

[7] HULL R. Matthew's Plant Virology. Academic Press，New York，NY，USA，2002.

[8] JIANG Y X，DE BLAS C，BARRIOS L，et al. Correlation between whitefly（Homoptera：Aleyrodidae）feeding behavior and transmission of tomato yellow leaf curl virus. *Ann. Entomol. Soc. Am.*，2000，93：573-579.

[9] JIMENEZ-MARTINEZ E S，BOSQUE-PÉREZ N A，BERGER P H，et al. Volatile cues influence the response of *Rhopalosiphum padi*（Homoptera：Aphididae）to barley yellow dwarf virus-infected transgenic and untransformed wheat. *Environ. Entomol.*，2004，33：1 207-1 216.

[10] JOHNSON D D，WALKER G P，CREAMER R. Stylet penetration behavior resulting in inoculation of a semipersistently transmitted closterovirus by the whitefly *Bemisia argentifolii*. *Entomol. Exp. Appl.*，2002，102：115-123.

[11] KAVALIERS M，CHOLERIS E，PFAFF D W. Genes，odours and the recognition of parasitized individuals by rodents. *Trends Parasitol.*，2005，21：423-429.

[12] MCLEAD G，GRIES R，VON REUβ S H，et al. The pathogen causing Dutch elm disease makes host trees attract insect vectors. *Proc. Biol. Sci.*，2005，272：2 499-2 503.

[13] MITCHELL P L. Heteroptera as vectors of plant pathogens. *Neotrop. Entom.*，2004，33：519-545.

[14] NAULT L R. Arthropod transmission of plant viruses：a new synthesis. *Ann. Entomol. Soc. Am.*，1997，90：521-541.

[15] NG JCK，FALK B W. Virus-vector interactions mediating nonpersistent and semipersistent transmission of plant viruses. *Ann. Rev. Phytopathol.*，2006，44：183-212.

[16] PALACIOS I，DRUCKER M，BLANC S，et al. Cauliflower mosaic virus is preferentially acquired from

the phloem by its aphid vectors. *J. Gen. Virol.*, 2002, 83: 3 163-3 171.

[17] PERRING T M, GRUENHAGEN N M, FARRAR C A. Management of plant viral diseases through chemical control of insect vectors. *Ann. Rev. Entomol.*, 1999, 44: 457-481.

[18] PRADO E, TJALLINGII W F. Aphid activities during sieve element punctures. *Entomol. Exp. Appl.*, 1994, 72: 157-165.

[19] RICHARDS O W, DAVIES R G. Imms's General Textbook of Entomology, vols. 1 - 2., 10th ed. Chapman & Hall, New York, 1977.

[20] STEWART L R, MEDINA V, TIAN T, et al. A mutation in the lettuce infectious yellows virus minor coat protein disrupts whitefly transmission but not in planta systemic movement. *J. Virol.*, 2010, 84: 12 165-12 173.

[21] TJALLINGII W F. Salivary secretions by aphids interacting with proteins of phloem wound responses. *J. Exp. Bot.*, 2006, 57: 739-745.

[22] TJALLINGII W F, Esch TH. Fine structure of aphid stylet routes in plant tissues in correlation with EPG signals, Physiological Entomology, 1993, 18: 317-328.

[23] TONKYN D W, WHITCOMB R F. Feeding strategies and the guild concept among vascular feeding insects and microorganisms. In: Current topics in vector research, Vol. 4. Springer, New York, 1987.

[24] WAYADANDE A C. Electrical monitoring of leafhoppers and planthoppers: feeding behavior and application in host - plant resistance studies. 86 - 105. In: M. M. Ellsbury, E. A. Backus, and D. L. Ulmann. History, development, and applications of AC electrical insect feeding monitors. Thomas Say Publications in Entomology, Entomological Society of America, Lanham. MD, 1994.

(闫凤鸣)

第三章　刺吸电位技术原理

取食是动物生长发育过程中的基本行为，取食行为是动物生态学研究的重要方面之一。取食行为不仅仅是指取食这一单一事件的发生，而是包括了食物选择、判断等复杂内容的广义事件。对于大型动物来说，取食行为的研究可以通过直观的观察获取大量信息；但是对于小型动物，尤其是具有特殊口部构造，如具有刺吸式口器的昆虫来说，取食行为的研究就必须借助一定的技术手段才能顺利进行。

刺吸式口器昆虫具有特殊的口器构造，整个或部分口器形成针状结构，以便吸食取食对象的汁液。对刺吸式口器昆虫的取食行为的研究包括了对取食位置的选择、口针的运动、昆虫与植物的其他相互作用等等。显然，刺吸式口器昆虫的取食过程大部分隐藏在取食对象的表层之下，很难直接观察到，更不用说刺吸式昆虫在取食过程中利用的物理化学信号。这就使得对于刺吸式昆虫的取食行为的研究落后于其他口器昆虫。对于刺吸式昆虫取食行为的研究具有突破性意义的事件是 McLean 和 Kinsey（1964）发明了监测蚜虫取食行为的电生理仪器。虽然由于种种原因，该仪器并没有大范围推广，但是将刺吸式口器昆虫的取食行为与电信号波形相联系的思路却确立了下来。后来的研究者通过不断地改进，改造设计并制造出了几种用于进行刺吸式昆虫取食行为监测的电生理仪器，并以其记录的图形为名，统称为刺吸电位记录仪（electrical penetration graph，EPG），其中以带有 $10^9\Omega$ 输入阻抗的直流系统（DC-EPG）应用较为广泛。但是需要说明的是，EPG 的结果也只能反映刺探行为的某些方面，需要和其他技术结合起来才能够获得取食行为的更多信息。

刺吸式昆虫的取食过程受其视觉和嗅觉的影响，但是最关键的过程却是个体在植株上降落之后通过口器对植物表面和内部的物理和化学因素进行评价，这个过程最终决定了个体是否在植株上持续取食。刺吸式口器是由昆虫的上颚、下颚和下唇形成的槽管状或针状的结构，可以刺入动植物的组织内吸食动植物体内的汁液。取食时，口针的不同部分协调刺入动植物表皮，唾液腺分泌凝胶型唾液和水溶性唾液，形成固定的通道，动植物汁液由于本身压力或者虫体的抽吸作用，沿着通道流入刺吸式昆虫体内。无论是主动吸食，还是被动取食，刺吸的过程都需要在一系列神经—肌肉系统的协调作用下进行，而这种神经—肌肉调节的昆虫体内电化学性质的变化和由于刺吸导致的植株—昆虫连通性的改变均是电信号的变化，可以被用来进行取食行为的监测。EPG 就是利用了这一点，将刺吸式昆虫的取食行为形象化了。最早的 EPG 只有 4 种不同的波形（McLean 和 Kinsey，1967），而目前 DC-EPG 至少已经鉴别出来 6 种波形，并且还有新的波形不断被认识。

一、EPG的电路原理

EPG技术电路原理如图3.1所示。昆虫通过导电银胶与直径10~20μm的金丝相连，进而接入电路中，这样，昆虫、植物、土壤以及电子元件就构成一个完整的回路。只有在昆虫的口针刺破植物表面，进入植物组织中时，该回路才被接通，口针到达不同组织层次会产生不同的电势，而昆虫以其他形式与植物的接触，由于过高阻抗的存在，并不会产生明显的EPG信号（Tjallingii，1995）。

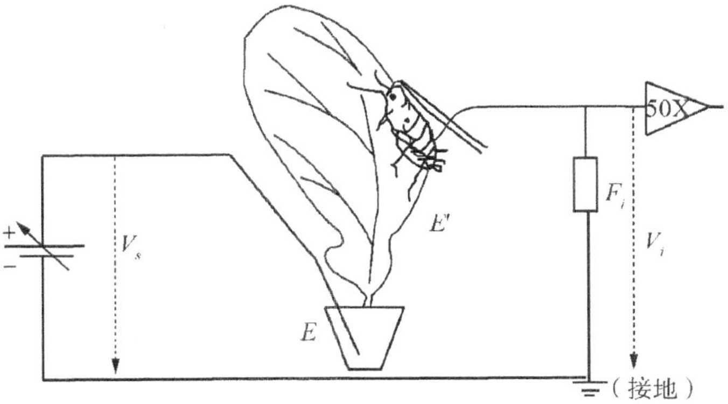

图3.1 刺吸电位技术原理（Tjallingii，1995）

在整个回路中，总电势V包括以下三个部分：电源电动势V_s，土壤与植物间的电势差E，昆虫与植物间的电势差E'。其中，电源电动势V_s是确定并且可调的，土壤与植物之间的电位差E对于每一组实验是固定的，而昆虫与植物之间的电位差E'则是不确定的，在口针刺探到植物不同的组织时，E'会有不同的值。在试验测量出的电位V_i的变化则是由于植物内部的组织导电性不均一引起的，即昆虫与植物的电阻R_a的变化，引起了记录值V_i的变化，两者关系如下式：

$$V_i=R_i \times V/(R_i+R_a)=R_i \times (V_s+E+E')/(R_i+R_a)$$

由于V_s是可调的，所以可以通过改变V_s而达到改变V_i的大小甚至极性的目的（雷宏和徐汝梅，1996）。当回路电压V为正时，V_i也为正；V为负时，记录显示图形处于0以下；而当V为0时，$V_s=-E-E'$，EPG信号完全消失。但是实际上，由于E和E'不能准确预测，所以EPG信号很难完全消失。由上式也可以看出，输入阻抗R_i一定时，记录信号完全是由R_a，也就是试虫—植物之间的阻抗决定的。这些电信号对我们理解刺探行为具有很重要的作用，不过这些信号在实际的记录中可能千差万别。如果要把这些信号都记录下来，我们就需要选择一个严格的输入阻抗R_i，使得大部分信号或者信号的有用部分能够被正确地记录下来，并可以进行后续的分析。最终记录的信号V_i实际是总电势V的一部分，其大小是由$R_i/(R_i+R_a)$决定的，选择合适的输入阻抗也就决定了最终记录信号的比例范围，当$R_i \gg R_a$时，$R_i/(R_i+R_a) \approx 1$，记录信号几乎就是回路总电势。在上式中，记录信号的变化由电阻成分［（R_i/（R_i+R_a）］和电势成分（V_s+E+E'）之分。在记录范围相对固定的情况下，当输入阻抗在试虫电阻这个数量级 $10^9\Omega$ 时，V_i/V 数值适当；当输入阻抗为 $10^8\Omega$ 时，信号中的电阻成分相对较高，即记录信号对试虫—植物间的电阻变化敏感度高，而电动势成分低，即记录信号对试虫—植物间的电势差不敏感；当输入阻抗为 $10^{10}\Omega$ 这个数量级时，则与低输入阻抗的情况相反，记录信号中电阻成分低，电势成分高。因此，在直流 EPG 设备中，输入阻抗一般选择 $10^9\Omega$ 这一数量级。

EPG 信号的记录同样需要进行优化，才能为后续的分析提供精准的分析材料。目前的 EPG 设备中使用的放大器也具有一定的特殊性。由于 EPG 回路的电阻高达 $10^9\Omega$ 这一数量级，所以回路中的电流值<1pA，记录设备中设置一个高值电阻器才能获得一个适当放大（一般为 50～100×）的电势值，方便记录。

二、影响 EPG 波形的因素

EPG 技术之所以可以作为研究刺吸式昆虫取食的重要实验方法，是因为 EPG 的波形与昆虫的刺探行为、唾液分泌、取食等生理过程相对应，可以通过对 EPG 波形的分析认识昆虫的取食，分析为害原因等。EPG 信号中包含了电阻成分和电势成分，也就是说，可以通过改变电阻或者电势而引起记录信号的变化。经过多年的试验和讨论（Kimmins，1986；Kimmins 和 Tjallingii，1985；Spiller，et al.，1990；Prado 和 Tjallingii，1994；Tjallingii 和 Hogen，1993；Tjallingii，1985），各种 EPG 波形的生物学意义已经形成并不断完善，在此基础上可以推断导致波形变化的因素在取食过程中的作用。

影响 EPG 记录波形的因素很多，可以按照不同的标准进行分类，如按照影响因素来源可以分为植物源因素、虫源因素等，按照影响对象可以分为影响电阻成分和影响电势成分。按照影响因素来源来介绍，可以较好地理解 EPG 记录波形变化与生物学意义之间的关系。

（一）植物源影响因素

不同植物组织具有不同的导电性，即便是同一组织细胞内外膜电位也存在差异，所以当试虫口针到达不同植物组织时，EPG 记录的波形会出现明显的变化，口针尖端与组织细胞的相对位置（细胞内外）也可以反映在 EPG 的波形当中。细胞膜电势的差别影响了整个回路的电势成分，口针进入到细胞内部，在记录波形上将产生一个明显的电势降落（potential drop，pd），由于整个回路的电势成分为电源电动势与各个部分自身电势的加和，各个部分的电势相比电源电动势虽然小，但是仍然占有一定比例，所以膜电势变化导致的记录波形改变的幅度较大。而不同组织的导电性差异则会影响回路的电阻成分，电阻成分的改变量相对于自身的电阻及输入阻抗来说比例较低，所以最终在记录波形上的表现较为微小。

（二）虫源影响因素

来源于试虫的某些因素也会对 EPG 系统的电阻和电势成分造成影响，例如蚜虫取食过程中食窦泵和唾液泵的活动会造成口针内液体组分和浓度的变化，导致该部分的导电性出现变化，进而反映在记录波形上。另外，带电粒子通过口针形成的局部电流也会影响回路的电势成分，试虫的神经、肌肉活动同样是由生物电信号驱动的，也可以造成电势变化，这些变化的电势成分也可以在记录波形上反映出来。

事实上，无论是植物源影响因素还是虫源影响因素，往往都是在一个生物过程中发生的，同一组织或者细胞同时影响了电势成分和电阻成分，而且对两种成分的影响往往发生相互叠加，最终得到的是一个综合的记录波形。要将电阻成分和电势成分区分开来，可以通过改变电源电动势的办法来实现。V_s 的变化不会对 E 和 E' 产生明显影响，但是 V_s 的改变影响了回路电流，记录结果 V_i 是电流和部分电阻成分的乘积。由此可以推断出：第一，电势成分不随 V_s 的变化而变化；第二，随着 V_s 的变化而变化的信号成分为电阻成分。记录波形既然是综合作用的结果，那么也就不难理解为什么 EPG 的记录波形千差万别，各种记录信号的频率、幅度等波形特征并不严格相同的事实。这样的分析也再一次印证了上述的记录波形不会完全消失的结论。

参考文献

[1] 雷宏，徐汝梅 . EPG——一种研究植食性刺吸式昆虫刺探行为的有效方法 . 昆虫知识，1996，33（2）：116-120.

[2] EVERT R F，ESCHRICH W，EICHHORN S E，et al. Observations on penetration of barley leaves by the aphid *Rhopalosiphum maidis*（Fitch）. *Protoplasma*，1973，77：95-110.

[3] HARRIS F K，JAMES E B. Reguraitation by *Myzus persicae* during membrane feeding：its likely funvtion in transmission of non-persistent plant viruses. *Annals of the Entomological Society of America*，1973，66：793-796.

[4] KIMMINS F M. Ultrastructure of the stylet pathway of *Brevicoryne brassicae* in host plant tissue，*Brassica oleracea*. *Entomologia Experimentalis et Applicata*，1986，41：283-290.

[5] KIMMINS F M，TJALLINGII W F. Ultrastructure of sieve element penetration by aphid stylets during electrical recording. *Entomologia Experimentalis et Applicata*，1985，39：135-143.

[6] MCLEAN D L，KINSEY M G. A technique for electronically recording aphid feeding and salivation. *Nature*，1964，202：1 358-1 359.

[7] MCLEAN D L，KINSEY M G. Identification of electrically recorded curve patterns associated with aphid salivation and ingestion. *Nature*，1965，205：1 130-1 131.

[8] MCLEAN D L，KINSEY M G. Probing behavior of the pea aphid，*Acyrthosiphon pisum*，I. Definitive correlation of electronically recorded waveforms with aphid probing activities. *Annals of the Entomological Society of America*，1967，60：400-406.

[9] MENTINK P J M，KIMMINS F M，HARREWJIN P，et al. Electrical penetration graphs combined with stylet cutting in a study of host plant resistance to aphids，*Entomologia Experimentalis et Applicata*，

1984, 35: 210-213.

[10] MITTLER T E. Studies on the feeding and nutrition of *Tuberolachmus salignus* (Gmelin) (Homoptera, Aphididae) . 1. The uptake of phloem sap. *Journal of Experimental Biology*, 1957, 34: 334-341.

[11] NAULT L R, STYER W E. Effects of sinigrin on host selection by aphids. *Entomologia Experimentalis et Applicata*, 1972, 15: 423-437.

[12] POLLARD D G. Plant penetration by feeding aphids (Hemiptera; Aphidoidea): a review. *Bulletin of Entomological Research*, 1973, 62: 631-714.

[13] PRADO E, TJALLINGII W F. Aphid activities during sieve element punctures. *Entomologia Experimentalis et Applicata*, 1994, 72: 157-165.

[14] ROHFRITSCH O. Ultrastructure of the nutritive tissue of *Chermus abietus* L. fundatrix of *Picea excelsa* L. *Marcellia*, 1977, 40: 135-149.

[15] SPILLER N J, KOENDERS L, TJALLINGII W F. Xylem ingestion by aphids – a strategy for maintaining water balance. *Entomologia Experimentalis et Applicata*, 1990, 55: 101-104.

[16] TJALLINGII W F. Membrane potentials as an indication for plant cell penetration by aphids. *Entomologia Experimentalis et Applicata*, 1985, 38: 187-193.

[17] TJALLINGII W F. Electrical signals from the depths of the plant tissues: the electrical penetration graph (EPG) . *In*: *Proceedings of an IFS workshop in chemical ecology* (held at the Universidad de Chile , Santiago Chile, 24-30 September 1995), 1995: 49-58.

[18] TJALLINGII W F, HOGEN E T. Fine structure of aphid stylet routes in plant tissues in correlation with EPG signals. *Physiological Entomology*, 1993, 18: 317-328.

[19] UNWIN F M. A versatile high frequency microcautery. *Physiological Entomology*, 1978, 3: 71-73.

(薛 堃 罗 晨)

第四章 EPG 波形的生物学意义

EPG 波形可以反映昆虫口针在植物特定组织的行为，进而揭示昆虫相关的生物学规律，如昆虫和植物之间的相互作用（Ellsbury et al.，1994；Walker & Backus，2000），昆虫传播病毒和其他病原体的机制（Martin et al.，1997；Backus et al.，2005），寄主抗性和抗性因素在植物组织中定位，农药、刺激剂和拒斥剂对刺吸式昆虫取食行为的影响（Jiang et al.，2001；Backus et al.，2005；Garzo et al.，2002）等。

EPG 波形的生物学意义是分析昆虫刺吸行为的基础（Tjallingii，1988）。利用 EPG 无论进行何种目的的研究，都要准确判别各类波形的生物学意义（Tjallingii，1990）。严格意义上来说，任何一种没有利用 EPG 研究过的昆虫，都要进行波形和生物学意义之间的关联研究。但在实际运用中，对于不同目的的研究，可以采取不同的策略。在利用 EPG 开展研究之前，首先应了解所研究昆虫的取食特点。不同的昆虫取食植物的部位是不一样的。根据吸食部位，植食性刺吸式昆虫可分为三类：①韧皮部吸食：胸喙亚目昆虫如粉虱、蚜虫、介壳虫，半翅目蝽类等；②木质部吸食：头喙亚目昆虫如叶蝉、蝉等；③叶肉吸食：缨翅目蓟马类锉吸植物叶肉细胞，部分同翅类叶蝉也属于这类（汤清波等，2011）。

一、蚜虫和粉虱 EPG 主要波形及其生物学意义

以蚜虫为例，昆虫口针在植物组织内所对应的 DC-EPG 波形分为三大类（图 4.1），其相对的取食行为生物学意义如表 4.1 所示。

（一）从表皮至吸食部位（韧皮部）前的 A 波、B 波和 C 波

A 波出现于蚜虫口针刚刚刺入植物的叶面组织，与其他波相比，幅度最大，历时很短，只有 5~10s，这种波形总是伴随着凝胶型唾液的分泌。

B 波紧随 A 波，与凝胶型唾液鞘的形成有关，蚜虫平均 5~10s 发生 1 次，其相对 A 波振幅由变化剧烈逐渐趋于平缓，随着唾液鞘的不断固化，导电性会起伏不定，所以波形会忽高忽低，同时慢慢过渡到 C 波，而通常在记录数据中，多为 B 波和 C 波叠加后的复合波形。

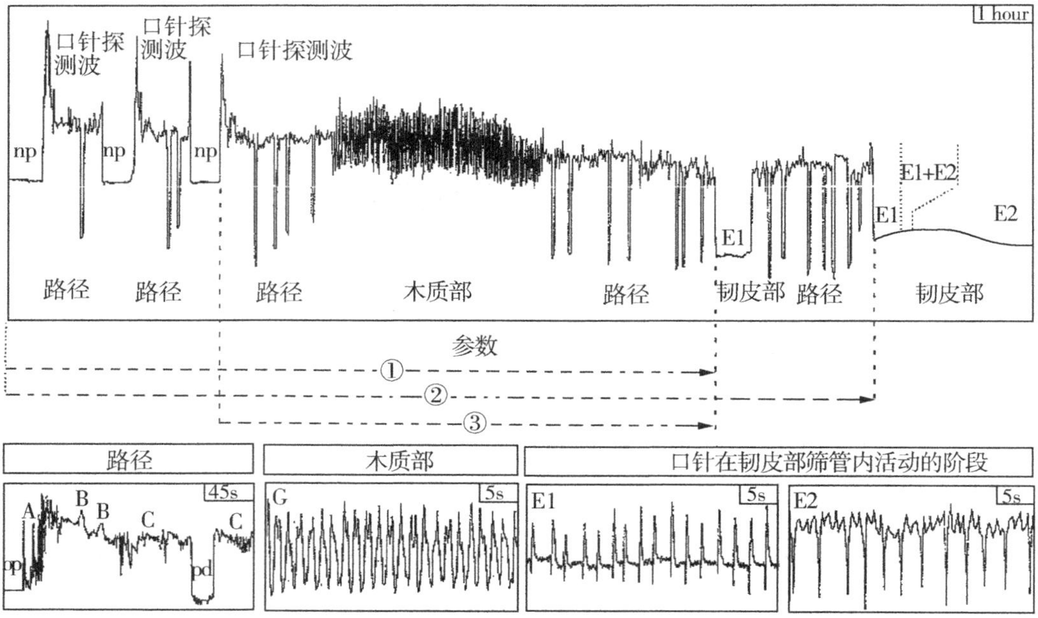

图 4.1 蚜虫 EPG 波形示意（Tjallingii，1978）

①从记录开始到第一次 E 波发生；②从记录开始到第一次韧皮部取食间的时间；③E 波发生前的刺探持续时间。

表 4.1 蚜虫类刺吸式昆虫 DC-EPG 波形的生物学意义

阶段	波形	刺吸式昆虫行为	口针端部所处植物组织和部位
刺探路径阶段	A	刺探，分泌凝胶型唾液	表皮
	B	分泌凝胶型唾液	表皮或叶肉组织
	C	刺探	叶肉细胞间隙或韧皮部筛分子外部
	pd	细胞内刺探	生活细胞内部
	F	口针遇到机械阻力	所有组织，细胞外部
韧皮部阶段	E1	分泌水溶性唾液	韧皮部筛管分子
	E2	被动吸食	韧皮部筛管分子
木质部阶段	G	主动吸食	木质部

C 波表明口针的位置在叶表皮与维管束之间的胞间游走。一般把 A 波、B 波、C 波统称为路径波（pathway waveform）。

当口针在细胞间时，会对周围细胞进行尝试性的刺探，一旦口针刺入细胞膜，由于细胞膜内电压为负，在波形上会出现一个电动势突然下降（potential drop，pd）的现象，这种尝试性的刺探时间一般持续 5～15s。如果昆虫在口针刺探过程中出现多次的 pd 波，表示该昆虫“品尝”行为较多。有些昆虫的 pd 波包含 3 个亚波，昆虫携带病毒会引起 pd 亚波持续的变化，在研究昆虫传播非持久性病毒时较有价值。

（二）G 波

G 波表明口针在木质部的主动取食，一般较少发生。

（三）韧皮部吸食对应的 E1 波和 E2 波

路径波比较乱，振幅大，持续时间相对较短，而韧皮部吸食波形（蚜虫中的 E1 和 E2）比较规律，振幅小，持续时间长。E1 表明唾液分泌，昆虫分泌唾液一般也与克服韧皮部筛管分子中的抗性因子相关；E2 表明韧皮部的被动吸食植物汁液过程。长于 10min 的 E2 波一般认为是蚜虫在韧皮部的持续吸食（Tjallingii，1988）。

根据昆虫蜜露的分泌来判断吸食波形：吸食开始后 1～3min，昆虫的腹部末端就有蜜露分泌出来，这时显示的波形肯定是吸食波 E2 波。

除此之外，在 EPG 波形分析时还经常涉及一系列详细的参数指标，这些参数指标也是上述波形指标的延伸，如蚜虫在棉花上的 EPG 指标及其对应的植物组织如表 4.2 所示。

表 4.2　棉蚜 EPG 指标及相应的寄主叶片层次（薛堃等，2008）

指标	棉花相关植物层次
总持续时间/s	叶表、表皮和叶肉
C 波持续时间总和	叶表、表皮和叶肉
C 波百分比/%	叶表、表皮和叶肉
非刺探持续时间总和	叶表、表皮和叶肉
np 百分比	叶表、表皮和叶肉
第一次刺探发生时间	叶表
刺探总次数	叶表
非刺探总次数	叶表
电势降落总次数	叶肉和细胞质
第一次电势降落发生时间	叶表、表皮和叶肉
第一次刺探的持续时间	表皮、叶肉和细胞质
第一次刺探到第一次电势降落的时间	叶肉
E 波发生前的刺探持续时间	叶肉
E 波发生前的刺探中电势降落次数	叶肉和细胞质
E 波发生前的刺探中电势降落间的平均间隔	叶肉和细胞

二、叶蝉等昆虫 EPG 波形及其生物学意义

对于叶蝉类等头喙亚目昆虫，由于其主要吸食的部位与蚜虫和粉虱等昆虫不同，所以其取食的 EPG 波也与上述昆虫不同。Backus 等（2005）利用AC-EPG 技术分析了北美大叶蝉 *Homalodisca coagulata* 成虫在葡萄上的取食选择行为，把 *H. coagulata* 的AC-EPG 波形分为 3 大类：路径波（pathway waveform）、吸食波（ingestion waveform）和取食中断波（interruption waveform），每个大类又可以分为不同类别和亚类，分别代表不同的生物学习性（表 4.3）。苗静和韩宝瑜（2007，2008）利用 DC-EPG 初步确定了叶蝉 7 种主要 DC EPG 波形，即 A 波、S 波、C 波、E 波、F 波和 R 波以及非刺探波 np 波，其中 A 波为刺探波，S 波为口针向韧皮部刺探和进入韧皮部中的分泌唾液波，C 波为口针到达韧皮部之前的主动取食波，E 波和 F 波为口针在韧皮部中吸食波，R 波为取食间歇波，并以 S 波、E 波和 F 波各波的平均持续时间以及含有各波形的单次刺探的平均持续时间为指标把 9 个茶树品种抗叶蝉的能力分为 3 个等级。

表 4.3　叶蝉 AC-EPG 波形的生物学意义（Backus et al.，2005）

阶段	大类	类别	亚类	口针端部所处植物组织	刺吸昆虫行为
路径波	A	A1		表皮，薄壁组织/叶肉	形成唾液鞘主体；口针束延伸/收缩，分泌水溶性唾液
		A2		薄壁组织/叶肉	唾液鞘延伸、打磨和加固，分泌部分水溶性唾液
	B	B1	B1s	薄壁组织/叶肉，髓，木质部，韧皮部	路径，取食中断，偶尔吸食；下颚口针活动
			B1w	同 B1s	分泌鞘状唾液；口针静止（待定）
		B2		薄壁组织/叶肉，髓	下颚口针活动以便离开唾液鞘或者切断组织；唾液鞘延伸/分支
			B2m	薄壁组织/叶肉，髓	同 B2，但是口针活动弱于 B2
吸食	C			木质部，韧皮部很少	吸食植物汁液
取食中断	N	非路径 N 波		木质部	分泌水溶性唾液（待定）
		路径 N 波		薄壁组织/叶肉	波形类似路径波，但是口针所在植物组织较深

三、其他昆虫的 EPG 波形

如果是 EPG 新记录的昆虫，而且没有现成的波形与生物学联系供参考，可以使用下面这些方式识别波形，进而判断其生物学意义（汤清波等，2011）：①根据可能出现的顺序进行推测：昆虫口针刺入植物表皮，EPG 波形会出现大的振幅，在蚜虫中称为 A 波，随即会分泌唾液鞘，随着唾液鞘的不断固化，导电性会起伏不定，所以波形会忽高忽低，这称为 B 波、C 波。一般把 A 波、B 波、C 波统称为路径波（pathway waveform）。韧皮部或木质部吸食的波形一般会先有短暂的水溶性唾液分泌的波形（蚜虫中为 E1），接着是振幅比较低但很规律的吸食波形（蚜虫中称为 E2），持续时间较长。②根据昆虫蜜露的分泌来判断吸食波形：吸食开始后 1~3min，昆虫的腹部末端就有蜜露分泌出来，这时显示的波形肯定是吸食波。③在应用中，如果不是特意研究波形的生物学意义，而是以研究昆虫取食或植物抗性为目的，只要识别 4 种波形就可以了：路径波、2 种吸食波、np 波（非取食波）、pd 波。

总之，EPG 技术在刺吸式昆虫与植物相互关系、植物抗性机制、昆虫传毒机制等研究中发挥着越来越大的作用，但是掌握其操作技巧及 EPG 波形分析需要一个过程，而且不同种类的昆虫可能在 EPG 波形分析上有所差异，其代表的生物学意义也不尽相同。因此，对于不同的试验材料和不同的试验目的，我们需要反复摸索，举一反三，才能达到熟练应用和解决实际问题的目的。

参考文献

[1] 苗静，韩宝瑜．假眼小绿叶蝉（*Empoasca vitis* Gothe）在不同品种茶树上的取食行为．生态学报，2007，27（10）：3 973-3 982.

[2] 苗静，韩宝瑜．MeSA 诱导茶树抗叶蝉取食效应的 DC-EPG 分析．植物保护学报，2008，35：143~147.

[3] 汤清波，张大山，姬琨，等．刺吸电位技术应用中的几个问题．应用昆虫学报，2011，48（5）：1 519~1 527.

[4] 薛堃，邓素，王戎疆，等．转 Bt 基因棉叶片表面特性与棉蚜取食行为的关系：非靶效应的个案研究．中国科学 C 辑：生命科学，2008，38（2）：188-198.

[5] BACKUS E A，HABIBI J，YAN F M，et al. Stylet penetration by adult *Homalodisca coagulataon* Grape：Electrical penetration graph waveform characterization，tissue correlation，and possible implications for transmission of *Xylella fastidiousa*. *Annals of the Entomological Society of America*，2005，98（6）：787-813.

[6] ELLSBURY M M，BACKUS E A，ULLMAN D L. History，development，and application of AC electrical insect feeding monitors. *Entomological Society of America*，Lanham，MD：Thomas Say Publications in Entomology，1994.

[7] GARZO E, SORIA C, GÓMEZ-GUILLAMÓN M L, et al. Feeding behavior of *Aphis gossypii* on resistant accessions of different melon genotypes (*Cucumis melo*). *Phytoparasitica*, 2002, 30 (2): 129-140.

[8] JIANG Y X, NOMBELA G. Analysis by DC-EPG of the resistance to *Bemisia tabaci* on an Mi-tomato line. *Entomologia Experimentalis et Applicata*, 2001, 99 (3) : 295-302.

[9] MARTIN B, COLLAR J L, TJALLINGII W F, et al. Intracellular ingestion and salivation by aphids may cause the acquisition and inoculation of non-persistently transmitted plant viruses. *Journal of General Virology*, 1997, 78 (10): 2 701-2 705.

[10] TJALLINGII W F. Electrical recording of stylet penetration activities. In: Minks A K, Harrewijn P, eds. Aphid, their Biology, *Nature Enemies and Control*, 1988, 2B: 95-108.

[11] TJALLINGII W F. Electronic recording of penetration behaviour by aphids. *Entomologia Experimentalis et Applicata*, 1978, 24: 521-530.

[12] TJALLINGII W F. Continuous recording of stylet penetration activities by aphids. In: Campbell RK, Eikenbary RD. Aphid-Plant Genotype Interactions. Amsterdam: Elsevier, 1990: 89-99.

[13] WALKER G P, BACKUS E A. Principles and applications of electronic monitoring and other techniques in the study of homopteran feeding behavior. *Entomological Society of America*, Lanham, MD: Thomas Say Publications in Entomology, 2000.

（汤清波　罗　晨）

第二部分

实验篇

第五章　刺吸电位技术基本操作与实验技巧

本章将以荷兰生产的直流刺吸电位仪为例，介绍 EPG 系统的组成、安装及基本实验操作技巧等。

一、EPG 系统的组成

（一）Giga 系列刺吸电位仪

荷兰科学家 Freddy Tjallingii 专门为记录刺吸电位图谱信号开发的直流刺吸电位仪（DC-EPG）被称为 Giga 系列放大器。“Giga”是由这个系列的直流刺吸电位仪有高达 1Giga Ω（10^9Ω）输入电阻而命名的。Giga 系列放大器分为三种即 Giga-4 型、Giga-8 型和其改进 Giga-8d 型，最多时它们分别能同时记录 4 只和 8 只昆虫的取食行为信号。

在 Giga EPG 系统中，通过将植物和昆虫并联接入电路，使得昆虫在植物上活动产生的电阻和电位，如生物电位都可记录。

在 Windows 系统下，使用者通过操作软件 $Stylet^+$ 1.1（早期使用软件 PROBE），利用一块独立的 AD 卡或者 USB 设备将采集的 EPG 信号数字化并存储在电脑中，并且在采集过程中通过软件实现对 EPG 信号实时监控。该软件还有一个独立的 EPG 分析模块用来初步分析和处理采集到的 EPG 信号。这套软件将在之后的章节中详细介绍（参见“常用软件使用指南”）。

（二）Giga 的主要部件

Giga 的主要部件如图 5.1 所示。

Giga 的主要部件分为三类：

控制盒：调节每个通道植物电压和增益，包括测试按钮。

EPG 电极：小型前置放大器，每个通道 1 个。

电源适配器：110V 或 220V AC ±8V DC。

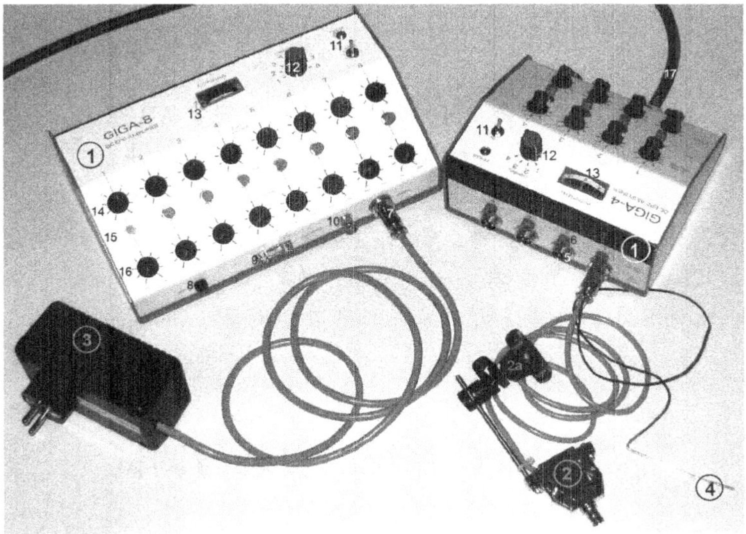

图 5.1　Giga 刺吸电位仪的主要部件（引自 Manual of Giga-4/8）

①——Giga-4（后视）和 Giga-8（前视）

②——EPG 电极：前置放大器（每个通道 1 个）

2a——金属棒转夹，直立固定附件

③——电源适配器，220V（110V）AC/±8V DC

④——植物电极（每个通道 1 个）

5——电极接入口（每个通道 1 个）

6 ——植物电极接出口（每个通道 1 个）

7——电源接入口

8——地线接口（接法拉第笼等）

9——4 或 8 通道输出连接器（接 17）

10——指定通道信号输出接口（由 12 指定）

11——电源开关（带指示灯）

12——通道选择旋钮

13——指定通道信号指示表（由 12 指定）

14 ——植物电极调节旋钮（每个通道 1 个）

15——校准脉冲按钮（每个通道 1 个）

16——增益调节旋钮（50~100 倍）（每个通道 1 个）

17——AD 连接线（从输出接口到 AD 装置或电脑）

小贴士

1. 老款 Giga 控制盒需要在信号输出接口（图 5.1-17）连接外置 AD 转化器，将输出的模拟信号转化为数字信号，由电脑接收。而新款 Giga 控制盒已经装有内置的 AD 转化模块，只需要将信号输出接口（图 5.1-17）与电脑直接相连即可。

2. 一般而言，如果电脑安装了 AD 转化器的驱动软件和软件 Stylet$^+$，该电脑能使用软件 Stylet$^+$的采集模块和分析模块，可以采集并分析 EPG 数据，这台

小贴士

电脑一般作为与 EPG 系统相连的操作电脑。而安装 AD 转化器驱动的电脑也可以安装软件 Stylet⁺，但是这些电脑只能使用其分析模块，方便后续 EPG 数据分析。

3. 采购仪器时需要注意，Giga 的电源适配器有两种型号，中国大陆地区使用电压 220V 的电源适配器。

4. 软件 Stylet⁺及其说明书等资料可以在网页 www.epgsystems.eu 下载。

二、EPG 系统的安装与调试

（一）数据采集部分

安装 AD 转换器和 Stylet⁺软件（参考本书后续章节“EPG 常用软件使用指南”）

（二）Giga EPG 系统的安装（以 Giga-4 为例）

Giga EPG 系统的安装（以 Giga-4 为例）步骤如下：

（1）在实验台等位置安装一个适合的法拉第笼（参考本章后续章节“法拉第笼的设计与制作”）。选址时注意环境，避免震动。注意不要在法拉第笼里安装电灯等电器，如需控制光周期，请将数控灯等物安装在法拉第笼外。

（2）如图 5.2 所示，安装 2 个普通铁架台，使用 2 个固定夹固定随机附带的支撑横梁（图 5.2，箭头所示），并把这些装置放在法拉第笼中，越靠里越好。

（3）将 Giga 放在法拉第笼里靠前的位置，将 4 个 EPG 电极的插头插入 Giga-4 仪器背部的电极接入口（图 5.1-5），只有一个方向能使之插入，旋转插头直到它们正确卡住。轻轻旋转插座的外圈，不要用太大力旋转。当其不好操作时，先逆时针方向转动外圈，直到螺纹适合。

（4）将植物电极（图 5.1-4）红色一端依次插入至 Giga 后面的相应接口（图5.1-6）上。

（5）同样，将电源适配器连接在 Giga 正面三孔插座上（图 5.1-7），请注意连接之前不要将电源适配器插入 220V 电源上，并确认 Giga 电源开关处于 OFF 状态。

（6）将 9 孔 AD 电缆连接在 Giga 前端 9 孔接口上（图 5.1-9），并将另一端接入 AD USB 设备，USB 暂时不要连接电脑，待电脑软件安装完成后，再连接 USB 电缆。如果是新款，就直接用连接线将接口（图 5.1-9）与电脑相连。连接时间同前。

（7）最后，通过接地线将仪器与法拉第笼相连。先将地线黑色插头插入 Giga 地线插孔（图 5.1-8）中，再将另一端的鳄鱼夹夹在法拉第笼的金属部件上。为了检测仪

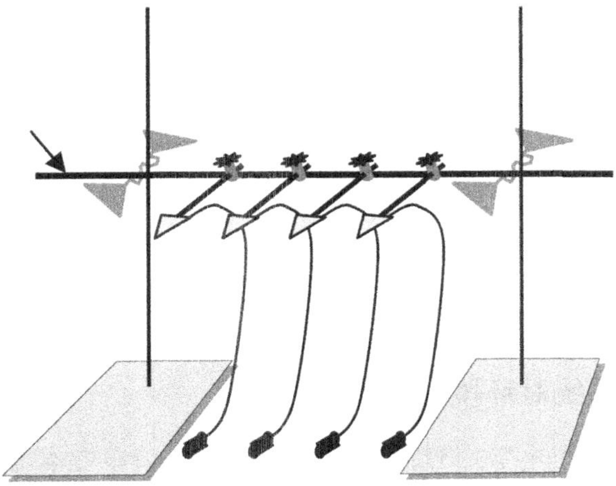

图 5.2　装有 4 个电极的铁架台（引自 Manual of Giga-4/8）

器安装好后的接地效果，可以使用万用表检测，如果接地良好，万用表应显示电压为 0V 时的电阻。

（三）仪器调试

当仪器安装完成、第一次使用前，或较长时间未使用仪器时，或记录到的 EPG 数据不正确时，就需要对设备进行调试，让整个设备处于良好的工作状态。对照说明书检查其连通性、各种接口是否有效等是设备开箱检查的一般步骤。对于 EPG 来说，还涉及一些特殊的调试程序。

这里主要介绍 EPG 电极测试。这个测试主要针对 EPG 电极的正确操作和初级放大器，并且不需要准备实验昆虫。

1. 准备工作

（1）EPG 电极既可安装在法拉第笼中，也可在外面进行安装测试。

（2）将电源适配器接入交流电源（110V 或 220V，中国大陆地区使用 220V），并确认 Giga 开关处于关闭状态。

（3）连接测试线，将其红色香蕉头一端插入通道一的植物电极接口。

（4）在进行下一步骤前，先用双手触摸法拉第笼的金属部件，释放身体携带的静电。

（5）为了避免损坏输入接口的零部件，小心将测试线的黄铜钉端插入电极的信号输入接口上（BNC 接口）。

（6）旋转 Giga 面板上植物电极调节旋钮（图 5.1-14）到 0 位置，如红色指针在 +/-符号之间。

（7）打开 Giga 开关，电源指示灯亮起。

（8）旋转通道选择旋钮（图 5.1-12）至位置 1，旋转植物电极调节旋钮，使信号指示表上（图 5.1-13）指针位于 0。

2. 测试步骤（如果 EPG 电极正常）

（1）轻轻正向或负向旋转植物电极调节旋钮，并检测信号指示表上红色指针是否同时正向或负向偏移。另外，长按校准按钮（图 5.1-15）大约 1s，并观察信号指示表上同时出现负向偏移，释放时指针回到原点。

（2）将信号指示表上红色指针留在正向位置，并将检测线的黄铜钉从 EPG 电极上移除。此时，观察信号指示器的指针是否能回到 0 位置（或非常接近 0）。

（3）如果在这种“开路”输入时，EPG 电极如果出现较大的正向或负向偏移，EPG 电极可能出现故障。

按上述步骤检查其他通道和电极。

（四）EPG 系统对环境的要求

EPG 的记录结果是电势信号，从 EPG 的原理（请参考本书第一章和第三章）可知，EPG 的记录信号与其电源电动势处于同一个或者是相近的数量级，EPG 系统是一套敏感的设备。实验环境中存在的电磁波会对 EPG 的记录产生干扰，形成记录波形中的噪声。一般环境中的电磁噪声的频率不高，表现在 EPG 记录波形上为周期性的基线变化，以及某些无规律的电势的突然变化。为了减小或完全消除这样的环境电磁噪声，EPG 设备安置的环境需要从几个层次加以注意：

（1）EPG 整个设备的安置需要选择一个相对封闭的环境，以尽量减少外界的干扰。相对封闭的环境中，无论是意料外的光干扰，还是环境电磁干扰，都会比开放空间更低一些。同时应该避免放置在窗口或门口等位置，以减少外界温度、光线、振动对其造成的影响。

（2）EPG 设备需要单独安置。EPG 设备按照其价格来说，并不算是大型（高值）仪器设备，所以在国内的绝大多数科研教学单位内并不会为其单独安排一个房间，经常会和其他设备共同放置。但是就 EPG 设备本身来说，最好是与其他设备，尤其是可以向外产生电磁辐射的设备隔离开来。

（3）EPG 设备可设立单独的屏蔽间以减小外界电磁干扰。在没有隔离条件的地方，EPG 设备必须放置在法拉第笼中来屏蔽噪声，减小电磁干扰（Jiang et al.，2007），以达到保持良好的记录波形的目的。

三、电极制作

（一）电极概述

前文提到 EPG 系统的电极分为两种：与植物连接的植物电极（plant electrode）和与实验昆虫连接的昆虫电极（insect electrode）。植物电极插在植物生长的土壤中，而昆虫电极末端金丝用银胶粘连在昆虫的前胸背板上。当 EPG 系统工作时，昆虫电极和植物电极将昆虫与植物分别连入生物电流放大器中，当昆虫口针刺入植物组织时，回路接通，回路电流经放大器放大后输出一系列电流波谱。EPG 系统中每个通道都有一个植物电极和昆虫电极与之配套。

（二）电极材料的选择与电极的制作

EPG 系统的植物电极和昆虫电极的主要部分都随设备提供，但考虑到实际情况，仪器的植物电极和昆虫电极都会出现损坏和遗失的情况，能自己制作电极是使实验能够顺利进行的保证。

电极的组成比较简单，理论上，只要是导体都可以作为备选的电极，但若电极制作不当，对实验结果还是有较大影响的。

选择电极最重要的标准是电导率。根据 EPG 的原理，记录信号实际上是植物—昆虫的电阻 R_a 两端的电势变化，在整个回路中 R_a 和 R_i（输入阻抗）是电阻的主要成分，其余均为导线或其他元件的电阻。如果加入的昆虫电极电阻较大则会增大记录电势，同时导致记录信号中的电阻成分与电势成分的微小变化更加不易分辨。因此，在选择电极材料时需要尽量选择电阻小的材料以保证记录信号的敏感，如铜、铝、银或金等。考虑到成本因素，一般选择铜作为电极的基本材料。

1. 植物电极 植物电极以选用长度 10cm 左右较粗铜棒为佳，若过短则不利于插入土壤，易受到土壤湿度上下不均的影响，若植物极铜棒过长，则容易与植物或外界物体接触，干扰实验。稍粗一些的直径，保证了植物电极可以轻易地插入土壤又不易弯曲。

2. 昆虫电极 昆虫电极的结构较植物电极稍微复杂，如图 5.3 所示，由三部分组成：铜钉（brass insect pin）、铜导线（copper extension wire）和金丝（dorsal gold wire）。

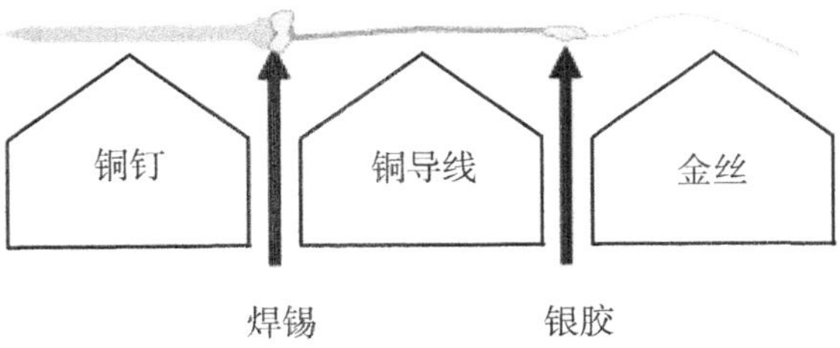

图 5.3 完整的昆虫电极（引自 Manual of Giga-4/8）

昆虫电极的第一部分为铜钉，可以插入 EPG 提供的昆虫电极探头中，保持持续的连通性。首先，铜钉的直径需要适合昆虫电极的接入口，过粗则不能插入放大器，且对放大器有一定的损伤；过细不能牢固地插入放大器，容易接触不良。其次是铜钉的长度要合适，以 2.5~3.0cm 为好，铜钉过短时，在插入或取下电极时容易触碰到铜钉和铜丝的连接处，使该处容易断裂；若铜钉过长，则容易触碰到其他的器件，不利于实验的顺利进行。

电极的第二部分为长度 5cm 左右的铜丝，与第一部分铜钉通过焊锡连接，具有一定的柔韧性，可以弯曲成不同的形状，保证那些在叶片远轴面取食的刺吸式昆虫能够正常连接。

电极第三部分为长度2~3cm的金丝，通过导电胶与铜丝相连。选择金丝的目的是为了保证电导、机械强度和柔韧性。整个电极具有几乎可以忽略的电阻值，并可以任意角度弯曲，适宜粘连实验昆虫（后续章节“金丝的选用和粘连昆虫技巧”将对其详细介绍）。

3. 昆虫电极制作步骤

（1）取2cm长的铜导线若干（直径0.2mm，已随仪器提供），使用电烙铁将铜导线焊在黄铜钉的顶端，制作成昆虫电极。如果铜导线用完，可使用普通110V或220V电线内未加涂层的内芯代替。

（2）如图5.4所示，非常小心地在体式显微镜下处理金线及缠绕金线的线圈。良好的光照条件非常重要，请打开体式显微的辅助光，或在操作台附近用较亮的台灯作为光源。金线轴两端分别有顶端标记，用镊子从一端截取金线，截取后，需要将剩余金线继续固定在顶端标记处，以便下次截取。建议首先截取实验所需长度的金线10~12cm，够五六根电极使用即可。不要提前截取成最终使用的长度（如2cm），这样非常容易遗失；也不要一次截取太长（如20cm），这样金丝非常容易打结。可以将截取的金线放置在培养皿中，操作时注意防风。

（3）将昆虫电极铜导线的末端插入银胶管液面下2~3mm蘸取银胶，将金线的一端粘好，干燥1~5min后截取适合实验的长度（如2cm）即可。继续操作直到制作好足够数量的昆虫电极。

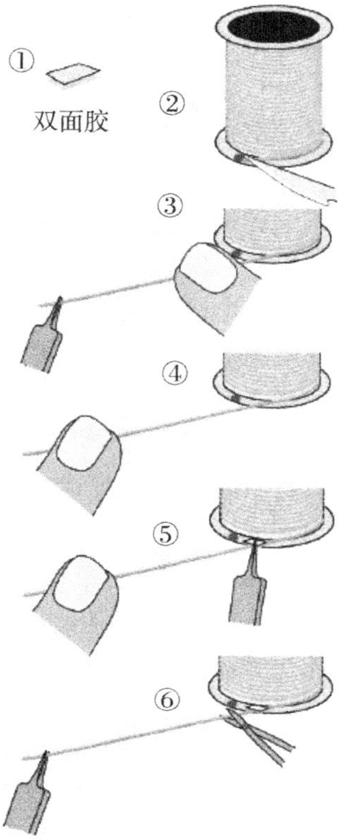

图5.4 金丝的取用
（引自Manual of Giga-8d）

小贴士

1. 金丝也可以先在铜导线上缠绕几圈固定后，再在连接处涂上银胶加固并增加导电性。

2. 昆虫电极属于非耗材，可以循环使用，金丝也不需要每次实验都更换。

3. 较多的昆虫电极放置在一起容易出现缠绕的现象，取用不方便且锡焊处容易断裂，最好将昆虫电极铜钉尾端插入固定好的泡沫板上用以存储。

4. 如果用镊子截取金丝不方便的话，也可以尝试用显微剪或者维纳斯剪（用作显微操作的小型剪刀）来截取。

5. 焊接铜导线和铜钉时需要使用到的实验仪器或耗材有：40W或以上的电烙铁、焊锡，焊锡膏（又名松香或助焊剂）。焊接时，局部温度较高，请勿直接

小贴士

在实验台或者塑料制品上操作，容易使之熔化而发生危险。另外，金属材料导热性好，请不要在金属台面上操作，容易烫伤且不利于热量在铜钉上聚集，不易焊接；最好使用废旧木块等物作为支撑进行焊接工作。

6. 由于昆虫电极制作不易，请每次实验时提前准备好富余的昆虫电极，以免影响实验进度。

四、导电银胶的配制和使用

（一）导电银胶概述

前文提到，在昆虫电极的制作过程中及昆虫电极与实验昆虫连接的时候均需要使用到导电银胶。顾名思义，导电银胶是具有导电功能的含银粉的胶水（后文简称银胶）。银胶是 EPG 实验中最常用的消耗品，除了随 EPG 设备购置的导电胶外，还可以用其他导电银胶代替，如电镜制片中使用的银胶等。不过，大多数成品导电胶含有大量有机溶剂，可能对实验昆虫的体壁或行为产生影响，故建议使用 EPG 系统配套银胶或自行配制银胶。

自制银胶时，胶水一般选择水溶性胶水，尽量避免有机溶剂的存在而导致意外效应的发生。其中导电效果的实现是通过向其中添加适量银粉，添加量一般为胶水体积的 1/3～1/2。

河南农业大学经过反复实验，研制了国产银胶，并取得了国家发明专利（专利号 ZL201010610889.7），经过不少单位试用，其导电性和黏性基本与进口银胶相当，可以满足国内科学工作者 EPG 研究中的银胶需要，且实验成本更低，购买周期更短。银胶大致由银粉、水基化液体纸胶、纯净水和聚乙二醇辛基苯基醚（TritonX-100）科学组配而成。其各成分的质量百分比范围分别为：银粉为 34.74%～44.36%，水基化液体纸胶为 31.20%～43.43%，纯净水为 18.48%～31.20%，TritonX-100 为 0.10%～0.20%。

（二）银胶的配制

1. 材料

（1）水。

（2）普通胶水。十分重要。一般使用普通透明的胶水（图 5.5），但不是所有胶水都适用。多尝试几个品牌。可以

图 5.5　适合配制银胶的胶水

（引自 Manual of Giga-8d）

先少量配制。制成后，在显微镜下观察银粉和胶水的混合情况。混合均匀者为佳。

（3）去垢剂。Triton×100 等（100mL 液体 1 滴足够）。

（4）银粉（细颗粒或片，小于 10μm）。银粉价格比较贵，且用量不多，请购买最小量。

2. 配制（10mL 银胶，使用电子天平）

（1）轮流将下述材料称量到一个容积大于 20mL 有盖小瓶（瓶盖可拧紧）中：4g 水，4g 胶水，4g 银粉。

（2）将材料充分混合（建议使用电动振荡混合器）。

3. 注意

将水和去垢剂配成储备溶液。100mL 水加上 1~2 滴去垢剂。

小贴士

1. 新配制的银胶性能最好，因此银胶要随用随买，一次性不要大量购买。每次配制 1~2 管（2mL 左右）可以使用半年左右。有效期 6 个月。

2. 购买到的银胶，可以分装在几个小型带盖的离心管中，盖紧并用封口膜密封。每次只取出一份（约 200μL）日常使用，其余部分 4℃ 下冷藏保存。

3. 银胶为水溶性的，在储存、静置中容易沉淀。使用前，请充分振摇或在振荡混合器上混匀。每次使用后，仍须密封低温保存。

4. 每次使用完银胶后，马上盖上容器盖子，避免水分蒸发。实验时，不要将装有银胶的 EP 管敞口放置管架上。

5. 使用较长时间的银胶，随着水分的不断蒸发，银粉会形成沉淀附着到管壁上，大大影响使用效果。为了节约，不用丢掉这些银胶，可以添加适量蒸馏水，在振荡器上混匀，还可以使用一段时间。注意：如果干透，即使再添加水分，也无法使成分和属性恢复。

6. 更换使用一种银胶时，如果金丝、银胶与昆虫粘贴稳定且牢固，但 EPG 系统中大部分通道的波形仍然显示接触不良，请将 2mL 银胶加入 1 滴洗洁精后充分混匀，之后波形会有所改善。

五、金丝的选用和粘连昆虫技巧

（一）金丝的选择

一般情况下，昆虫电极的端部为金（铂）丝，现在也有实验室尝试用镀银碳纤维（卢少华，2015）。选择金丝作为电极的原因在于其良好的导电性、延展性和较低的价格。适用于 EPG 实验的市售金丝一般为直径 12.5μm 和 18μm 两个规格。金丝会在一定程度上影响昆虫的活动，有时也可能影响昆虫的取食，所以应尽量减少连接金丝对昆虫取食行为的干扰。故选用金丝应遵守如下原则：①电极金丝与实验昆虫的连接牢靠，以保证实验的顺利进行；②金丝附着于实验昆虫背板不至于明显影响实验昆虫的活动；③控制实验成本。一般情况下，EPG 实验中选用金丝的基本原则是金丝的直径与实验昆虫的大小相匹配，即实验昆虫个体和力量较大时，如盲蝽、桃蚜等，选择粗一些的金丝；而实验昆虫较小时，如棉蚜、粉虱等，可选用细金丝。虽然选用直径 10μm 左右的细金丝在与烟粉虱类较小昆虫连接过程中有一定的难度，但是在使用这种金丝后，实验的成功率由原来的 40% 提高到将近 100%（汤清波等，2011）。金丝长度一般应该在 0.5~5cm，一般而言，蚜虫等金丝长度为 2cm，叶蝉或飞虱则需要 5cm，而体形较小的粉虱只需要 0.5~1cm 长的金丝就足够了。

（二）金丝与实验昆虫的连接

金丝与昆虫的连接，就是将昆虫电极金丝的末端蘸取少量银胶，利用银胶将金丝固定在实验昆虫的背板上（图 5.6）。昆虫电极与实验昆虫的连接是 EPG 实验操作的难点之一，初学者很难高质量地完成（岳梅等，2005），金丝表面比较光滑，横截面较小，很难聚集银胶，而实验昆虫的连接需要一定的接触面积才能保证其牢固性。有研究证明，金丝附着在实验昆虫背板上会影响其活动和取食行为（Prado and Tjallingii, 1999），但是精准的操作会尽量降低这种影响，对实验昆虫取食行为进行记录。昆虫电极的充分准备是提高昆虫电极与实验昆虫的连接成功率和质量的保证。

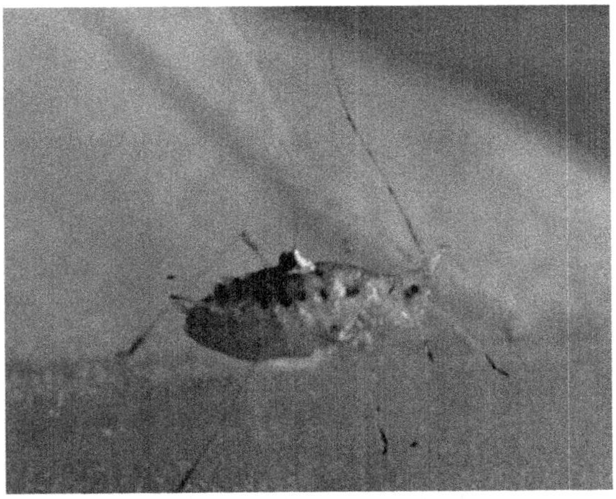

图 5.6　金丝和银胶连接到蚜虫的背部

（引自 www.epgsystems.eu）

对于电极的处理主要是如何在有限的接触面积上更多地黏附导电胶，以便更好地粘连实验昆虫。将金线端部浸入银胶中，抽出后让附着的银胶在空气中干燥 1~2min，再将金丝端部浸入银胶，如此反复至金丝端部形成一个银胶球。对于不同的实验昆虫，银胶球的大小也应不同，像粉虱等小型昆虫，银胶球就需要尽可能小，不然不仅会对昆虫造成负担，影响其取食，还有可能在粉虱活动的过程中使银胶球接触到植物叶面造成实验失败（汤清波等，2011）。

除了上述多次蘸取导电胶形成胶球的办法外，也可将金丝端部弯折成钩状，直接扩大与实验昆虫背板的接触面积，或利用导电胶的表面张力形成的膜使得粘连更加简便、牢靠。对于体形较大的昆虫，也可以直接使用解剖针或铜导线（也可使用移液器）蘸取 1 滴银胶，滴在昆虫背部，将昆虫电极的金线端插入到银胶里面，等待其干燥。

（三）昆虫与金丝电极连接后的适应问题

金丝粘连后的昆虫不要直接放在EPG实验所用植物上，而应该另备一株植物用于让昆虫适应背部粘有银胶和金丝的情况，等所有实验昆虫都与金丝连接完毕后，再进行饥饿处理。这个过程需要30~60min，视不同昆虫情况而定，虫体大的、生命力强的时间相对长一些；虫体小的、生命力弱的，时间相对短一些。实验前的饥饿处理对于实验的成功与否至关重要，饥饿处理的时间过长，会造成实验过程中G波出现频繁，饥饿时间过短会使第一个刺探前的时间太长。经过饥饿处理后的昆虫，应在EPG仪器开启前同时放在记录植物上。

有的学者不赞成在连好金丝后进行饥饿或适应处理，而认为应该把连好电极的昆虫直接进行记录实验。这样在分析这些记录数据时，应该把记录开始到开始刺探前的这段时间去掉，因为先后连接金丝的昆虫所受的干扰时间和程度不同，可能使得在记录植物上的刺探开始前的适应期不同。

不管如何选择，一定要保持对照组和处理组之间和组内处理一致。

六、EPG记录

一般情况下，刺吸式昆虫都是在有光照的条件下进行取食的，因此，EPG实验需要在有足够光照的条件下进行，如果室内光源不足，可以在法拉第笼外增加辅助光源。而且进行EPG实验的时间应尽可能和所饲养昆虫的光周期保持一致。

EPG仪器与制备好的昆虫电极连接并记录的步骤如下：

（1）将实验植物放置在法拉第笼内的合适位置。

（2）注意开始连接前，将双手接触法拉第笼的金属部分，释放静电。

（3）确定Giga的电源开关处于“关闭”状态。

（4）将已经连接实验昆虫的昆虫电极接入到EPG昆虫电极接口（图5.1-②）内，如图5.7所示，调整转夹（图5.1-2a）使其位于植物附近一个合适的位置，使实验昆虫可以小范围在植物上面爬动。

图5.7　将昆虫电极插入接口

（引自Manual of Giga-8d）

（5）将对应植物电极插入至该植物培养盆的土壤中。

（6）重复该操作直至所有通道都装好昆虫电极和植物电极。当然，如果不需要使用全部通道也可以，但最好使用标号靠前的通道，如只有3只昆虫，使用通道1至通道3。

（7）打开Giga的电源开关。打开软件开始数据采集。

（8）调整植物调节旋钮使记录到的 EPG 波形处于合适的状态。

（9）关闭法拉第笼。

（10）记录结束后，关闭软件的采集模块。打开法拉第笼。关闭 Giga 的电源开关。取下昆虫电极和植物电极等配件。取出实验植物，清理实验台。

连接好后的 EPG 系统如图 5.8 所示。为了展示清楚，法拉第笼未关闭。

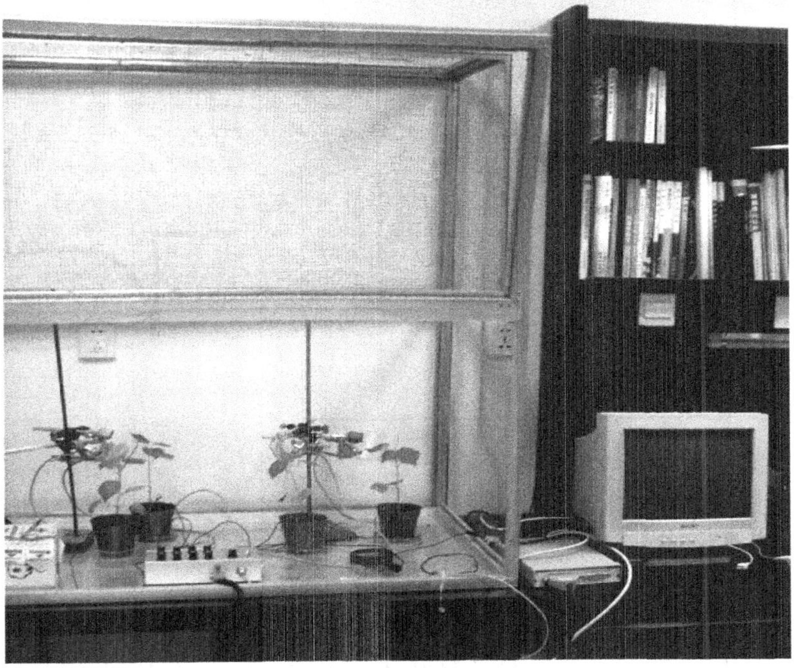

图 5.8 连接好后的 EPG 系统

小贴士

1. 要想获取比较完美的波形，还要对植物电压和放大系数进行调校。一般而言，应该使植物电压的指针处于中间稍偏向正的位置，放大系数为 50 倍比较适宜。

2. 记录 EPG 波形时，当昆虫发生刺探行为时，观察 pd 波的波形，如图 5.9-A 所示，pd 波应该是比较规则的类似正方形脉冲，则说明银胶连接正常。如果出现类似电容干扰，则说明接触不良。如果短时间没有记录到 pd 波，在昆虫发生刺穿行为时，按下 Giga 上的脉冲校准按钮时，如图 5.9-B 所示，输出信号脉冲应该是规则的正方形脉冲，则说明银胶连接正常。如果只能看见脉冲的边缘内，或正方向脉冲边缘不规则，说明接触不良。

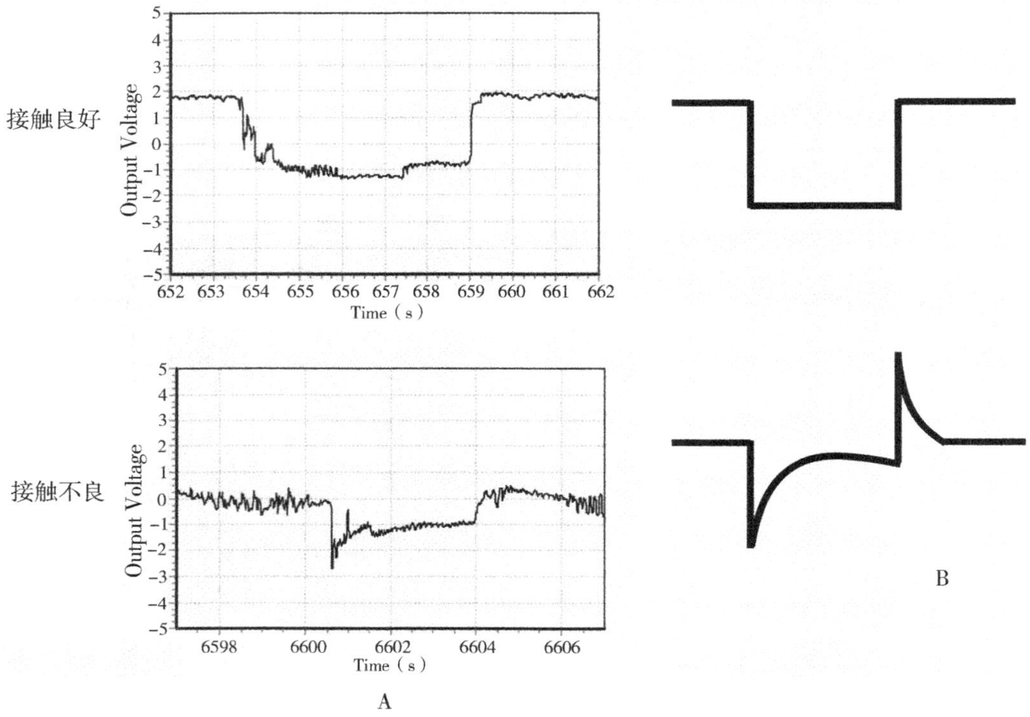

图 5.9　接触良好和接触不良的 EPG 波形

七、实验昆虫的固定与负吸装置的设计

（一）实验昆虫的固定

在连接实验昆虫与昆虫电极连接时，像蚜虫这类动作缓慢、活动能力较弱的昆虫可以直接用小毛笔固定，但像飞虱类体形较大、活动能力强的昆虫，往往不易固定，导致连接效果不好。除了之前章节提到的昆虫电极的准备，对实验昆虫处理也能提高昆虫电极与实验昆虫的连接成功率和质量。对实验昆虫的处理的主要目的是降低实验昆虫的活动性，确保电极与实验昆虫粘连的顺利进行。具体的方法主要有三种，即低温、CO_2 麻醉和负压固定等。

低温法可以降低实验昆虫的活动性，其主要原理是昆虫在环境温度低时其活动性下降，甚至会出现冷昏迷的情况，在温度回升后可以顺利复苏或恢复活动性。这类方法使用较为简便，对于蚜虫或粉虱，可以将培养皿置于−20℃冰箱预冷或将其置于预冷的冰袋上，将实验昆虫置于培养皿中即可实现降低其活动性的目的。对于中大型实验昆虫，预冷培养皿或将培养皿置于冰上可以提供的低温时间较为短暂或者温度不够低，则可以将装有实验昆虫的培养皿直接置于冰箱冷冻室中数十秒以达到同样的目的。使用这类方法需要注意低温的程度和时间，温度稍高或时间过短不能有效降低实验昆虫活动性，温度过低或时间过长则会造成实验昆虫的死亡或昆虫无法恢复正常活力。所以必须进行预实验以确定实验昆虫的冷冻温度范围和时间。另外，冷冻过程会妨碍取食活动的实验昆虫不宜采用这类方法。例如，烟粉虱在冷冻室中经过 20~30s 即不能活动，取出置于操作台上即可进行电极连接；褐飞虱不宜采用冷冻法，因为经过冷冻的褐飞虱，恢复活动后取食能力较差，对实验结果影响较大（闫凤鸣，2003）。

CO_2 麻醉法需要营造一个高浓度 CO_2 的局部环境，在持续一定时间后，实验昆虫活动性减弱，局部高 CO_2 浓度环境消失后，实验昆虫会逐渐恢复活性。这种方法对实验设备有一定要求，实际使用范围有限。

负压固定一般是采用某些可以产生负压的装置，如真空泵、气流抽气泵（水老鼠）等，产生一个适当的负压，将实验昆虫腹部置于负压装置的吸头处固定，方便蘸有导电胶的电极金丝粘连实验昆虫，如图 5.10 所示。利用负压装置进行实验昆虫固定必须调节负压大小到适当范围，压力过小则不能固定实验昆虫，压力过大则可能造成实验昆虫死亡。合适的负压标准是既可以固定实验昆虫，又能够使得实验昆虫在粘连完成后可以被电极顺利拖离负压吸头。这种负压装置一方面可以帮助限制实验昆虫的活动，另一方面也可以检验金丝粘连的牢固程度。可以在粘连完成后，稍等几分钟，银胶凝固后，关上负压，将昆虫拖离负压吸头。

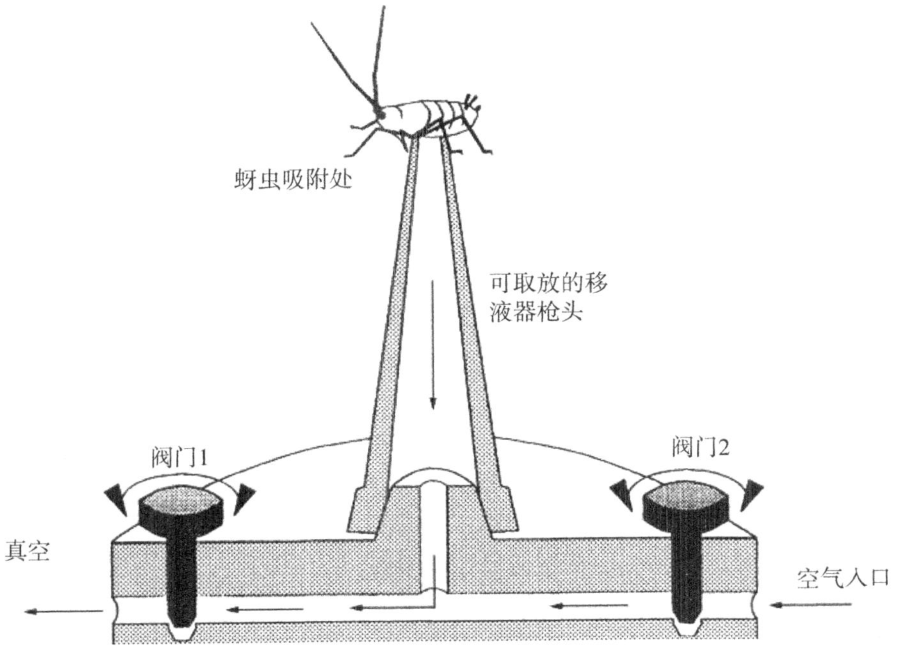

图 5.10 用于固定蚜虫的真空装置

（引自 Van Helden and Tjallingii，2000）

综上所述，固定实验昆虫时，推荐使用负吸法，尽量不要用 CO_2 麻醉法和低温冷却法。

（二）负吸装置的设计与制作

一般而言，真空泵、硅胶管、流量计、泡沫、移液枪头等即可组成一套负压装置，如图 5.11 所示，在粘连昆虫前，打开真空泵，将昆虫放到有泡沫板固定的、带有海绵的移液枪头端部，调整好昆虫的姿态，尽量使昆虫背板正对着操作者，这样便于快速、准确地粘连昆虫。昆虫种类、体形大小的不同，需要的负吸压力也不同，可通过调节流量计的空气流量来实现，如图 5.12 所示。

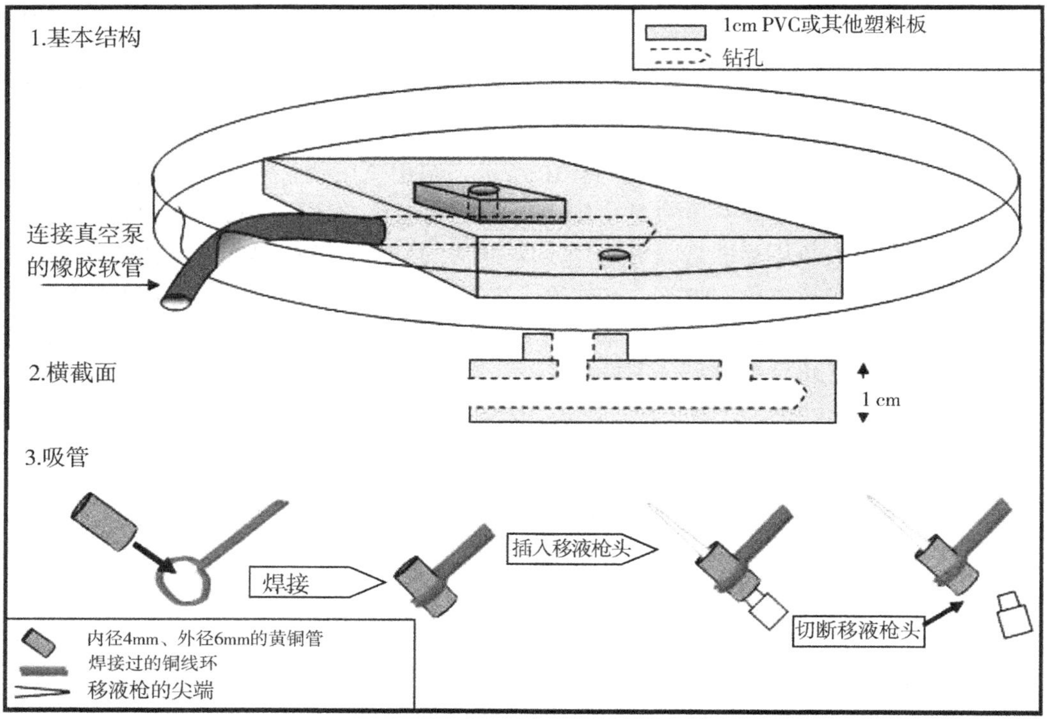

图 5.11　负吸装置制作示意

（引自 Manual of Giga-8d）

利用负吸装置粘贴昆虫操作步骤：

（1）将实验昆虫置于移液枪枪头的顶部，用剪刀或刀片去除枪头顶端的一部分，使枪头顶部大小正好适合昆虫体形大小。

（2）通过滑动纸片，调整吸力大小。

（3）转动把手将其转到一个合适操作的位置。

（4）将银胶粘到昆虫背部。

（5）将金丝插入银胶中。

（6）等待银胶干透。

（7）移开纸片，移走昆虫。

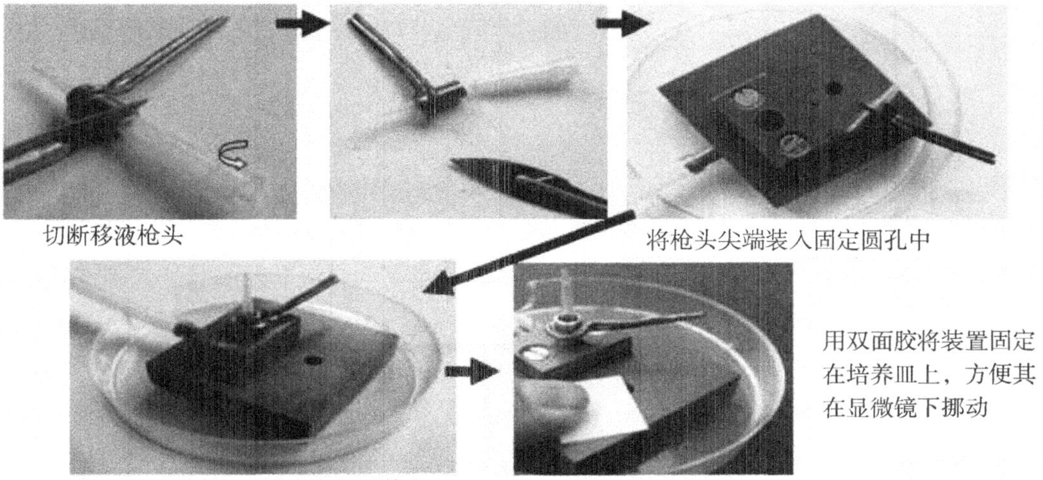

图 5. 12 负吸装置制作实物及操作图

（引自 Manual of Giga-8d）

八、法拉第笼的设计与制作

法拉第笼（Faraday Cage）又称为金属屏蔽网，是一个由金属或者良导体形成的笼子，是以电磁学的奠基人、英国物理学家法拉第命名的一种用于静电屏蔽的设备。通常来说 EPG 实验要求较低，只需要在平整的实验台上记录，并避免放置于振动的环境即可。法拉第笼则可以自行设计制作或委托承接金属架制作的小型加工点代为制作，价格比直接购买公司成品更为低廉。另外，为了良好可控的光照并避免自然光线的影响，可以在法拉第笼外侧上方安装数控电灯。

简单的法拉第笼可以使用普通金属网来制作。金属网选择镀锌或无包裹层的品种，不要使用有塑料涂层或有涂漆层的品种。如果受条件所限，只有涂漆层的金属网，可以用砂纸将金丝网边缘的部分涂层除去，这样可以确保笼面电接触点接触良好。一般而言，国内普遍采用铜网。金属网可以使用木头或金属的框架来支撑，如果是用木头框架支撑，每一面的金属网都应该互相接触，使法拉第笼形成一个整体。制作好的如图 5. 13 所示法拉第笼可以直接固定到实验台面上。还有一种更低成本的制作方法，如图 5. 14 所示。在天花板上安装滑轮组，用绳索和平衡锤（counter weight）将法拉第笼固定在天花板上。这样可以快速将法拉第笼拉起，之后形成开放的台面，方便实验者从各个方面接触到实验植物和昆虫等。但是需要注意的是，实验时放下法拉第笼后，底部的接地需要整理。另外，还可以直接以坚固的大网眼金属网制作成较大的桶，再在上面覆盖上合适的金属网，固定好。使用时，将金属桶倒扣在 EPG 系统上，代替法拉第笼的作用。这种方式成本低廉，制作简单，但是坚固性和实验稳定性差，可以作为预实验或应急使用。

图 5.13　法拉第笼

（引自东乐公司网页 www.bio-equip.com）

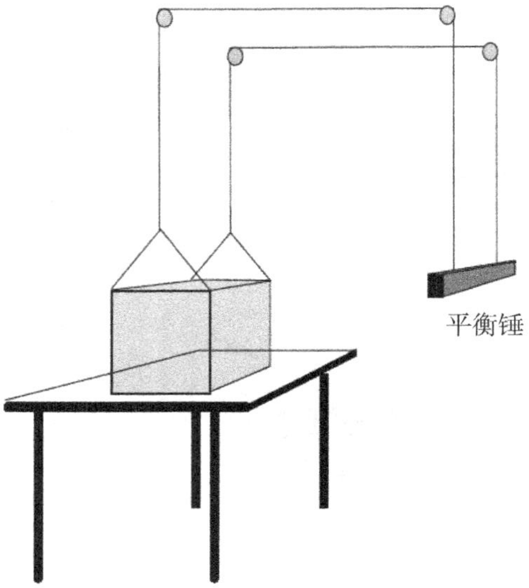

图 5.14　一种法拉第笼的制作方法

（Manual of Giga-4/8）

不论上述哪种方法，金属网的选择都是一致的。为了较好地屏蔽电磁干扰，一般选用网眼为 1cm 以下的金属网。另外不要选择网眼小于 2mm 的金属网，这样金属网成本较高且光线不易进入到法拉第笼内，影响实验植物的光合作用。

法拉第笼的大小需要将实验植物和昆虫电极以及支持用的铁架台全部包括入内。EPG 系统的控制单元也可以放置在笼外，这样可以在 EPG 记录时方便地操作、调节植物电位。

底部也是法拉第笼的一部分，不能简单以实验台面代替，否则会导致电磁干扰从底部引入，它必须与笼子的其他金属部分互相接触，形成一个整体。底部可以用放置同样的金属网，也可以直接用一块没有涂层的厚铁板（0.8～1cm）代替。这样固定和支撑昆虫电极时，可以用稳定性和接地性更好的磁力表座代替普通的铁架台。另外还可直接在底部铺上一层铝箔，这样更方便、便宜，但是铝箔很容易被破坏，需要不断更换。

小贴士

1. 法拉第笼的前部可做成可移除式，打开时可以方便实验操作，关闭时能减少噪声。可以安装窗帘一样的拉索，上下拉动柔软的金属网或者直接用钩子固定。如果框架材料采用铁质的，也可以直接用吸铁石固定。不管哪种方法都要注意与笼体其他部分的接触。

2. 在 EPG 记录时，不要让植物或土壤直接与法拉第笼底部电接触。最好将花盆放置在塑料或陶瓷材料的绝缘小碟子上。

3. 在 EPG 记录时，如果实验需要用体式显微镜观察植物或昆虫，法拉第笼要么足够大，将显微镜放置在笼内观察；要么足够小，将显微镜放在笼外观察。

（罗　岚）

第六章　EPG 实验设计、指标选择及数据分析

一、EPG 实验设计的一般原则

刺吸电位（EPG）技术是研究刺吸式昆虫取食行为的一种常用技术，它可以记录刺吸式昆虫的口针插入植物内之后的所有活动。现在的 EPG 实验已不仅仅停留在简单记录这些昆虫在植物上的取食行为，而是以 EPG 技术为依托，通过这些取食行为，深入到筛选抗性植物品系，探索刺吸式昆虫传播病害的机制和途径控制，研究这些昆虫寄主转化的原理和过程等研究中了。EPG 实验需要的生物重复数多，处理时数据量大，分析各种指标多，总之，EPG 实验需要大量的时间成本。所以在进行 EPG 实验时，最重要的是：明确实验目的，了解 EPG 实验能解决的问题，并以此为基础进行实验设计。

EPG 实验虽然记录的是试虫在取食过程中的电位变化，属于电生理方面的内容，但是其目的是为了研究刺吸式口器昆虫的取食活动，本质上是行为学实验，所以从实验设计到数据分析需要遵从行为学研究的基本范式。行为学研究不同于严格的实验室微观实验，主要特点是个体差异大，实验重复性差，随机因素和系统差异在结果中共存，数据表现为方差大。因此，EPG 实验的总体原则就是在实验材料尽量均一的情况下，通过大样本量重复来减小大方差造成的影响。

根据实际操作经验，我们可以总结 6 条 EPG 实验设计中的原则：

1. 实验材料均一　人工饲料的均一性没有问题，当以生活植株作为昆虫食物的时候，植株的生长环境等外界因素，生长阶段、生长状况等植株本身的特征均要保持一致。试虫的均一性即要保持所用昆虫的饲喂方式一致，生长阶段和生长状况一致，昆虫电极粘连之后做相对一致的处理。总之，就是在尽可能的情况下避免差异的存在，减小随机因素对结果造成的干扰。

2. 对照的选择需要以实验目的为基础　EPG 实验毕竟不是纯粹的行为学实验，其具有一定的可控性，EPG 实验的对照选择是一个关键环节。通常情况下，可控实验中选择对照的目的是为了证明实验系统的有效性（阳性对照、空白对照），也是为了凸显实验结果（阳性对照和阴性对照）。在行为学研究中，因为条件问题，通常可能只有实验处理而没有严格的对照。EPG 实验中，我们一般也不设置阳性对照和空白对照，只是根据实验的不同对实验组进行不同处理以达到对比的目的。例如，在研究试虫在不同处理的植株上的取食活动差异的时候，除了植株的差异外，尽量避免其他可能的差

异来源，这样得到的结果才能够说明问题。

3. 昆虫电极随机使用，记录通道安排不固定 在实验过程中，昆虫电极粘取试虫会影响到试虫的取食活动，每个自制的昆虫电极都有其特殊性，对试虫的影响也可能存在差异，固定使用某个昆虫电极可能会造成结果中包含昆虫电极的系统差异。同样，记录通道也可能存在同样的问题，如植物放置位置、室内光线、靠近窗户或门口导致的温度差异和干扰等造成的误差。因此，在实验中，随机使用各个电极，不固定安排通道，在样本量足够大的情况下可以在一定程度上避免电极和通道的差异导致的结果偏差。如果需要更加严格的实验设计，则可以将昆虫电极和通道编号，以随机数表或者计算机程序产生随机数为基础进行排列。例如，要利用四通道的 EPG 装置（如Giga-4）记录一种昆虫在 4 种寄主植物上的取食行为，就是 4 个重复，分别编号为 A、B、C、D，利用数理统计随机表（表 6.1），将 4 株植物在 EPG 记录通道的安排列成表格，记录在实验记录本上，每次实验都要按随机顺序安排记录通道，而不是在实验的时候任意放置。

表 6.1　4 株植物在 EPG 记录通道的随机排列

重复	通道			
	1	2	3	4
1	A	B	C	D
2	C	B	D	A
3	B	D	A	C
4	D	C	B	A
5	A	C	D	B
…	…	…	…	…

4. 具有一定的样本数量 由于行为学研究一定程度上比分子实验中处理组内数据方差较大，因此行为学实验对于处理组内的数量是有要求的。一般的实验室微观实验要求至少 3 个重复才能达到数据统计的基本要求。在 EPG 实验中，“昆虫—植物”组合之间所产生的波形早晚和持续时间长短等指标数据，会有比较大的方差。由于 EPG 实验要求每次重复都应该使用新的昆虫个体和新的植物个体，因此，方差的产生既来源于植物个体之间的不同，也来自于不同昆虫个体的表现。EPG 实验则需要更多的样本量才可能得到可信的结果，样本量越大，结果越准确可信，但是考虑到实验成本和劳动量的大小，一般经验认为，约 20 个可用结果即可通过数据统计对实验目的进行有效说明。在实际操作中，有许多实验记录由于各种原因不能使用，建议每个处理记录多次，从中挑选出大约 20 个成功的记录进行统计分析（闫凤鸣，2003）。当然，在考虑样本数量的同时，样本的独立性必须保证。

5. 选择合理的记录时间 EPG 实验记录时间的长短，主要取决于实验的目的。如果只是研究寄主植物表面或叶肉层次物理化学性质对昆虫取食行为的影响，一般 30min 到 3h 的记录时间就足够了；而研究昆虫传播非持久性病毒的机制，最初几个 pd 波是十分重要的，记录几分钟就可以了，而研究传播半持久性或持久性病毒的机制，则需要几个小时（昆虫传毒机制请参考本书第二章）。一般昆虫与植物关系的研究，目的在于探讨植物次生物质在植物的哪个层次对昆虫的取食行为有影响，建议记录6h。如果想要得到昆虫取食的全部波形，就需要给昆虫足够的时间去完成这些取食行为。在蚜虫EPG实验中，从实验记录开始到第1个E2波的出现是一个较长的过程，通常情况下需要几小时。一般情况下，第一次韧皮部取食的时间较短，后来的韧皮部取食时间都会较长。一个理想的决定EPG记录时间的方法是，通过观察大量昆虫在寄主植物上的第一滴蜜露的分泌，即第一次到达韧皮部的取食时间，然后重复这个时间用来进行EPG实验。有研究证明，记录16h和8h，在结果上并没有显著的差异。8h大约相当于从实验开始到第一个E2波出现时间的2倍。一般的研究，6~8h的记录是比较合适的。当刚刚开展实验时，由于不太了解昆虫和植物之间的关系，可以稍微延长记录时间。例如，EPG实验记录了9h，但实际需要分析前8h的数据，这些EPG数据仍然属于有效数据。但是相反，EPG记录了6h的数据，之后发现应该分析8h的数据，那么之前记录的6h的数据由于时长不够不属于有效数据，白白浪费了工作量。

6. 记录结果的选择遵循统计学原则　生物学实验，尤其是行为学实验，由于存在着较大的方差，经常会被认为存在异常值而舍弃本来可用的实验结果，进而造成实验结果因为人为挑选而不可信。在EPG实验中，我们一般遵循统计学上的原则来对实验值是否异常进行判断。一个简单的标准是实验值与均值之差和3倍标准差（SD）进行比较，当$|S|<3SD$时，该值不能舍弃，当$|S|>3SD$时，该实验值可以认为是异常值。

二、EPG实验数据分析

EPG实验得到的结果是整个记录过程中的电势变化，是以波形的形式呈现的。获得EPG数据后，将EPG波形的内容“读”出来，才能够分析EPG记录的取食活动。一般来说，整个EPG记录波形是一个连续的过程，一个波形的结束就是下一个新波形的出现（图6.1）。即使昆虫未进行刺探或取食，其EPG波形也显示为np波。利用$Stylet^{+}$软件的分析模块，可以将EPG波形进行标注（具体方法参见第七章“EPG常用软件使用指南”）。

实际分析中，有以下几点需要注意：

如图6.2所示，从np波到A波的转变十分明显，但是有的波形的识别有时并不清楚，尤其是某些连续发生的波，如A波、B波和C波，且pd波发生在C波范围内，所以上述波形可统称为路径波（广义C波）；从广义C波到其他的波形转化一般比较明显。

某些波形之间存在过渡状态，如图6.3所示，大部分E1波和E2波都有明显特征，E1波和E2波之间往往存在过渡阶段（图中灰色部分），不好分辨和划分，比较简单的做法是取其中的中间点作为E1波和E2波的分界。

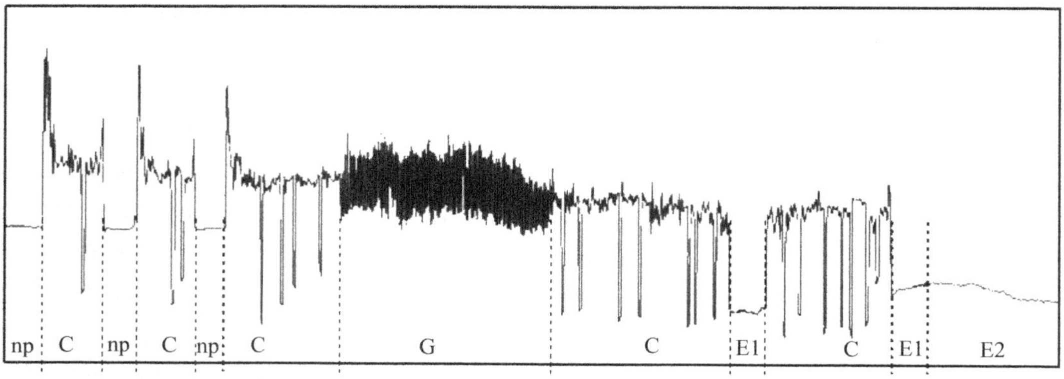

图 6. 1 一段 EPG 波形（W. F. Tjallingii，2010 - 3rd Chinese EPG course）（1）

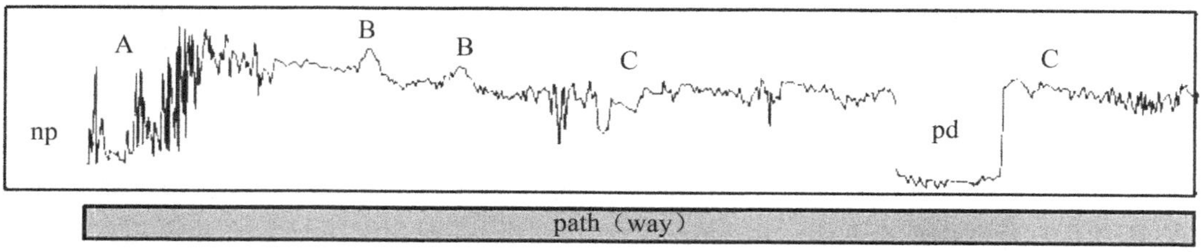

图 6. 2 一段 EPG 波形（W. F. Tjallingii，2010 - 3rd Chinese EPG course）（2）

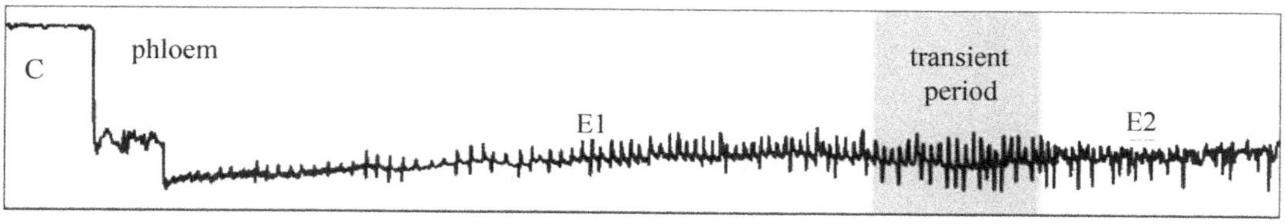

图 6. 3 一段 EPG 波形（W. F. Tjallingii，2010 - 3rd Chinese EPG course）（3）

pd 波在一般的实验中分析起来生物学意义不大，但是 pd 波大量充斥在 C 波中，如果没有特殊的实验要求，或分析结果需要针对某个问题（如传播非持久性病毒机制），可以将其直接并入 C 波计算，不需要单独标出，这样可以节省大量的分析时间。其他某些与最终目的无关的波形可以忽略，如 F 波。当 EPG 波形分析标注完毕后，就可以开始选择实验指标了。

充分了解各个波形的特征和生物学意义才能更好地对 EPG 波形进行具体分析（主要昆虫的 EPG 波形的生物学意义，请参考本书第四章）。

三、EPG 实验指标选择

将 EPG 波形的内容“读”出来后，需要考虑选择合适的指标（parameter 或 variable）来解读 EPG 记录结果。

EPG的指标分为两类：第一类为“与取食时间顺序无关的指标”（non-sequential variable），也可以称为通用指标（general variable），如某种波发生的总次数或平均持续时间、非刺探时间占总记录时间的百分比等，主要用于考察昆虫在某种植物上总的取食行为表现，或能够得到植物抗性方面的总的信息。第二类是“顺序性指标”（sequential variable），即以波形发生的时间顺序，把整个记录过程分为三个阶段：第一阶段是从刺吸开始到第一个E1波出现（反映叶表面性质，叶表皮和叶肉部分的性质），该阶段试虫寻找合适的刺吸位置，在整个阶段，EPG波形以路径波为主；第二个阶段是从第一个E1波出现到第一次韧皮部持续取食（E2波长于10min）（反映叶肉部分的性质），这个阶段的波形还是以路径波为主，间或出现短暂的取食波（以E1波为主）；此后至记录结束为第三阶段（反映韧皮部的性质）。本阶段情况相对复杂，可能出现全部为E波的情况，也可能是路径波与取食波交替出现，甚至是路径波占优势的情况。另外，在记录过程中可能出现代表取食木质部的F波，可以根据实验的需要计算或者排除该段波的影响。

各个阶段的不同指标，可以更细致地获得植物在不同层次（表皮、叶肉、韧皮部）对昆虫的抗性或刺激因素的信息，如“从记录开始到发生第一次E波的时间”“第三阶段内E1波发生的总时间”等。EPG指标选择的例子如图6.4和表6.2所示。

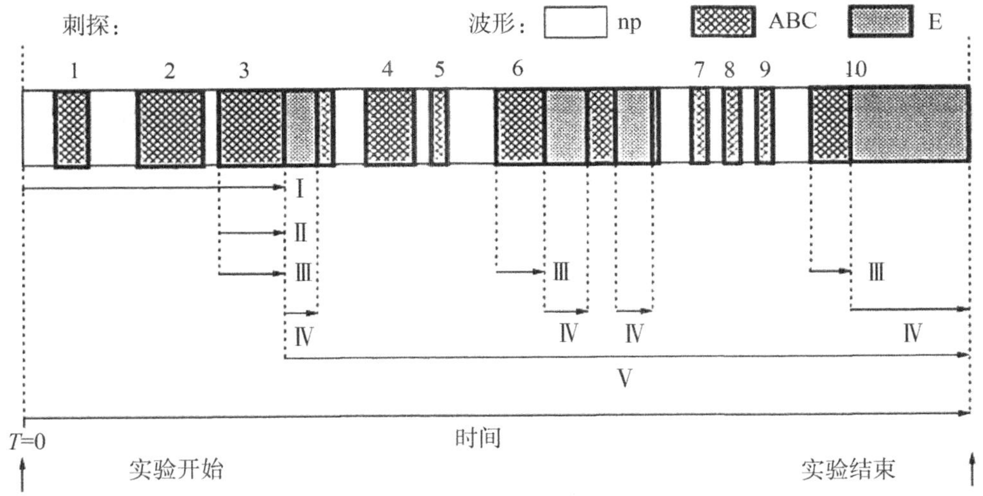

图6.4 DC-EPG分析中常用指标的含义示例（引自Helden et al.，2000）

将蚜虫放在植物叶片上以后，记录就开始了（$T=0$）。一段时间以后（第一个非刺探时间，1st np），刺探开始。通常先有几个没有E波的刺探（1和2），然后，开始有E波的刺探（3）。从记录开始（$T=0$）到第一个E波（1st E）的时间（箭头Ⅰ，指标Ⅰ）一般包括几个刺探。从包含第一个E波的那次刺探（3）开始到E波出现的时间为第二个指标Ⅱ（箭头Ⅱ）。可能有好几个刺探都含有E波，时间各不相同（箭头Ⅲ，指标Ⅲ）。通常用E波的总时间（箭头Ⅳ，指标Ⅳ）占E波出现后直到记录结束的总时间（箭头Ⅴ，指标Ⅴ）的百分比来表示出现的频率。

EPG实验指标需要考虑其所在阶段所代表的生物学意义。一般可供选择的指标如表6.2所示。需要说明的是，作为行为学实验，单一的实验指标对问题的说明是有限的，只有将一系列反映相关情况的指标结合起来，才能对结果进行正确地解读。具体到每个实验要解决的问题，并不需要对全部指标进行考量，根据需求自行选择能够说明问题的指标进行综合分析即可。

表 6.2　可供选择的 EPG 实验指标（Xue et al.，2008；2009）

阶段	指标	相关的植物组织层次
通用指标	刺探的总次数	表皮、叶肉和实质
	pd 的总次数	细胞内
	E1 波的总持续时间（s）	韧皮部
	E1 波的百分比（%）	韧皮部
	E2 波的总持续时间（s）	韧皮部
	E2 波的百分比（%）	韧皮部
	np 的总持续时间（s）	表面
	np 的百分比（%）	表面
	np 的总次数	表面
	刺探的总持续时间（s）	表面、表皮和叶肉
	刺探的百分比（%）	表面、表皮和叶肉
	G 波的总持续时间（s）	木质部
	G 波的百分比（%）	木质部
	发生 F 波的蚜虫比率（%）	木质部
第一阶段	第一次 E 波发生的时间（s）	除韧皮部外的其他植物组织
	第一阶段刺探次数	表皮、叶肉和实质
	第一阶段刺探次数占总刺探次数的比率（%）	表皮、叶肉和实质
	第一阶段刺探的百分比（%）	表皮、叶肉和实质
	第一阶段的 pd 降落数	细胞内
	第一阶段 pd 数占总 pd 数的比率（%）	细胞内
	第一次刺探的发生时间（s）	表面
	第一阶段 np 的百分比（%）	表面
	第一次刺探的持续时间（s）	表面、表皮和叶肉
	第一次 pd 的发生时间（s）	表面、表皮、叶肉和实质
	第一次刺探即可到达韧皮部的蚜虫比率（%）	表面、表皮、叶肉和实质
第二阶段	第一次持续取食的发生时间（s）	所有植物组织
	从第一次刺探到第一次持续取食的时间（s）	所有植物组织

续表

阶段	指标	相关的植物组织层次
第二阶段	第一次持续取食前路径波的持续时间（s）	除表面以外的所有植物组织
	第二阶段刺探的百分比（%）	表皮、叶肉和实质
	第一次持续取食前路径波当中的 pd 数	细胞内
	第二阶段 pd 数	细胞内
	第二阶段 E 波发生次数	韧皮部
	第二阶段 E 波百分比（%）	韧皮部
	第二阶段 np 的百分比（%）	表面
	第二阶段的持续时间（s）	所有植物组织
第三阶段	第三阶段持续时间（s）	所有植物组织
	第一次持续取食之后 E 波的次数	所有植物组织
	第三阶段 C 波百分比（%）	表皮、叶肉和实质
	第三阶段 pd 数量	细胞内
	持续取食前的路径波中 pd 的平均数	细胞内
	第一次持续取食的持续时间（s）	韧皮部
	持续取食的次数	韧皮部
	第三阶段 E1 波的百分比（%）	韧皮部
	第三阶段 E2 波的百分比（%）	韧皮部
	第三阶段 E1 波和 E2 波总和的百分比（%）	韧皮部
	第三阶段刺探的次数	韧皮部
	第三阶段 np 的百分比（%）	表面

（罗　岚　薛　堃　卢少华）

第七章　EPG 常用软件使用指南

一、EPG 常用软件介绍

EPG 波形不但在不同物种和不同个体间差异很大，同一个体不同刺探，甚至是一次刺探中都有可能出现明显的变化。EPG 记录的各类典型波形的频率是 0.1~75Hz，相差很大。对特定波形的识别依靠的就是频率和振幅两个特征，使用傅里叶分析（Fourier analysis）和幅值条带分析（amplitude histogram analysis）可以实现对记录结果的波形整合。以此为基础的各类实验记录和分析软件为实验的自动化和定量分析提供了很好的平台。

EPG 应用之初，在还没有记录与分析软件时，波形的记录和波谱分析完全依靠手工来进行，效率低，且具有个人偏好性；随着计算机的广泛应用，可用于自动记录的软件的出现，使得 EPG 的应用进一步扩大。

几种 EPG 常用软件包括：

1. 基于 DOS 系统的 Stylet 程序和用于 Macintosh 的 MacStylet 软件，可以完成 EPG 的记录与分析。Stylet 程序是最早的 EPG 自动记录和分析程序，但是也有一个明显的问题，就是不能实时显示记录结果。

2. Windows 系统的出现催生了界面更友好的 Probe 程序，该程序作为专门的 EPG 程序，前后有过多个版本。

3. 目前 EPG 设备推荐的软件是 Stylet$^+$，也是 DC-EPG 实验专用软件，可以实时显示波形。它是 Giga 系列刺吸电位仪的配套软件。它分为两个部分：Stylet$^+$d 模块可以记录并采集波形，Stylet$^+$a 模块可以对记录波形进行分析。该软件的一个问题是波形显示为 1h，每小时波形单独存为一个文件。

此外，还有一些通用软件可以进行 EPG 波形的记录与分析，如 Dataq Instruments Inc. 的 WinDaq Lite 采集软件和 WinDaq Waveform Browser 分析软件。

此外，还有一些使用者设计编写的程序，是研究刺吸式昆虫取食活动的有效工具，例如南京大学陈建群组开发出的 Realdisplay 程序（张闲等，2001）。

二、Stylet⁺软件及其安装

自1965年EPG技术诞生以来，先后经过多次改进，EPG技术在一定程度上已经相当成熟。而与EPG技术配套使用的电脑分析软件也几经更新，越来越方便快捷，这里我们就以目前Giga-4/8刺吸电位仪推荐软件的stylet⁺为例进行介绍。

（一）Stylet⁺软件

Stylet⁺软件是基于Windows系统（2000，XP，Me，Vista和Win7）的EPG信号采集分析软件。Stylet⁺软件不需要许可证直接从网站（www. epgsystems. eu）上下载即可。Stylet⁺的采集模块需要和Giga硬件和AD转化器（模拟—数字转化器）一起工作，并将EPG波形转化为数字信号存储在电脑硬盘中。市面上AD转化器多种多样，Giga-4和Giga-8分别配套的是两种低成本但是高品质的USB接口设备，Models Di185U和Models Di710（DATAQ仪器有限公司，DATAQ Instruments，Inc.，OH，USA）。Stylet⁺软件分析模块和PROBE分析模块十分类似，并可兼容用于分析PROBE和STYLET的数据，且不需要进行任何格式转化。信号图片能直接粘贴到PowerPoint或Word上作为出版和展示用。

（二）Stylet⁺软件及AD转换器的安装（以Giga-4/8为例）

1. 下载

（1）首先将电脑上有关EPG运行的软件删除干净。在Windows程序中找到Windaq710和Stylet⁺，删除掉C盘中的所有与EPG相关的程序，一定要删除干净（包括Dataq、EPG-2007、Probe、Stylet⁺和Windaq）。如果是首次安装，则跳过第一步，直接进行第二步。

（2）可以从www. epgsystems. eu网站（图7.1）直接下载（免费）所需安装的软件和硬件驱动old EPG install 2014. exe，保存到电脑的合适位置。

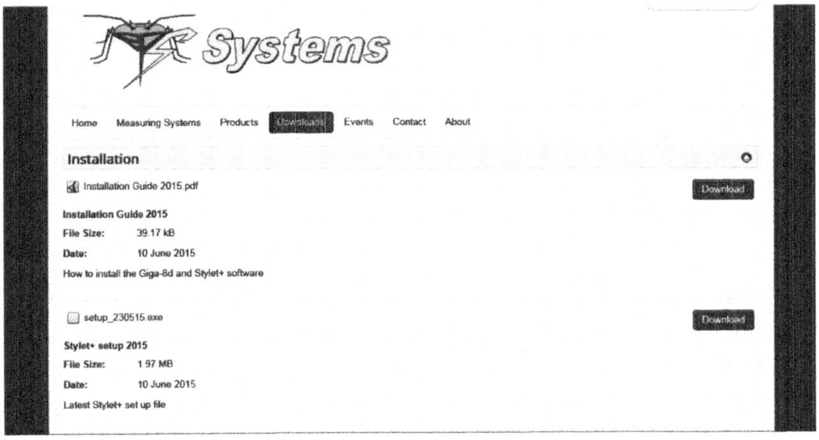

图7.1 网站www. epgsystems. eu的下载页面

（3）找到下载的程序，双击该程序。

（4）四个文件将被解压缩到 C：\ Program Files \ EPG Install 2014，如下：

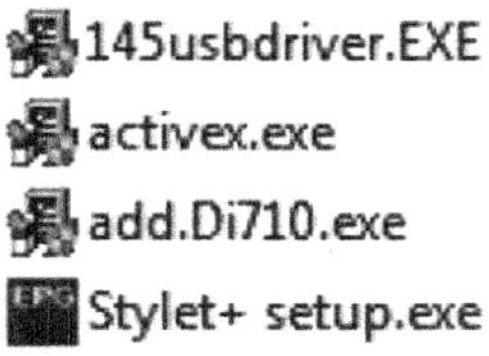

（5）如果有提示，请选择“Program is correctly installed”。

2. 安装 AD 转化器驱动程序

（1）请不要连接与 Giga 相连的 USB 线。

（2）Giga-8 请安装 add. vDi710. exe。这时会同时安装 activex. exe，先请根据提示安装 activex. exe。

（3）安装 add. vDi710. exe 请根据如下步骤进行：

接受版权声明并继续；

填写注册信息（姓名及单位，随意填写）；

按默认路径（C：\ DATAQ）安装；

选择仪器接口为 USB；

单击 OK 创建开始菜单组 WinDAQ；

同意所有用户使用权限；

不需要安装帮助软件；

点击 OK 结束安装。

（4）第（2）、（3）步骤之后，Giga-4 需另外安装 145usbdriber. exe。除了 AD 转化器的驱动外，商业 Dataq 软件 Windaqlite（免费版）也会被安装。

3. 安装 Stylet^{+}EPG 软件（包括采集模块和分析模块）

（1）运行 C：\ Program Files \ EPG Install 2014 \ Stylet^{+} Setup. exe，并根据提示安装该软件。

选择安装软件的语言（English）；

同意软件版权声明；

选择安装路径（C：\ Stylet^{+}）；

创建开始菜单程序组；

创建桌面和快速启动栏图标；

安装完成。

（2）安装该软件后所有文件都在文件夹 C：\ Stylet^{+}中，并且自己建立一个文件夹 C：\ EPG_ data 用于存放默认路径 EPG 数据。记录软件在记录结束后会生产以 . aq4，或 . aq8. 为扩展名的文件。

（3）软件安装完成后，在电脑桌面或程序栏里会出现软件数据采集和分析模块的快捷方式。

（4）将 Giga 和 AD 转化器连接至电脑，电脑桌面右下角提示发现新硬件，待自动安装驱动完成后，将提示新设备可以正常使用，软件安装完毕。

stylet⁺a

stylet⁺d

4. 更新

可以到网站上下载最新的软件。软件基本可以向下兼容。

小贴士

如果不安装 AD 转化器的驱动而直接安装软件 Stylet⁺，也能成功安装，但是软件的采集模块将无法打开，分析模块能正常使用。

三、数据采集软件（Stylet⁺d）

（1）连接好 Giga 和 AD 转化器后，双击 Stylet⁺d 图标，就可以进入数据采集模块。如图 7.2 所示。

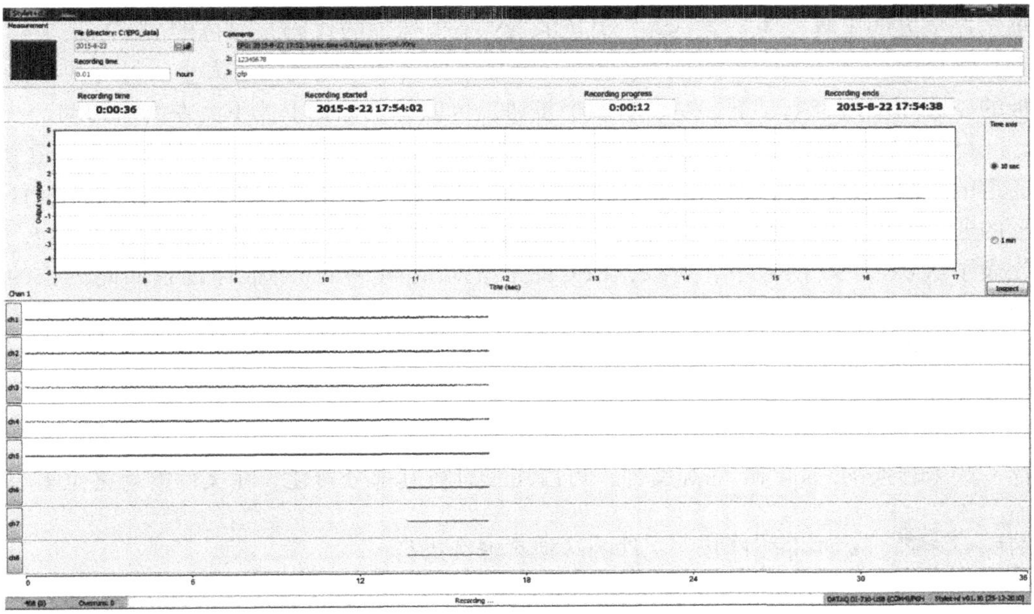

图 7.2　Stylet⁺软件的采集模块页面

（2）对记录进行命名。“

”文件夹默认保存在 C：\ EPG_data 中，不过我们可以通过单击界面上的“”，选择将记录结果存放在电脑中的其

他位置，记录后我们将获得以 . aq4（或 . aq8，8 通道）结尾的原始波形数据。

（3）记录时间选择。直接在“”中输入所要记录的时间。

（4）波形备注：在“”文档栏中有 3 个标注行，但只有第 2 个和第 3 个行能进行编辑，第 1 个能自动进行调整。一般在第 2 个标注行编辑实验或处理的条件、植物、昆虫等。第 3 个标注行能标注特殊处理的通道。

（5）单击“”，软件开始运行，同时变成红色“”，EPG 波形将会开始并出现在操作界面的空白面板上。当单击通道左边的时，主通道会显示选中的通道。

（6）通过右侧“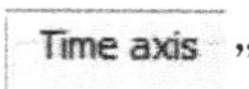”的时间选项可以选择主控界面上显示波形的时间为“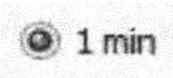”和“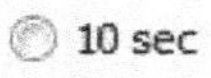”之间切换。

（7）从“”中我们可以知道记录数据的记录时间（时间）、记录开始的日期和时间、开始记录以来时间（完成）和记录结束日期和时间。

（8）单击按钮可以静止显示正在记录的通道波形，但不会打断记录。此功能可以用来在记录时观察 EPG 波形的细节。Inspect 显示窗口可以通过鼠标调整窗口大小。左上方的工具栏，黑色的 XY 是坐标轴的代码，绿色的“+”是坐标放大，红色的“-”是坐标缩小，蓝色的圆圈是坐标自动调整到合适大小。另外独立显示板的左边和下边有上、下、左、右四个按钮，可以将坐标轴上、下、左、右移动。

（9）当记录结束后，数据会存储到一个包括所有取样点的媒介文件中，4 通道和 8 通道的后缀名分别为 . aq4 和 . aq8。这些文件非常大，往往也包括了为成功记录 EPG 波形的通道数据。紧接着，它们将被转换为按小时存储的数据文件。文件名为 D##，## 代表了从 01 到 99 的数字，所以，EPG 最长记录时间 99h（不推荐记录如此之长的时间）。在进行转换的时候弹框（图 7. 3）下部会出现绿色的进度条。当转换结束时，弹框的上部将出现黄色部分。注意：在信息出现前，不要进行任何操作。转换好的文件将储存在指定的文件夹中。

（10）单独转换。如果在记录过程中出现电脑死机、断电等情况，电脑中可能会保存一些未转换的 . aq4 和 . aq8 文件，可以用工具将其部分转化。在文件的最终位置单击图标。文件将被打开，文件的转换将继续进行。

注意：

实验电脑不要连接网络；屏幕保护程序关闭，否则一旦进入屏幕保护程序，采集模块可能报告中断或关闭；在控制面板中设置电源为永远不关闭，防止电脑进入休眠状态而中断或关闭。

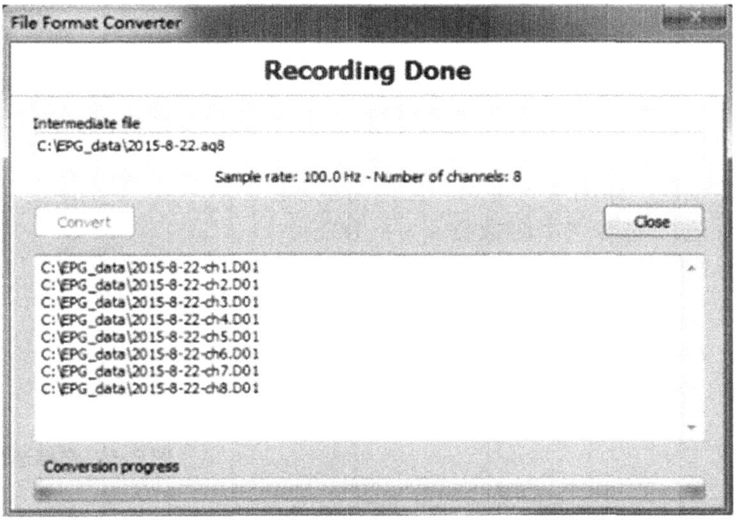

图 7.3　记录结束的提示框

四、数据分析软件（Stylet⁺ a）

双击“Stylet⁺a”EPG ana的图标后，Stylet⁺软件将打开 EPG 分析面板，当载入文档后，将出现如图 7.4 所示面板。

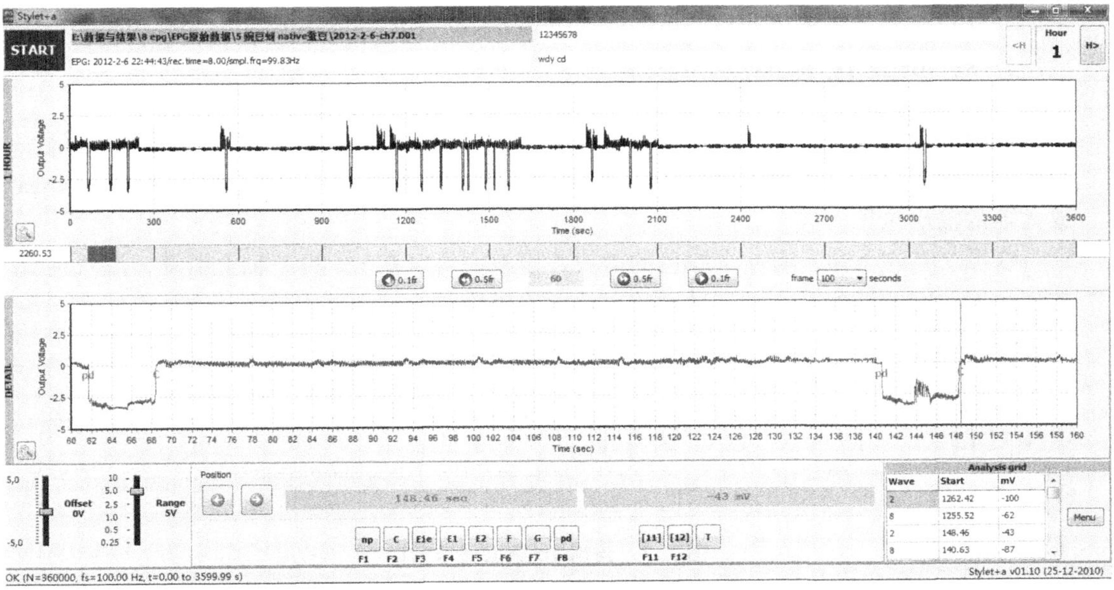

图 7.4　Stylet⁺软件的分析模块页面

（一）文档载入和面板信息部分

1. 文档载入

单击面板上的“START”按钮，可以进入文档选择面板，如图 7.5 所示。里面将显示该文件夹下所记录的第一小时数据（--. D01）。选择所需要的文件并单击“打开”打开文件。面板将显示该记录的第一个小时的 EPG 数据。

图 7.5　文档选择面板

面板将分为几个部分，上部是信息栏，其信息分布和功能如图 7.6 所示。

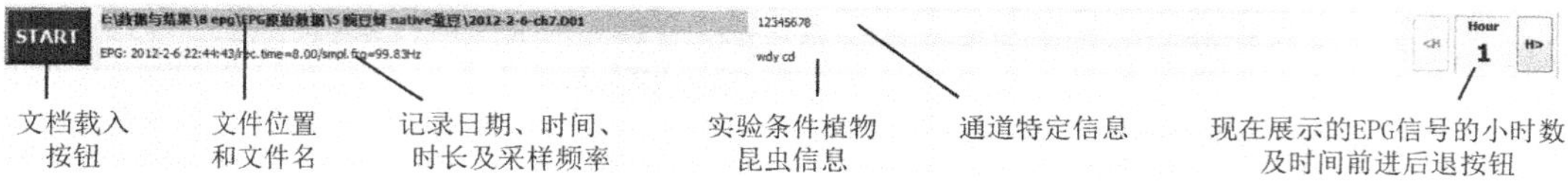

图 7.6　信息栏

2. 前进后退

展示的是第一小时的 EPG 数据，文件为--. D01。如果有后续记录，单击“H>”按钮将载入下 1h 的数据，如果没有将会出现“file does not exist”提示。同样，后退将单击“<H”。

（二）当前 1h 数据显示窗口

如图 7.7 所示，当前 1h（3 600s）数据在该显示窗口里展示。图中选择的是第一个小时的记录，所以 X 轴是从 0~3 600s 输出时间。Y 轴是从-5V 到+5V，代表了 Giga 放大器的输出电压。

在图 7.7 左下角，有一个绿色的滑块，该滑块可以在 1 h 的 Y 轴上滑动，滑块指示的一小段时间的波形将在细节展示的显示窗口里展示。

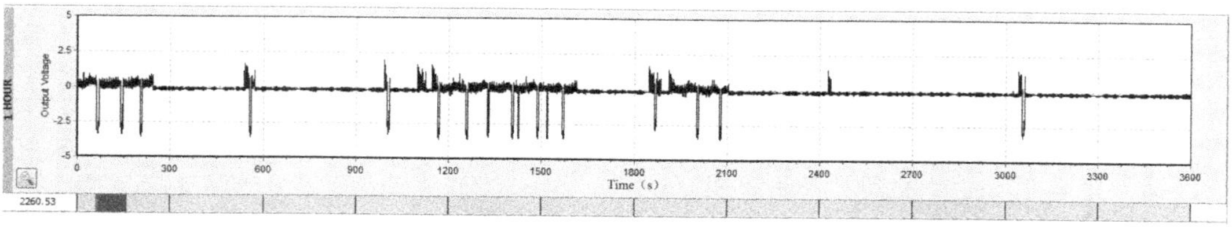

图 7.7　当前 1h 数据显示窗口

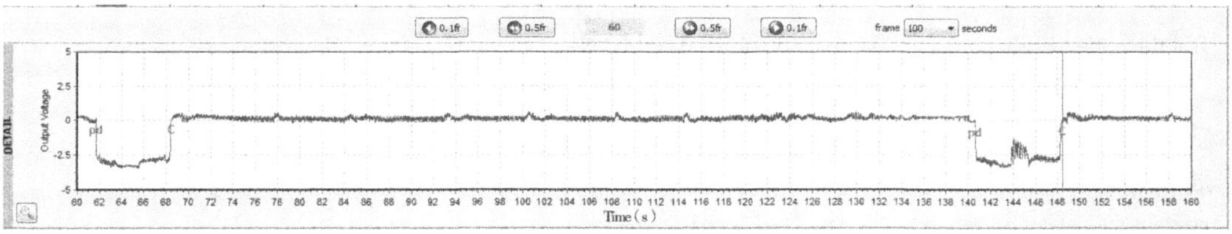

图 7.8　细节数据显示窗口

（三）细节展示窗口

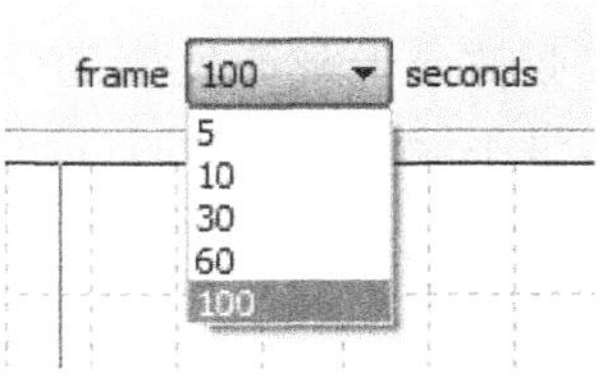

每帧尺寸：如图 7.8 所示细节展示图，每帧显示 100s 的波形，单击下拉选框（靠右黑色的倒三角）即可从选框中选择 5s、10s、30s、60s、100s 的不同每帧尺寸。同时，当前 1h 数据显示窗口中绿色指示滑块的大小也会相应改变。

图 7.9　帧幅移动按钮

图 7.9 的 4 个帧幅移动按钮可以向左或向右以 0.1 帧和 0.5 帧幅的幅度移动细节展示栏中的波形。

（四）信号判断区

在 Stylet$^+$软件分析模块的下部，图 7.10 为信号判断区的显示面板。

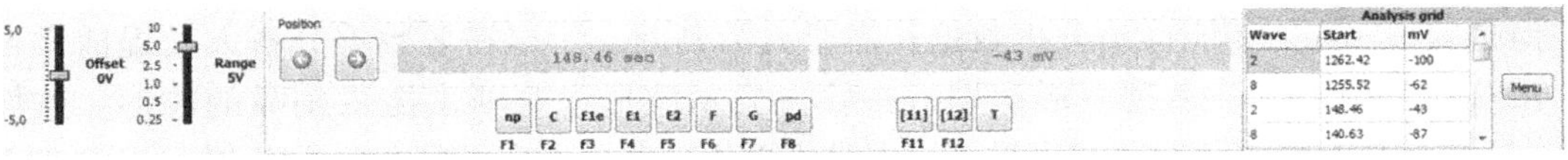

图 7.10　信号判断区

1. 电压比例和偏移　如图 7.11 所示，调节信号偏移（offset）设置可以垂直平移信号曲线，默认为 0，即不偏移。通过拖动滑块或直接单击相应位置，设置可以更改纵坐标的显示范围，默认是-5～+5V。Range 的设置，可以放大或缩小电压的振幅，通过拖动滑块使稍微超出量程或太小的 EPG 信号调节到合适的大小，在细节展示栏显示，方便分析。

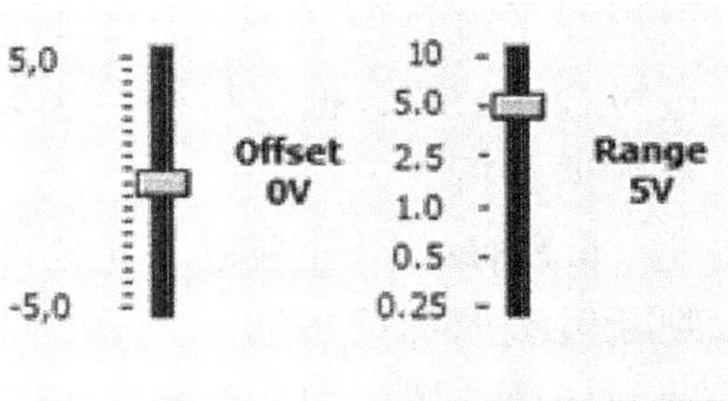

图 7.11　信号偏移设置

2. 信号特征观察菜单工具栏　单击图 7.10 右侧的“Menu”，即可显示信号特征观察的菜单工具栏，如图 7.12 所示。

（1）Smooth：平滑曲线功能。可以选择性去除 50/60Hz 的噪声，但该功能有时也会影响 EPG 信号。最好参考《EPG 常见故障与处理》中减少噪声的干扰，而不是通过平滑曲线的功能来处理噪声。过滤后的信号会呈红色，但是细节展示的帧幅一旦发生变化，都需要进行新的平滑。且显示过滤后的信号时，红色的光标将无法移动。该功能可根据实际情况使用。

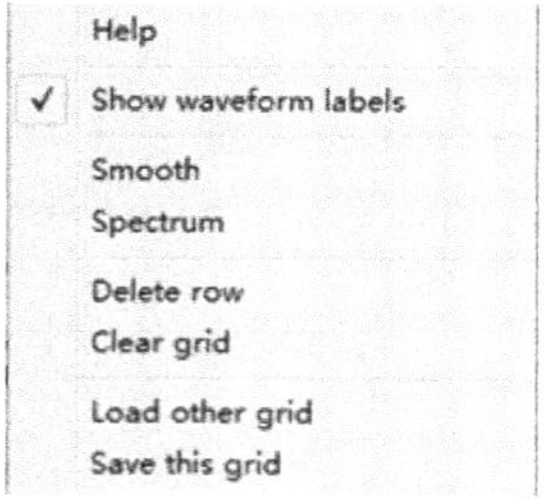

图 7.12　菜单工具栏

（2）Spectrum：信号频谱功能。单击该按钮后，出现一个新窗口（图 7.13），显示细节展示栏信号开始后 10s 信号的自动功率谱（如果信号展示栏的帧幅大于或等于 10s 时）。自动功率谱通过快速傅里叶变换，X 轴显示频率，Y 轴显示量级，即这 10s 内的发生信号每个频率的发生量。这个频谱有助于分析噪声的来源或判断 EPG 信号。注意：频谱分析只对类似正弦波或者规律发生的信号有效。不规律的信号不能生成清晰的自动功率谱。

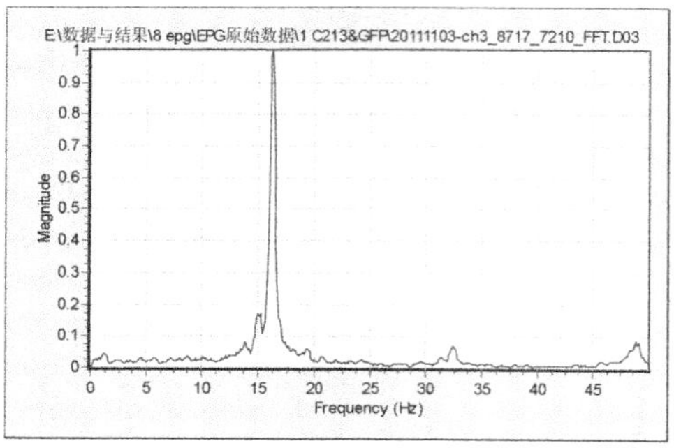

图 7.13　信号频谱

3. 细节展示栏中光标显示面板　当在细节展示栏中单击鼠标，将出现光标（红色直线），光标指示的时间和输出电压将如下所示。

4. 光标移动　光标可以通过鼠标单击波形或者单击  来移动。

5. 通过 EPG 信号判定数据　目前，软件的分析功能仅限于获得信号的细节，展示频率特征，通过平滑功能去除噪声，将光标挪动到合适的位置（时间）。通过这些细节我们能了解并判断到原本的生物学波形，但是软件 Stylet⁺不能自动分析数据，需要人为判断，所以需要数据分析人员提前学习 EPG 信号的识别，通过波形识别，在软件显示波形中判断出昆虫刺吸的波形。软件 Stylet⁺是基于蚜虫的取食波形设计，但是也能用于其他刺吸式昆虫。分析时可以重定义波形按钮的名称，也可以忽略其意义，直接用快捷键标注波形。软件 Stylet⁺是为了识别波形并了解每种波形从什么时候开始，持续时间和发生顺序而设计。这些数据必须通过对 Stylet⁺的数据进行进一步处理才能得到。（参考第六章“EPG 实验设计、指标选择及数据分析”）

6. Stylet⁺信号判定　在软件 Stylet⁺中，波形标签被分配给如图 7.14 上部所示信号判断区波形标记按钮（下面的 F 数字键为对应的快捷键）。在细节显示窗口，将光标挪动到任意位置，单击波形标记按钮，即可在右边分析栏（Analysis grid）获得该光标位置的基本信息（时间和电压）的数值和单击到的波形信息。分析数据时，将光标移动到每个波形的开头（波形分界的位置），分别单击该波形的波形标记按钮。从数据开始到结束，对波形进行这样的识别，即可在分析栏中获得一系列的数据，如图 7.15 所示。另外，根据假设，昆虫的取食行为是互相排斥的，即波形不会同时发生，一个波形的开始就是另一种波形的结束。所以尽管这种假设不是完全确实，但是十分有用。主要是 A 波、B 波、C 波不能严格区分，不过它们一起代表了口针在刺探的过程（广义称为路径波，pathway)。为了“解决”这个波形重叠的难题，将这些波形统称为 C 波（广义 C 波），这样它就可以和其他波形区分开。

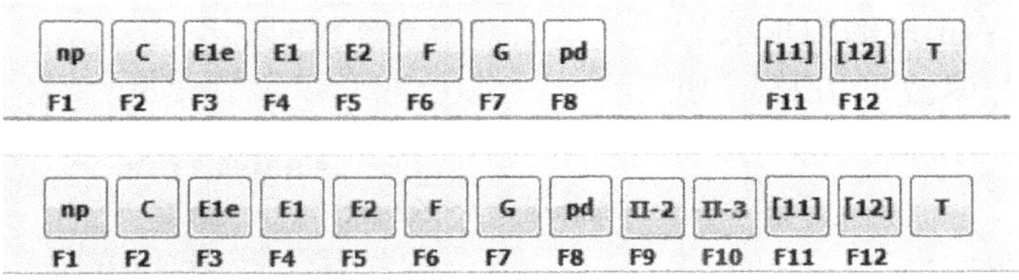

图 7.14　波形标记按钮

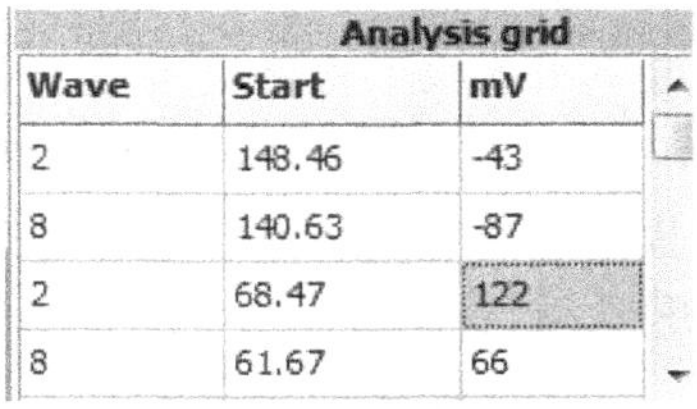

图 7.15　数据分析栏

7. 波形标记按钮　标记是根据键盘的功能键按数字排列的。功能键可以作为快捷方式使用。对于蚜虫来说，按键 F9 和 F10 只在单击了按键 F8 之后出现（图 7.8），在单击了按键 F2（C）后消失。这个功能主要选择性分析 pd 波的亚波形（最难区分）。尽管按键 F9 和 F10 只在单击了按键 F8 之后出现，但是快捷功能键 F9 和 F10 却一直有效。波形标记按钮及其功能如表 7.1 所示。

另外，对于其他种类的昆虫，请忽略蚜虫波形标记的提示，直接使用功能键标记波形。

表 7.1　蚜虫波形及其按钮

按钮	代码/F 功能键	蚜虫波形
np	1	非刺探
C	2	路径波，包括 A 波、B 波、C 波和 pd 波
E1e	3	细胞外 E1 波（不一定出现）
E1	4	E1，正常韧皮部唾液分泌
E2	5	E2，韧皮部取食
F	6	F，口针穿刺困难
G	7	G，木质部摄取水分
pd	8	pd，包括在 C 波中，作为额外分析之用（不推荐）
Ⅱ-2	9	pd 波亚波形Ⅱ-2（如果分析）
Ⅱ-3	10	pd 波亚波形Ⅱ-3（如果分析）
11	11	额外补充按钮，新的取食行为或者作为标记
12	12	同 11
T	99	结束。用于标记记录的数据末尾。数字 99 用于之后，是数据处理惯例

8. 分析栏　单击波形标记按钮（或者快捷键）后，将会有一行数字存储到分析栏中。第一列为波形代号（Wave），第二列为波形开始时间（s），第三列为输出电压（mV）。每次新单击波形按钮，将会有一行新的数字存储到分析栏中。由于显示篇幅大小，只有最后 4 行能显示出来，可拖动旁边的滑块查看其他数据。每一行新的数据，

会自动根据开始时间的顺序排列，插入到数据中，由最大值开始向下排列。

9. 更正 当分析错误或操作失误时，可通过单击该波形方格（变蓝）输入正确的波形值后单击其他方格即可，或者单击该波形方格（变蓝）后，单击“Menu”选择“Delete row”，来删除整行。

10. 重复判定警告 如图 7.16 所示，如果不小心 2 次单击相同的波形判定按钮，会有两列相邻的数据有相同的波形代码，这会导致警告的提示栏出现。同时分析栏会变成红色直到其被纠正。另外，如果不小心删除某行，导致相邻两行波形代码一致，也会出现类似情况。只要按照正确的波形顺序在两行之间插入新的波形，警告提示即可停止。

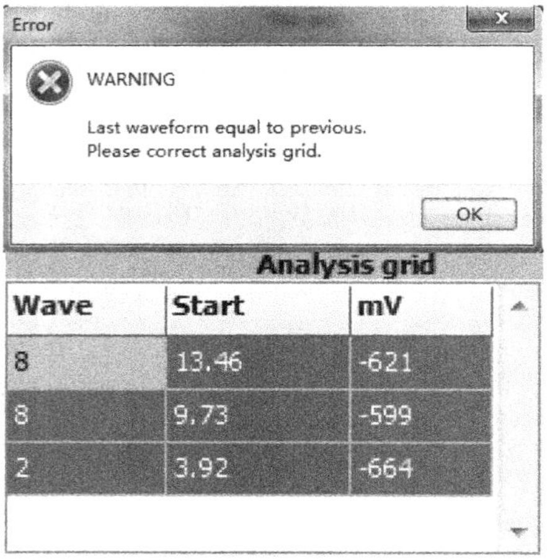

图 7.16 警告提示

11. 分析结束 当所有的波形改变都记录到分析栏中后，单击按钮 T 并加上完成的记录时间（在记录波形末端自动显示）。这个值将用来进行后续软件宏命令的数据分析。

12. 保存分析数据 当分析结束后，分析栏中的信息可以通过单击“Menu”按钮后弹出菜单“Save this grid”选项，保存为 .ana 文件，其文件名由之前的 EPG 波形的文件名自动生成。可以重命名或者改变保存位置。

13. 中断或重新开始分析 EPG 分析可以随时中断，并在之后重新载入。需要中断分析时，按上一条方法对分析后的信息进行保存。需要重新开始时，先打开 EPG 分析模块，重新载入需要分析的 EPG 波形，并在单击“Menu”按钮后，弹出菜单“Load other grid”选项后选择被打断的 .ana 文件。该文件之前分析得数据会在分析栏中重新载入，即可继续进行分析。

14. 同一个页面的下一次分析 为了数据的安全，在分析栏载入另一个文件时，之前的文件并不会自动删除，而是出现提示栏（图 7.17）。并在单击“Menu”按钮后弹出菜单“Clear grid”选项后，先清空分析栏才能载入或开始下一次分析。

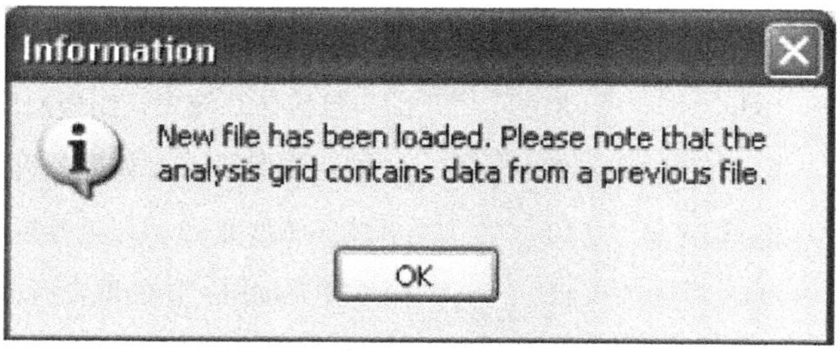

图 7.17 提示栏

15. 显示波形标签 单击该按钮后，根据之前分析的数据会在细节展示栏中出现红色波形标注，如图 7.6 中红色字母所示。由于分析时可能存在误操作，或者新手不熟练时，可以打开帮助判断波形。

16. 所有 Menu 选项

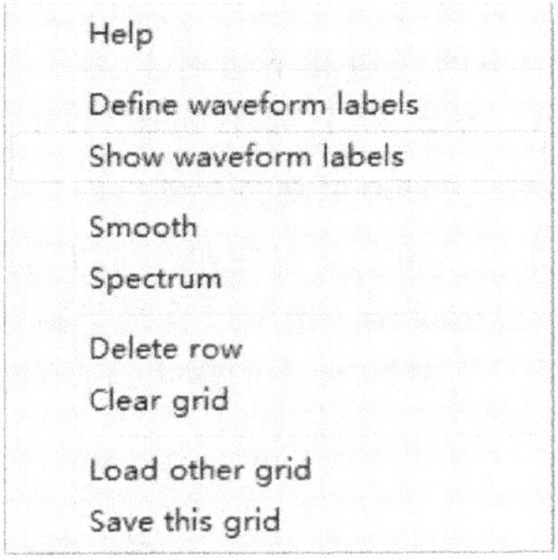

帮助（单击后有各选项的简单说明）
定义波形标签（旧版本软件无此功能）
显示波形标签
平滑曲线
频谱
删除行
清空表格
载入其他表格
保存此表格

图 7.18　菜单

(五) 展示板中的工具

在当前一小时数据显示板和细节显示板左下角中，都有一个放大镜工具。单击该按钮后，会出现一个独立的显示板（图 7.19），里面的数据和原显示板

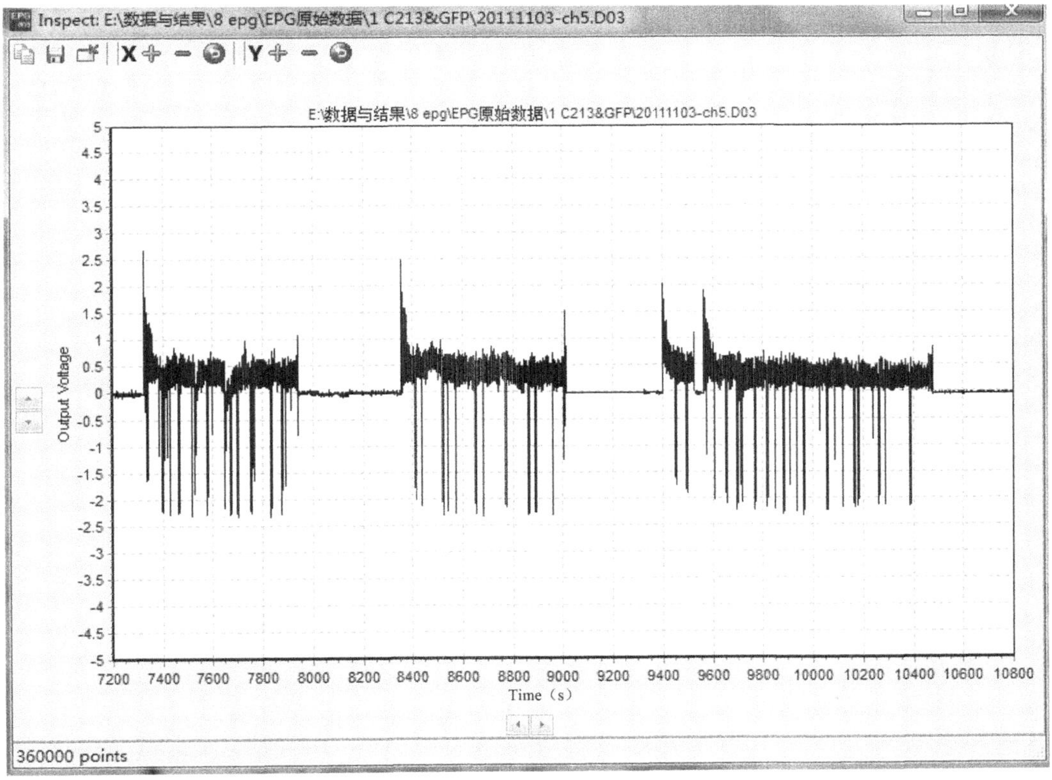

图 7.19　放大镜工具单击后出现的独立显示窗口

的数据一致（类似记录模块的［inspect］功能）。左上方的工具栏，黑色的 XY 是坐标轴的代码，绿色的“+”是坐标放大，红色的“-”是坐标缩小，蓝色的圆圈是坐标自动调整到合适大小。另外独立显示板的左边和下边有上、下、左、右四个按钮，可以将坐标轴上、下、左、右移动。显示板上部中间蓝色的文字 E:\数据与结果\8 epg\EPG原始数据\1 C213&GFP\20111103-ch5.D03，是波形来源文件。

工具按钮可以复制、保存和关闭该文本框。单击复制按钮，可以将该窗口的内容直接复制到 Word 或 PowerPoint 中作为展示或记录用。也可以直接复制到 AI 等软件中对图片进行波形拼接或提取等做进一步处理。复制到 AI 中的图片是非编组模式，该模式下坐标轴、标尺背景等都可以编辑。

（罗　岚）

第八章　人工饲料及其上的 EPG 记录

一、昆虫人工饲料

（一）概述

昆虫在自然界取食的饲料，一般称为天然食料。所谓人工饲料（artificial diets），通常指人为配制的饲料。完全由已知的纯粹化学物质配成的饲料，称为全纯饲料；如果至少含有一种不知化学结构物质（如蛋白质）或不知纯度的饲料，称为半纯饲料（忻介六，1976）。

自 1908 年 Bog Danow 首次用肉汁、淀粉和蛋白胨等制成人工饲料对黑颊丽蝇 *Calliphora vomitoria* 进行饲喂以来，昆虫人工饲料的研究逐步发展起来。截至目前，可用人工饲料饲养的昆虫已超过 1 400 多种，包括直翅目、半翅目、鞘翅目、鳞翅目等多个目的昆虫。半翅目昆虫人工饲料的研究始于 20 世纪 20~30 年代对叶蝉、粉虱和蚜虫传播病毒机制的研究。至 20 世纪 60 年代初才由 Mittler 和 Dadd 及 Auclair 和 Cartier 取得突破。他们建立的 Parafilm 膜饲喂技术和全纯饲料成为此后同翅类昆虫人工饲料饲养技术研究的蓝本，推动了该领域的迅速发展。至 1985 年，可用人工饲料饲养的同翅类昆虫种类已达 71 种，其中包括 50 种蚜虫、8 种叶蝉、7 种飞虱、3 种蜡蚧、2 种粉蚧和 1 种沫蝉（傅强等，2001）。

近年来，由于昆虫学研究的不断深入及害虫防治新技术的发展，需要供应成万上亿的生理标准整齐划一的昆虫，促使人工饲料的研究工作迅速发展，并成为昆虫学研究和害虫防治新技术中的基本技术之一。人工饲料不仅可以使昆虫发育整齐，生理一致，而且在很多情况下是解决季节性饲料短缺的主要途径，如冬季由于大多数自然食料不可利用，只有依靠人工饲料才能使研究继续下去（方杰等，2003）。

（二）人工饲料的成分

尽管不同昆虫之间对个别营养物质的需求略有差异，但几乎所有昆虫的营养需求基本相似。包括蛋白质、糖类、脂类、维生素、脂原物质、无机盐和水分等。

（1）蛋白质和氨基酸：除少数昆虫依靠体内共生物的帮助能以纤维素为食外，绝大部分昆虫都必须从食物中摄取蛋白质，通过消化系统将蛋白质分解成氨基酸，并在体内合成用于本身生命活动所需要的蛋白质。昆虫和哺乳动物一样，必需氨基酸共有

10 种，只有少数种类需要 10 种以外的氨基酸。因此从本质上说，昆虫对蛋白质的需求就是对氨基酸的需求。一种饲料其蛋白质价值如何，主要看它的氨基酸组成，以及在消化过程中所分解出来的各种氨基酸的量能否满足昆虫体内合成的需要（Mattson，1980；White，1984）。

（2）糖类：糖类也叫碳水化合物，是生物体内能量的最主要来源。同时糖类也是脂类和非必需氨基酸的前体物质。昆虫人工饲料中常用的糖类主要有葡萄糖、麦芽糖、蔗糖和淀粉，少数也用果糖、半乳糖和乳糖（王延年等，1984）。植食性昆虫一般都有极强的蔗糖酶活性，能很好地分解蔗糖，但不能很好地利用淀粉。故现在人工饲料中大多数使用蔗糖（忻介六等，1979）。糖类不仅对许多昆虫的取食有促进作用（Stadler et al.，1978；Albert et al.，1982），而且对它们的发育和繁殖也有重要影响（Abushama et al.，1977；Koyama，1985）。若干糖类有促进取食活动的作用，也就是说是一种取食刺激剂。

（3）脂肪酸和固醇：脂肪酸和固醇是昆虫人工饲料中常用的脂类物质。昆虫需要的脂肪酸主要分为两类：一是饱和脂肪酸与含一个不饱和双键的脂肪酸（油酸）；二是含有二或三个不饱和双键的不饱和脂肪酸（亚油酸和亚麻酸），目前研究表明，不饱和脂肪酸是许多昆虫的必需营养成分（王延年等，1984），一般饲料都用植物油作为脂肪酸的来源（还含有少量的生育酚和固醇等），替代纯的不饱和脂肪酸。固醇又叫甾醇，昆虫体内不能合成固醇，必须从食物中获得才能维持正常的生长发育。饲料中多使用胆固醇（王延年等，1984）。大豆卵磷脂作为营养的增补物，也能刺激某些昆虫的取食。胆固醇与卵和卵巢的发育有关（忻介六等，1979）。

（4）维生素：维生素是一种机体内无法合成的起调节及管制代谢作用的物质（忻介六等，1979）。分为水溶性和脂溶性两大类，必须从食物中摄取。研究表明，倘若昆虫完全缺乏某种维生素或某种维生素供给不足，代谢就会失调，生长停滞；饲料中如果没有硫胺素、吡哆醇、核黄素、烟酸和泛酸中的任何一种，都会导致低龄幼虫的大量死亡。叶酸和生物素也是必需物质，会影响高龄幼虫和蛹的发育。抗坏血酸既是很多植食性昆虫必需的营养成分，还是一种取食刺激物。维生素 E 对很多昆虫的生殖等都有重要影响。

（5）无机盐：无机物质在昆虫的生理活动和组织构成上起着非常重要的作用，而所有的无机物都需要通过食物获得，因此人工饲料中也必须含有无机物（王延年等，1984）。不同昆虫对无机盐的需要有一定的差异性，多数昆虫需要钾、磷、镁，不同程度需要钙、铁、锌、钠、锰、铜，植食性昆虫需要的钾比钠高。

（三）人工饲料的种类

根据保存状态，人工饲料分为三类：液体人工饲料、固体人工饲料和半固体人工饲料。一般而言，刺吸式昆虫主要取食液体人工饲料，而鳞翅目昆虫则主要取食固体人工饲料。

根据饲喂方式，用于刺吸式昆虫的人工饲料包括以下三类：

（1）液体人工饲料囊：两层封口膜包被 1~2mL 的人工饲料。先在瓶口覆上一层封口膜，接着在膜上滴上一定数量的人工饲料，然后再覆上另一层膜，把周围封严。

（2）液体人工饲料盒：可以将载玻片大小、厚约 5mm 的聚酯从侧面挖个凹槽，在槽内加液体人工饲料，用封口膜密封。也可以使用实验室现成的带凹槽的载玻片。这种人工饲料便于从侧面观察昆虫口针的运动。

（3）半液体人工饲料：根据不同昆虫的需要，可以配合琼脂制作成半固体（半液体）人工饲料，但要注意其导电性（闫凤鸣，2011）。

根据培养时间和营养成分，刺吸式昆虫的人工饲料分两类：一类是用于长期饲养的，需满足昆虫生长发育繁殖的需要，成分复杂，包含蛋白质、糖类、脂类、维生素、脂原物质、无机盐和水分等；另一类是用于特殊实验目的，一般为体积百分比为 5%～10%的蔗糖溶液，能够维持昆虫几小时或几天的需要即可。

二、人工饲料上的 EPG 记录

由于植物成分多样且复杂，所以植物上的 EPG 记录很难评价单一或已知成分对昆虫取食行为的影响。而人工饲料成分已知，添加的成分可以定量，在人工饲料进行 EPG 记录可以定性或定量某类物质对刺吸式口器昆虫取食行为的影响，而且重复性较好（李晓敏等，2013）。人工饲料经常选择蔗糖溶液，因为它成分单一，配制方便，能基本满足短期内昆虫的营养需要，而且在蔗糖溶液中添加某些植物次生物或农药等物质，可以避免植物复杂的结构和反应对实验结果的干扰，方便研究单一物质对刺吸式昆虫取食行为的影响。

进行人工饲料上的 EPG 记录，目的主要有：一是进行波形与生物学意义的关联研究；二是进行植物次生物质、毒蛋白或农药等对昆虫取食行为影响机制的研究；三是进行昆虫传播植物病毒机制的研究。此外，人工饲料还可以不与 EPG 研究结合，单独用于昆虫唾液成分的收集（闫凤鸣，2011）。

（一）EPG 液体饲料囊装置的制作

（1）在超净工作台上进行液体饲料囊的制备。

（2）在透明开口圆柱形容器（高 5cm，直径 4cm）口上覆上一层封口膜（parafilm），这一层可以稍微厚一点，但是注意不要将封口膜绷得太紧，防止封口膜破裂，导致人工饲料的溢出。

（3）在膜上加入 2～3mL 5%～10%蔗糖溶液（或加上含有植物次生物质等需要检测物质的蔗糖溶液）。

（4）在人工饲料上轻轻快速覆盖上一层封口膜（parafilm）。这一层尽可能薄，也不要太松，必须提供一定压强，方便昆虫取食。但也不要将封口膜绷得太紧，防止封口膜破裂，导致人工饲料的溢出。

（5）将封口膜在容器周围压紧，使之形成液体饲料囊。如图 8.1 所示，将长为 5～8cm 铜丝的一端浸入饲料囊中的液体中，另一端连接到 EPG 电路系统植物电极上，作为输入电极。

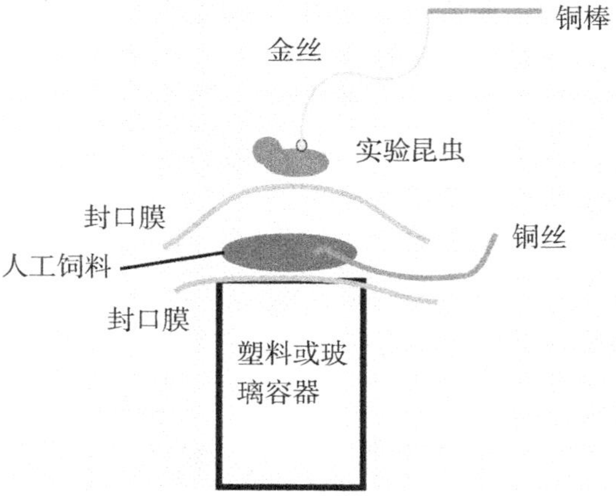

图 8.1　EPG 人工饲料囊示意图

(6) 将粘好金丝的昆虫轻轻放置在饲料囊上面。其余参考普通 EPG 操作即可。

(7) 也可将饲料囊的外层轻轻涂上少许蔗糖溶液，吸引昆虫取食。

如果要进行波形与生物学意义的关联研究，可以用显微摄像机观察，记录昆虫口针的运动，通过图像混合器，将 EPG 波形和昆虫口针运动情况同步记录到计算机中并显示在屏幕上（Joost et al.，2006；闫凤鸣，2011）。

注意事项

利用人工饲料作为介质进行昆虫 EPG 记录，应该注意以下几点：

(1) 不管使用哪一种人工饲料，都需要把植物电极连接到人工饲料中，不然无法形成回路。

(2) 由于昆虫的取食需要许多刺激因素，因此记录其在人工饲料上的取食行为，成功率会比较低。解决这个问题，可以采取两个措施：一是适当延长昆虫在记录前的饥饿时间，二是可以在人工膜的表面适当涂些植物汁液以刺激昆虫刺探。

(3) 人工饲料毕竟不能和植物一样有那么多的结构层次，所以昆虫在人工饲料结构膜上比在植物上产生的 EPG 波形类型要少，但可以在一定程度上揭示昆虫的取食行为。

(4) 由于蔗糖溶液是透明的，不容易观察到昆虫分泌唾液和取食的情况，可以在蔗糖溶液中加一些微小颗粒的碳粉，以便于通过碳粒的微小运动，来证明昆虫的取食活动（闫凤鸣，2011）。

(二) 昆虫在人工饲料上取食与植物上取食的差别

昆虫在液体饲料囊上取食与在植物上取食有很大不同：

（1）液体饲料囊缺乏植物表面的物理化学环境，也没有植物内部那么多层次（闫凤鸣，2011）。一般来说，EPG波形也没有在植物上取食的波形复杂，只有简单的几类波形，如非刺探波、口针插入封口膜的路径波和取食波等几种波形，且波形与植物上的略有差别。

（2）液体饲料囊内只有蔗糖溶液而没有其他营养物质，昆虫的持续取食会受到影响（闫凤鸣，2011）。

（3）植物维管束内韧皮部的液体是正压，昆虫是被动吸食，而液体饲料囊内的压力有限，昆虫需要调整取食策略，如利用肌肉抽吸等（闫凤鸣，2011）。所以，人工饲料上的取食是由昆虫主动吸食来完成的，其波形类似昆虫在植物木质部主动摄取水分的波形。

三、示例：烟粉虱在人工饲料囊上的EPG记录

（一）实验步骤

1. 实验前材料准备　实验前需准备的材料有烟粉虱成虫若干头、冰袋、人工饲料、封口膜、银胶、振荡器、连接有金丝的铜棒、金丝（直径10μm左右）、铜线、剪刀。

实验前需做的几个步骤：

（1）先用剪刀将封口膜剪成1cm见方的正方形或宽度0.5cm、长1cm左右的长条形，用来制作人工饲料囊。

（2）将银胶从冰箱取出，在振荡器上震荡，直到在EP管管口形成黏稠度合适的银胶溶液，以备使用。

（3）检查黏结在铜棒上的金丝长度是否合适，长度1.5～3cm为宜。

2. 烟粉虱的固定及与金丝的连接　烟粉虱的固定。将1.5mL EP管内若干头烟粉虱放在冰袋上冰冻或者直接在冰袋上划动几秒，烟粉虱即不能活动。烟粉虱活动停止后即可进行电极的连接。将金线端部浸入银胶中，抽出后让附着的银胶在空气中干燥1～2min，再将金线端部浸入银胶，如此反复至金线端部形成一个尽可能小的银胶球，连接在烟粉虱的前胸背板上，连接好烟粉虱之后，进行饥饿处理20min。在实验昆虫饥饿期间可进行人工饲料囊的制作。

3. 人工饲料囊的制作　作为EPG记录人工饲料一般均是在蔗糖溶液的基础上添加活性物质。不同蔗糖浓度时产生的EPG波形存在一定差异，研究表明：10%的蔗糖溶液更适用于不同生物型烟粉虱取食行为的EPG研究，相较于5%及15%的蔗糖溶液，10%浓度下EPG波形的相对振幅、频率与在植物上的行为最为接近，因此液体饲料选择10%的蔗糖溶液（李晓敏，2013）。

将实验所需要的成分配置成人工饲料，用两层封口膜包被300～400μL的饲料，制成人工饲料囊。如图8.2所示，在瓶口覆上一层封口膜，再在周围缠绕几圈封口膜，形成一个凹槽，在缠绕封口膜的同时将导电铜丝缠绕进去，以连接植物电极，然后在

膜上滴上一定数量的饲料，然后再覆上另一层膜，把周围封严即可。

图 8.2　人工饲料囊的制作

4. EPG 记录　将制作好的人工饲料中的铜线连接到植物电极上，将饥饿之后粘连有试虫的铜棒插入到昆虫电极的探头中，虫子口针向下平放到饲料膜上。当每个通道的人工饲料囊和烟粉虱都接入调试校正好的 EPG 记录仪后，单击软件上 start 进行 EPG 记录。一般每个处理需要有效数据为 20 个以上，记录时间为 6h。

（二）波形识别和分析

在人工饲料上的波形与在植物上的波形相似，而且相对于植物来说比较简单，人工饲料毕竟不能和植物一样有那么多的结构层次，但可以在一定程度上揭示昆虫的取食行为（Joost et al.，2006）。人工饲料上的波形一般只有 C 波、E1 波、E2 波、np 波。np 波（non-probing 或 non-penetration）表示昆虫的口针尚未刺穿液体饲料囊，基本是一条直线；路径波（pathway waveforms）包括 A 波、B 波、C 波三种波，统称为路径波 C 波，表示昆虫在吸食之前的刺探行为，代表口针刺破囊膜、鞘唾液分泌及逐渐凝固等行为，只是与取食植物相比，在液体饲囊中波形相对较为单一；E1 波表示在吸食汁液前水溶性唾液的分泌（类似于在植物韧皮部、筛管中水溶性唾液的分泌）；E2 波表示昆虫吸食液体的行为（类似于被动吸食植物韧皮部汁液的行为）；E 波是 E1 波、E2 波的总和，总体表示昆虫吸食汁液的行为。在 EPG 波形指标上，C 波持续总时间、E1 波持续总时间及 E2 波持续总时间等是 EPG 和人工饲料技术体系较为重要的行为指标，具体分析指标可参照李晓敏等（2013）。图 8.3 为烟粉虱在烟草叶片与人工饲料上取食对比的 EPG 波形图。

（三）常见问题及解决办法

利用人工饲料进行 EPG 记录有一定的优势，但也会出现一些除了常见的银胶黏性不好、噪声等问题外的其他特殊问题。当烟粉虱不刺探或刺探不出现 E 波可能是因为如下原因：与在植物上相比，人工饲料对烟粉虱的吸引力较低，人工饲料里只有蔗糖溶液而没有其他营养物质，烟粉虱刺探饲料薄膜的概率较低，导致成功率较小。可以在封口膜上涂上少量蔗糖溶液引诱其取食。烟粉虱不取食还可能有如下原因：①银胶没有粘在烟粉虱前胸背板上，粘在烟粉虱翅膀上或头部等，影响烟粉虱活动；②烟粉虱放置在人工饲料上时，烟粉虱口针及腹板没有放正；③饥饿时间不当，过度的饥饿致使其活性下降。

另外，还可能是人工饲料囊制作不当导致的。人工饲料囊是由两层封口膜中间夹

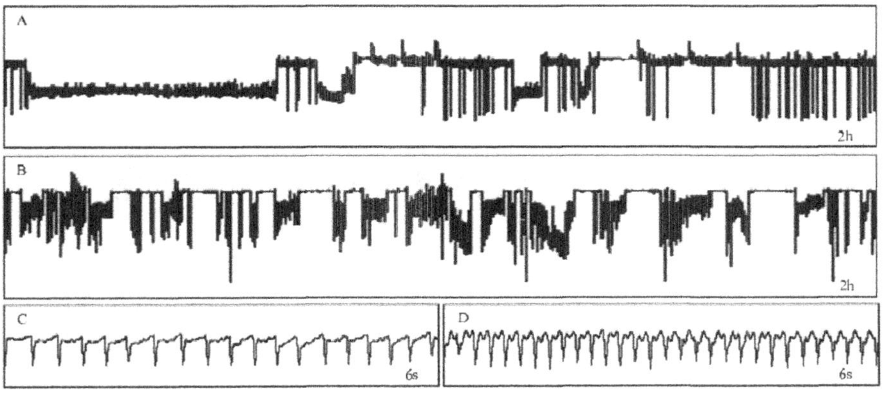

图 8.3 烟粉虱在烟草叶片和人工饲料囊上取食典型波形（李晓敏等，2013）

A. 烟粉虱在烟草上的取食波形；B. 烟粉虱在 10%蔗糖溶液的人工饲料囊上的取食波形；C. 烟粉虱在烟草上取食的 E2 波；D. 烟粉虱在 10%蔗糖溶液的人工饲料囊上取食的 E2 波

层蔗糖溶液配制而成的，而最上面一层封口膜的厚度决定了烟粉虱取食的效果，如果封口膜太厚，烟粉虱口针刺探不透，会影响烟粉虱的取食。另外，植物维管束内韧皮部的液体是正压，昆虫是被动吸食，而液体饲料囊内的压力有限，昆虫需要调整取食策略，如利用肌肉抽吸等，因此两层封口膜夹层里的液体饲料量太少，也会影响烟粉虱的取食，甚至导致 E2 波不出现。将与昆虫接触的封口膜拉制得足够薄，并紧紧封住足量的液体饲料可改善该状态。

（王雪丽 李静静 黄翠虹 罗 岚）

第九章　新波形的研究方法

自EPG技术出现后，刺吸式口器昆虫的取食行为研究得到了长足发展，特别是有关蚜虫的EPG波形和行为的关系也被研究得相当透彻，便利于研究者们使用现有的波形推测蚜虫的行为。同时，EPG技术也被应用于其他刺吸式口器昆虫的行为研究，如叶蝉（Miranda et al.，2009）、粉虱（Janssen et al.，1989）、褐飞虱（Seo et al.，2009）、粉蚧（Calatayud et al，1994；Cid and Fereres，2010；Huang et al.，2012）等。然而，如果要研究一类前人没有研究过的刺吸式昆虫的取食行为，就必须首先建立昆虫行为和EPG波形之间的关系的模式（闫凤鸣，2002）。本章介绍几种取食行为和波形相关性研究的常用技术手段。

一、组织切片的显微观察

（一）电镜观察

透射电镜的应用：对取食处进行标记，在所需检测的波形发生时，在尽量避免植物组织内口针发生位移的前提下，将虫体快速切离植物。按照标记的位置，切取适当大小（一般为5mm×5mm×5mm，若昆虫个体较大，则适当放大以确保口针被完整切取）的植物组织为供试样品。样品被迅速浸入2.5%的戊二醛，4℃下固定过夜；次日，按照以下步骤进行处理：①用移液枪小心移去戊二醛，用pH值为7.4的磷酸缓冲液（PBS）冲洗样品，静置或缓慢晃动15min，重复3次；②小心挑取样品放入2%锇酸，氧化1~2h；③用PBS冲洗样品，静置或缓慢晃动15min，重复3次；④用酒精由低到高（40%、60%、70%、80%、90%、100%）进行梯度脱水，每次脱水静置或缓慢晃动15min，之后用丙酮浸泡样品20min；⑤挑取样品进入丙酮树脂（树脂为Epon 812）的混合液（$V_{丙酮}$：V树脂包埋液=1：1），浸泡1h；⑥挑取样品进入丙酮树脂的混合液（$V_{丙酮}$：V树脂包埋液=1：3），浸泡3h；⑦挑取样品浸入树脂包埋液，室温过夜；⑧次日，挑取样品浸入树脂包埋液，70℃烘箱过夜。

样品被固定后，在显微镜下进行定位切片，制作成厚度为2~3μm的超薄切片，在透射电镜下进行观察。

（二）光镜观察

切取昆虫取食位点的植物组织（叶片为5mm×5mm或长度为2~3mm的茎）；用1%

酸性品红染色 3min，清水洗净，在滤纸上吸干多余水分，在 Karnovsky 固着液进行固定；然后室温下用梯度浓度酒精进行脱水（10%～100%）；脱水后用 Historesin 试剂盒进行渗透（Leica，Heidelberg，Germany）；用软树脂（Leica Historesin，Heidelberg，Germany）进行包埋，连续切片，制作 10μm 的半薄切片，用亚甲基蓝进行染色后进行显微观察。方法参考 Miranda 等（2009）和 Bonani 等（2010）。

昆虫取食 1～3h 后，切取取食位点的植物组织，浸入 CRAF Ⅲ固着剂 24h（最初的 2h 在真空中进行）；用 HEPES 清洗样品，梯度浓度酒精脱水，于 52℃ 用 Paraplast X-tra进行渗透，之后于 56℃用 Paraplast Ultra 进行包埋。包埋后的样品进行连续切片，厚度为 10μm，用 5%（*W/V*）氨水进行封片，过夜烘干后，用 100%的二甲苯脱蜡，用 0.5%番红进行染色，然后用 0.01%固绿复染，最后用中性树胶封片剂（Fisher）封片后观察。方法参考 Backus 等（2009）和 Backus 等（2005）。

二、液体食物法

为了清楚直观地观察刺吸式昆虫的取食行为及其和 EPG 波形的联系，可以让连接好电极的昆虫在人工液体饲料上取食，同时在显微镜下观察昆虫的取食情况。液体食物法的实验装置主要包括四部分，如图 9.1 所示：饲料盒、显微摄像仪、EPG、数据同步处理设备（一般电脑即可）。其中饲料盒有两种，可供昆虫进行不同角度的取食，如图 9.2 所示。本方法可参考 Joost 等（2006）。

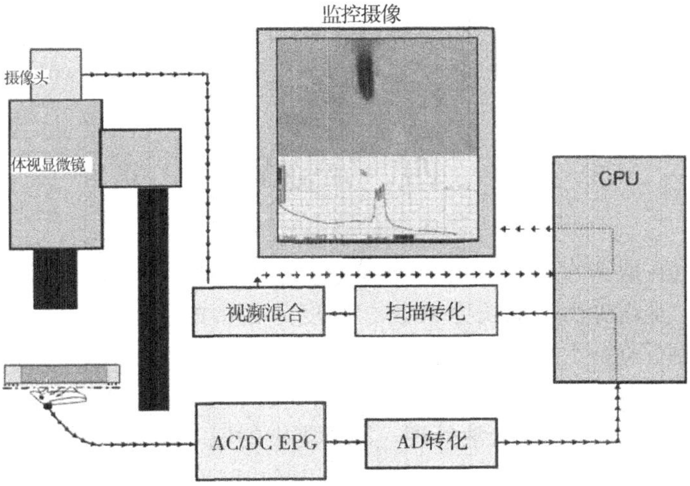

图 9.1　液体食物法实验装置

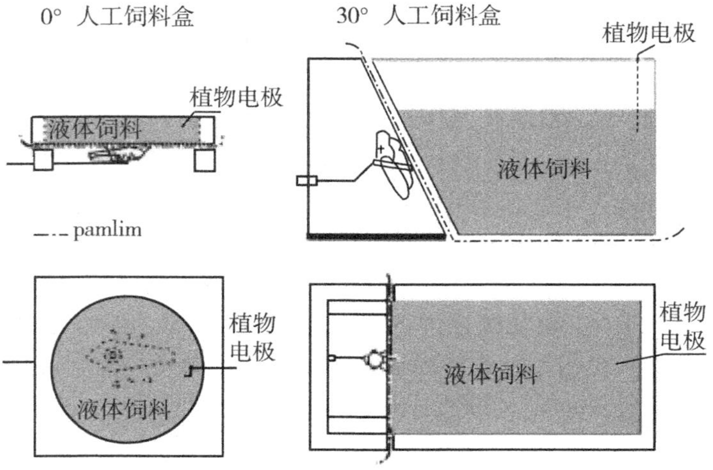

图 9.2 液体人工饲料盒

三、DC-EPG 和 AC-EPG 联用

DC-EPG 和 AC-EPG 系统在识别蚜虫的行为上各有优劣，可以将这两种系统结合起来使用，让两种系统同时显示同一头蚜虫产生的波形，借此进行对比，能更客观、更全面地观察昆虫的取食行为。Backus 和 Bennett 在前人的基础上设计了一种 AC-DC 联用监测仪，对比之前的 AC 和 DC 的简单联用，具有设置更细致、对比更明显等的特点。该方法可参考 Backus 和 Bennett（2009）。

四、蜜露钟

刺吸式昆虫在取食过程中，通常会分泌蜜露。因此，蜜露分泌是用来判断昆虫取食的重要依据。蜜露钟由一周长为 20cm、高度为 2cm 的中空铝质圆柱和一节 5 号电池驱动的 24h 制的石英钟组成。蜜露钟装置放在已用导电胶黏合好的虫体的正下方。圆柱的表面环绕一条 20cm 长、2cm 宽的薄层色谱用的硅胶纸（Merck，60F254）。硅胶纸事先用 0.2%正丁醇茚三酮混合液处理。因蜜露中的氨基酸成分作用，滴落在硅胶纸上的蜜露显示出紫红色。每次蜜露分泌的起始和终止时间被准确标记，并与同步记录的 EPG 波形相对照，就可以初步判断该波形和昆虫吸食是否有关，且可以知道取食时间的长短。

五、其他技术

（一）口针截断技术

口针截断技术是电子显微镜和EPG结合的一个必要补充。当口针刺入植物组织时，快速截断口针，再经过处理即可通过电镜观察定位口针位置，将某波形与其发生时口针的精确位置联系起来。一般使用的口针截断技术分为机械方法（Mittler，1957；Nault and Styer，1972）和其他方法，如激光法和高频波法（Unwin，1978）。机械方法相对粗糙，适用范围有限，最大问题是不能保证在截断的同时口针前端的位置不发生变化。激光或高频波法使用相应的激光发生器或高频波发生器，快速准确地截断口针，能够最大程度保证口针位置不会因截断的过程而发生变化。

使用口针截断技术除了能够准确定位口针在植物组织中的位置外，还能够进行植物韧皮部液体样品的收集。通过EPG实验和电镜观察发现，E波发生时，试虫口针的尖端位于筛管细胞之中（Kimmins and Tjallingii，1985），其他的波形发生时，并没有汁液溢出（Mentink et al.，1984）。通过这个系列的研究可以发现，通过EPG监测，当E波发生时，口针尖部进入植物筛管细胞，截断口针，植物韧皮部汁液就会从口针的断口处溢出。收集溢出的液体即可用于化学分析，确定植物韧皮部汁液的基本成分。

（二）放射性示踪技术

放射性示踪技术与EPG结合的基本思路是在人工饲料或植物中添加放射性标记物，如^{14}C或^{32}P，通过监测其放射性来观察试虫的取食情况。在人工饲料中加入^{32}P标记物，当试虫取食时即可将其吸入试虫体内，人为中断取食，测定试虫体内的放射性就能够反映出之前取食的累积量。如果使用EPG进行记录，每个取食阶段的物质摄入量即可被准确推算出来。使用该方法证明，桃蚜的最大取食率为6.7pL/s，平均取食率为2.5pL/s。如果在人工饲料中使用该技术可以进行定量研究，那么使用放射性标记技术于生活植物，则可以进行半定量研究，主要原因在于生活植物中放射性同位素的分布是不均衡的，所以难以对每个部分和整体进行精确定量。

（三）肌电图（Electromyography，EMG）

将两个玻璃毛细电极插入试虫的唇基部，试虫仍然会保持刺探的行为。这时，两个电极可以记录下口针刺探过程中肌肉电势的变化，即为试虫的肌电图。肌电图和EPG具有一定程度的相似性，记录内容可以相互印证。但是实践发现，很多时候肌电图记录与EPG并不相符，只有在电极靠近试虫唾液泵时，肌电图记录和EPG记录才能很好地相互印证。

（黄　芳　薛　堃）

第十章　EPG 常见问题及处理方法

一、软件、记录及实验设计问题

（一）为什么双击 Stylet⁺程序图标，却无法启动软件？

这并不是软件与系统之间的兼容问题，而是 Stylet⁺软件本身的 bug，它会经常无法启动软件，甚至几天一次，只能通过重装 Stylet⁺软件来解决，具体步骤如下：

（1）在“程序”中卸载 Stylet⁺软件。重启电脑。

（2）再删除 C：\ Stylet⁺ \ 下的所有文件，重启电脑。请注意不要误删之前的 EPG 数据。

（3）再次安装 Stylet⁺软件。重启电脑。

（4）双击 Stylet⁺程序图标，即可启动软件。

（二）为什么双击 Stylet⁺a 程序图标（分析模块）可以启动软件，但是双击 Stylet⁺d 程序图标（采集模块）却不能启动软件（图 10.1）

这是由于该电脑上没有安装与 Giga 硬件和 AD 转化器驱动程序，使得该电脑只能使用软件的分析模块而不能使用采集模块。解决方法：安装 AD 转化器驱动程序即可。如果不是实验用采集数据电脑，采集模块可以忽略，只使用分析模块。

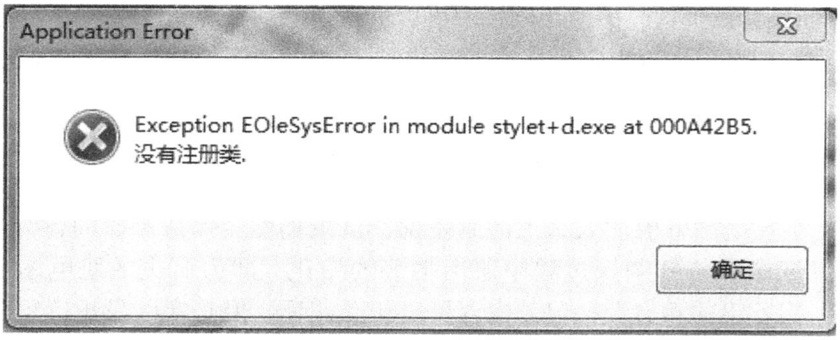

图 10.1　双击 Stylet⁺d 程序错误提示

（三）为什么单击“开始”记录按钮后，EPG波形不是正常的波形偶数通道显示正弦波，奇数通道显示为噪声？

电脑和Giga之间的USB连接出现问题。请观察软件右下角是否写有“Demo device”，如果有则表示 $Stylet^+$ 程序未检测到Giga的硬件设备，$Stylet^+$ 程序进入演示版（即Demo版）模拟运行。解决方法如下：

（1）关闭Giga电源和 $Stylet^+$ 程序。关机。

（2）检查电脑和Giga之间的USB连接接口有没有连接或接触不良。

（3）电脑开机。打开Giga和 $Stylet^+$ 程序。

（四）为什么打开软件后，软件显示Undefined？

这可能是电脑软件无法正确驱动Giga。电脑和Giga之间的USB连接出现问题或软件出现错误。解决方法如下：

（1）先检查电脑和Giga之间的USB连接是否出现问题。

（2）如果连接没有出现问题，重新安装软件即可。

（五）为什么单击“开始”记录按钮后，EPG波形不是正常的波形所有通道均显示较粗的直线条？

通道显示较粗的直线条，表示放大器的回路没有接通，无法进行EPG波形的记录。如果所有的通道都显示为较粗的直线条，可能是电脑和Giga之间的USB连接出现问题，或电脑检测到了Giga，但Giga没有正常工作。解决方法如下：

（1）观察Giga的电源开关是否在“ON”的位置。如果没有，打开Giga的电源开关即可。

（2）如果Giga的电源开关在“ON”的位置，请观察电源指示灯是否亮起。如果指示灯没有亮起，则表示Giga的电源适配器没有连接到室内电源。检查电源适配器是否接好，且室内电源的插座是否正常工作。

（3）如果以上都正常，则根据上一问题的操作，重新进行电脑和Giga之间的USB连接。

（六）为什么单击“开始”记录按钮后个别通道EPG波形显示较粗的线条？

通道显示较粗的直线条，表示放大器的回路没有接通，无法进行EPG波形的记录。如果仅有通道显示为较粗的直线条，其余通道显示出正确的EPG波形，可能是由以下原因导致的：

（1）可能是显示较粗直线条的通道的实验昆虫与金丝之间的连接出现问题，导致昆虫已经在植物上取食，但由于昆虫与金丝连接出现问题，波形没有被Giga放大器记录到。解决方法：每次粘好昆虫后，在体式显微镜下观察昆虫与金丝的连接状态，避免出现接触不良的问题。

（2）还有可能是实验昆虫放置的问题，比如说金丝较短，限制了其活动；或金丝被叶片缠绕，使昆虫无法够到叶片；或昆虫从叶片上落下之类的问题导致其无法取食。解决方法：注意金丝长短，放置昆虫时，尽量放置在它喜欢取食的部位，并固定植物，方便其取食，减少昆虫因寻找取食部位带来的活动对 EPG 波形造成的影响。

（3）如果开始新的实验，使用新的昆虫或植物，或长期未开展实验后重新开始实验，则可能是个别通道植物电极增益值没有正确调节。解决方法请参考后面“增益调节”。

（七）为什么单击“开始”记录按钮后，无法记录到正常的波形?

这可能是因为 EPG 电极没有正常工作。检查方法请参考本书第五章中的章节“EPG 系统安装与调试”中 EPG 电极的检测。如出现问题，请及时与维修工程师联系。

（八）为什么单击“开始”记录按钮后记录到 EPG 波形是正常的波形，但是却不能正常显示（例如波形太小了，不好分辨，或者波形太大了，超出量程）?

这可能是波形的增益出现问题，解决方法请参考本章后续“信号调节问题”部分。

如果增益已经调到极限小，如 50 倍，但是记录到的波形还是超出量程，可能是 Giga 放大器内部出现问题，请参考问题（十五）；如还不能解决，请及时联系维修工程师。

（九）为什么单击“开始”记录按钮后，EPG 波形较为正常但是信号曲线噪声太大?

这可能是 Giga 仪器记录 EPG 波形时引入了噪声，解决方法请参考后面“噪声检测与控制问题”。

（十）为什么点击“开始”记录按钮后，EPG 波形较为正常但是波形振幅整体偏小，像 pd 波或者 E 波这样的负向波形总有恢复到 C 波水平的趋势（图 10.2），但是观看细节波形却的确是该波形的特点?

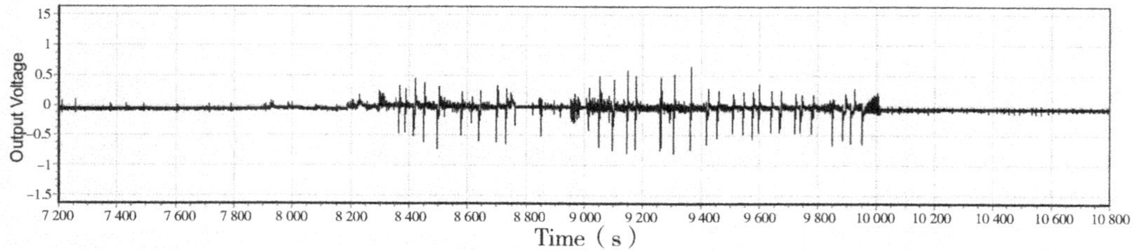

图 10.2　一段整体偏小且连接不好的 EPG 波形

这可能是实验昆虫与金丝之间的连接出现了问题（参考本书第五章“EPG 记录”

中小贴士 2）。解决方法：提高银胶的质量和昆虫与金丝连接的质量。

（十一）为什么单击“开始”记录按钮后，EPG 波形是正常的但是记录一段时间后 EPG 波形的基线整体出现了偏移，有的甚至超出量程（图 10.3），或者波形振幅突然变小，或者出现混乱的波形（但不是噪声，如图 10.4 所示）？

这两种情况一般是在记录 EPG 波形时，昆虫在植物上爬行，寻找合适取食的位置时，其活动影响了金丝与昆虫的连接，导致金丝与昆虫之间的电阻出现变化，进而影响了 EPG 波形。出现这种情况，如果 EPG 波形还在量程内，数据还是可用的；如果超出了量程，只能放弃数据。

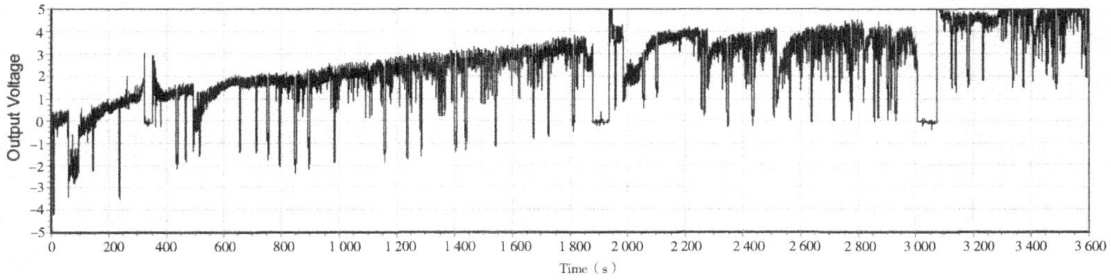

图 10.3　一段基线漂移且超出正向量程的 EPG 波形

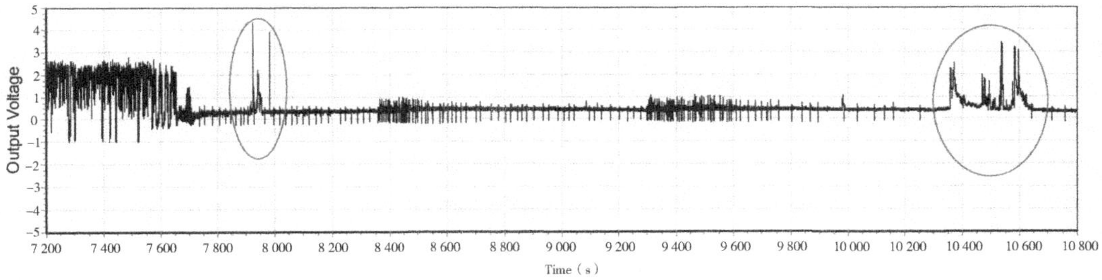

图 10.4　一段 E2 波内有干扰信号的 EPG 波形

而出现混乱的波形可能是金丝碰到了叶片或者种植植物的泥土带来的噪声。

想要避免这些情况，只能注意金丝与昆虫的连接，使其更紧密（但是要注意过多的银胶会给昆虫的行动带来负担）；将昆虫放置在植物适合取食的部位；处理植物如翻转并固定叶片，使昆虫易于取食等。

（十二）为什么设定记录 8h，但是记录结束时数据量不够 8h 或者稍微多点数据？

一般来说，每个通道每小时产生 360 000 个数据点（每秒 100 个数据点）。但是最后 1 h 记录时，有时会少一些数据点或者多记录到一些数据点并存到下 1 h 的文件中。

这是由实际取样频率和准确 100Hz 取样的差异引起的。对于长达几小时的记录时间而言，几十秒的差异并不会带来太大的影响，可直接忽略。

（十三）为什么有时候会出现溢出错误？

在采集模块页面的左下角有一个溢出（overruns）的指标栏，该数值应该尽可能小。如果在进行数据采集时，使用或开打了其他的程序或者联机到局域网或者是互联网会导致溢出值得增加。请不要在进行 EPG 数据采集的时候进行这些操作。

（十四）为什么有时候会出现“103 错误”（Error 103）？

如果数据文件（. D01）是只读文件，那么在单击分析模块的［Start］按钮时，就会出现该错误提醒。在文件进行转化或者拷贝的时候，数据文件可能会出现这种只读特性。如果数据是保存在 CDRom 里（用光盘保存），那么它一直都是只读文件。单击鼠标右键，在文件属性选项里改变只读状态即可。如果是光盘保存的数据，必须先拷贝到电脑硬盘才能进行分析。

（十五）为什么波形振幅太大以至于超出量程（此时增益已调节到最小）？

Giga-8 放大器一般和 DI-710 AD 转换器连用。在旧型号的 Giga-8 放大器，有时会输出错误模式（UH 代替 UL），这时信号数据错误，可能不适用于正负 5V 显示的量程。这个可以通过软件进行调整。在采集页面调出设置（Option）菜单，选择所有的 DI-710 设备一直使用 UL 模式，并单击 OK 即可，如图 10. 5 所示：

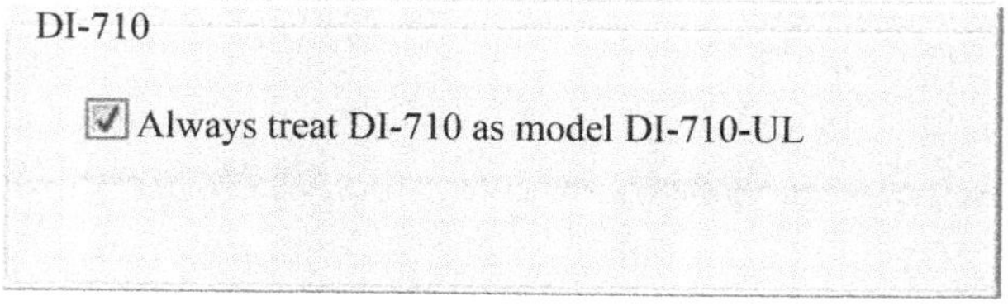

图 10. 5 选择 UL 模式

（十六）为什么做 EPG 记录时，每次都要实验不同的植物和不同的昆虫？

昆虫与植物的关系是很复杂微妙的，昆虫对于植物的任何伤害，都可能使植物产生诱导反应。也就是说，前一个昆虫取食所诱导的植物抗性的增强，就有可能对下一头昆虫的取食行为造成影响。这样，如果用一株植物做多头昆虫的实验，先后实验的两头昆虫所处的植物是不同的，这就增加了实验的方差。因此，每次都用新的植物和新的昆虫。

（十七）如果EPG实验中，植物不易获得或者植物间差异较大（如带毒量不同）怎么办？

对于植物获取难度较大（如体形较大、生长周期长或者需野外采集等）的情况，也可以考虑采用次选方案，不同的昆虫在同一株植物上记录（EPG实验时每次更换昆虫，但不更换植物），但是一定要考虑到之前取食的昆虫对植物的影响。对于生长快速的植物，一定要保持植物状态一致；否则，植物间生长状态差异带来的影响可能覆盖昆虫处理或植物处理之间带来的影响。

如果植物数量少或差异大，甚至不能保证每个通道一株植物，还可以在一株植物上同时记录2只昆虫，如果你的EPG是4个频道的，1和2频道两只昆虫用一株植物，用频道1的植物电极插到植物花盆土壤里；同理，3和4频道用同一株植物，用频道3的植物电极插到土壤里。另外，也可以尝试用4只昆虫在一株植物上进行记录（原来没人做过），只用频道1的植物电极，看是否可以。

二、信号调节问题

想要正确记录和采集数据，并且得到最佳的输出信息，需要分别调整每个操作通道的EPG信号。一般来说，EPG信号调节应该在仪器开始记录且昆虫开始刺探植物之后马上开始。如果在调整后才进行EPG的记录，那么调整之前的昆虫和植物之间相互作用的EPG波形（可能很重要的）可能就错过了。所以调整到最优的分辨率之后，调整每一个通道，将会得到最好的信噪比。一次调整一个通道内所有的信号，然后就保持不变。

EPG信号一般都会改变，会和自己理想的调整效果不符。但是一次记录时，最多调整一次增益，过多地调整经常会造成EPG波形注释困难，如信号不停地变化，不好判断是由昆虫造成的，还是人为调整增益的结果。

每个通道的调节步骤如下（图10.6）：

（1）连接一只蚜虫在某一个通道上，在等待其进行第一次刺吸前，把植物电压调整旋钮在中间的位置时，此时GAIN旋钮应当在逆时针方向旋转的位置（50×）。观察板上仪表，在没有刺探时，板上仪表的指针会在大约中间的位置（0），屏上显示为0基线。

（2）第一个EPG波形时调整植物电压，使板上仪表的电压峰值到达+4V或+5V。稍等一会儿，当EPG连续到达B波或C波形时，信号的电压值应当到达最大值，大约4V。

（3）接着按下校准按钮，看脉冲电流的底部。如果电压水平与0参考线（基线）重合，说明调整成功。但是，大多数时候不是这样。所以按下校准按钮，重新调整植物电压旋钮，直到合适为止。脉冲按钮和pd波形的电压相同（-50mV），所以当标准脉冲降低到0参考电压时，pd电压也会为0。

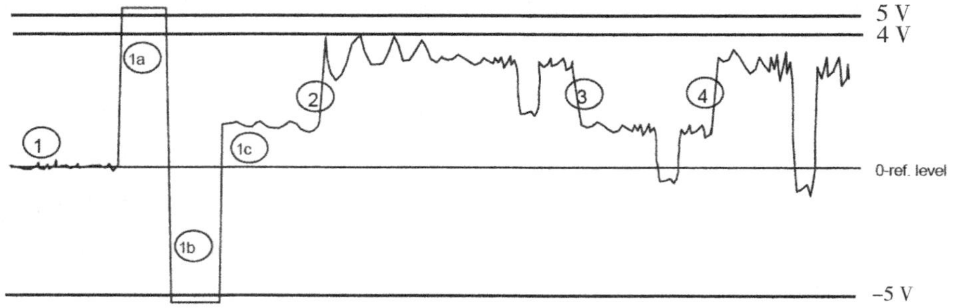

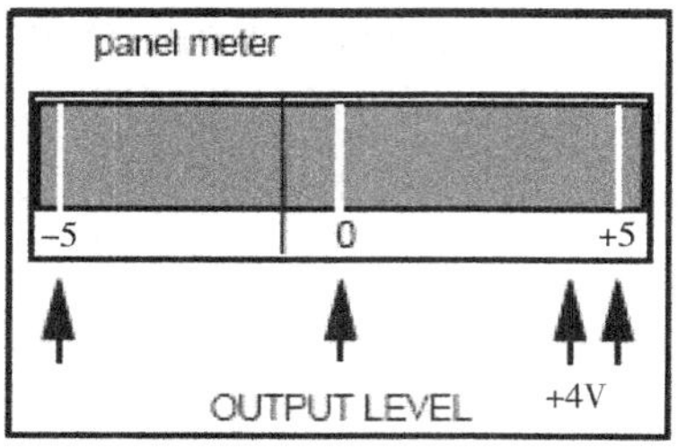

1.等待昆虫刺吸开始时，信号可能是以下三种情况：
1a.正向超出量程信号
1b.负向超出量程信号
1c.信号较弱，接近0
2.调整植物电压调节旋钮直到信号输出为最大值，即C波时为+4V。
3.当按下脉冲校准按钮时，重新调整植物电压调节旋钮，使信号输出降到0以下。
4.松开脉冲校准按钮，调节增益旋钮，使最大信号输出为4V。

图 10.6　植物电压及增益的调整

（4）为了回到 4V 的最大输出电压水平，按照第（2）条调整，现在不是用植物电压而是用增益按钮，这样 pd 波或者标准脉冲电压会继续保持在 0 线。

不适当的调整可能会使波形变坏，同样，虽然在不需要放大时是最佳记录，但当分析时放大信号查找详细资料，噪声（50Hz 或 60Hz）也会放大，因而会扰乱信号图。

三、噪声检测和控制问题

噪声一直是困扰电生理记录实验的一大难题。电生理记录仪高放大倍数可以将微小的电信号放大并记录，环境中微小的噪声也同时被放大。通常使用的 DC-EPG 对外界干扰十分敏感，外界电波的干扰会影响记录的信号，使有意义的信号变得难以辨认，而且外界嘈杂的环境会影响昆虫的取食活动。所以整个实验装置应置于一个电压比较稳定、无外界干扰的环境中，因此，DC-EPG 必须使用接地的法拉第笼来屏蔽噪声。

一般而言，电子噪声又称工频干扰，它是我们日常电气电子设备等带来的 50Hz 的噪声（美国是 60Hz），它会以电磁波的辐射形式，对电气设备和电子设备造成干扰，导致设备运行异常。EPG 记录时出现的噪声可能来自于环境或系统，我们可以在实验过程中分辨这两种噪声，但有时候两者却很难分辨。

（一）环境噪声

环境噪声主要来自于实验室所处的环境，例如其他仪器的电磁干扰，通常使用法拉第笼可以减少环境噪声对 EPG 记录的影响。

1. 准备工作

（1）将EPG电极安装在法拉第笼内，Giga最好安装在法拉第笼之外，以避免对EPG测量的影响，法拉第笼前面保持打开状态。

（2）连接Giga和AD转换器到电脑，在软件中设置1h数据采集（详见第七章EPG常用软件使用指南中的“EPG记录仪及软件的安装”）。

（3）将一个没有连接昆虫的黄铜钉安装在EPG电极上。

（4）将Giga电源开关打开。

2. 测试步骤

（1）如果测量受环境噪声影响，在软件上将会有规律地显示正弦波或粗线。

（2）关闭法拉第笼前面纱网，查看软件中噪声振幅是否减小，如果振幅较大，请参见第（6）条。

（3）用手接近EPG电极，但不要接触，观察软件中显示曲线是否有变化；注意此时身体不要碰到任何仪器或者法拉第笼的金属部分。

（4）正常情况手接近EPG电极时，振幅增大，离开后振幅减小。这时产生的噪声会叠加在第（1）条产生的噪声上。

（5）当手靠近EPG电极，仔细观察振幅，同时另一只手接触法拉第笼，此时可以观察到噪声恢复到初始水平［即第（1）条观察到的］。

（6）法拉第笼屏蔽和接地可以有效地减小环境噪声。检查法拉第笼是否工作的步骤如下：

a. 使用万用表检查法拉第笼和Giga系统的接地情况（万用表电阻应显示连通或阻值较小），再观察显示的噪声水平。

b. 检测法拉第笼每个面纱网的电气连接情况，包括法拉第笼的底部和可移除的前面部分。正常应为一个低阻值或连通状态。检查时，用万用表的一个接头接触法拉第笼的一个表面，再用另一个接头接触另一个表面。另外，EPG电极间和外部的BNC输入连接器也需要检测。

c. 所用的架子和部件需要与法拉第笼接地。

d. 禁止在法拉第笼内安装任何交流电设备，可以把它们安装在法拉第笼的外部。

e. 当噪声水平较高时，关闭法拉第笼前面的纱网，看有无改善。

（7）当笼体前面开放时，噪声足够低的话可以不影响EPG记录。如果不是这样，那么就需要改进法拉第笼，比如说安装一个可移除的法拉第笼前部。

（二）系统噪声

系统噪声通常是指设备内部或电脑产生的50Hz声波。它非常难以去除，而且上一节第（6）条的a~e措施都不起作用。它是由EPG记录仪本身产生的噪声。它可以通过电源、回路或者安全接地等的插座将法拉第笼外甚至是实验室外或者实验楼外的电气设备产生的噪声导入到仪器中。不同于环境噪声，减小系统噪声，请按以下步骤操作：

（1）检查电脑使用的电源是否为三相插头，如果不是，请更换。

（2）将电脑（笔记本）的一颗螺钉与法拉第笼连接。

（3）法拉第笼需要外部接地，可以连接在金属暖气管或水管上。

（4）察看接地是否有接触不良的情况。

（5）如果上述步骤都完成且都无效，可以尝试相反的操作。如将三相插头换成两相的，将地线从水管或暖气系统上移除。有时候“接地”也会引入部分噪声。

（6）如果有间歇性的系统噪声，请尝试关闭附近的其他用电设备，以查找噪声源。如果有这样的设备，可以用金属网将其包裹并接入其他地线，或者调整实验室的设备将其搬离。

（7）如果无法做到，就把 EPG 系统更换地线或者更换放置地点。

（罗　岚　闫凤鸣）

第十一章　EPG 系统的改进和原创设计

EPG 技术问世已经 50 多年了，在我国应用也已有二三十年之久。随着信息技术、生物技术的发展，EPG 技术的水平也应该有所提高和改进，以适应新时期科学研究的需要。我们通过长期应用，对 EPG 的优势和不足有着深刻体会，开展了对 EPG 系统的一些改进工作，同时对 EPG 的未来应用进行了展望。

一、EPG 系统的改进

目前的刺吸电位仪存在很多不足，主要表现在：①对环境干扰敏感；②信号获取软件无法得到数字化的 Excel 文件；③通道少：目前市场上均是 4 个和 8 个通道的仪器，无法满足大规模、高通量检测样品的需要；④EPG 的波形分析软件无法自动识别波形。

（一）硬件方面的不足及改进

1. 硬件的不足

（1）信号获取和分析软件的不足：目前 EPG 信号获取和波形分析采用 A/D 采集卡商家的配套软件（如 Windaq）或专门开发的软件（如 Stylet 或 Probe），功能不够全面，目前尚无一套中文或中英文对照的软件，给初学者学习造成很大困难，不利于 EPG 技术在我国的普及和应用。

（2）早期设计普遍采用 12 位或 14 位 A/D 采集卡，信号分辨率不够高，影响波形细节判断。

（3）使用材料的问题：在进行 EPG 实验前，需使用银胶将金属丝粘在昆虫身体背板上，虽然金（或银、铂）丝十分纤细、有一定的柔韧度，但对昆虫的束缚作用也很明显，在一定程度上其记录结果会与自由状态下的昆虫行为有差异。由于金属丝限制了昆虫的活动，有时会导致 EPG 波形不能正确反映自然状态下昆虫与寄主植物的相互关系的实际状况。

2. 对 EPG 的改进计划

（1）抗干扰设计：采用新型硬件设计方法全面提高仪器的抗干扰性能，将各模块单独分布在独立的腔体内；对电源采取隔离和滤波设计，严格限制其纹波电压；模块间信号隔离及抗干扰性等。

（2）全面提高波形分辨率：将输入阻抗由 10^9 Ω 提高到 10^{10} Ω，使仪器的输出波形的分辨率比现有仪器提高 10 倍。

（3）采用高分辨率 A/D 采集卡：目前荷兰生产的直流 Giga-4/8 EPG 仪采用 14 位 A/D 采集卡（早期还有采用 12 位 A/D 采集卡的）。计划采用 18 位 A/D 采集卡，全面提高 EPG 波形细节分辨能力。比如 14 位 A/D 采集卡相当于 10inch（inch 为非法定计量单位，1inch=2.54cm）电视观看图像，则 18 位 A/D 采集卡相当于 40inch 电视观看图像。

（4）使用碳纤维代替细金（或银、铂）丝，节约实验费用和提高实验的准确性：碳纤维是一种具有特殊结构的材料，巨大的长径比使其强度为钢的 100 倍，重量则只有钢的 1/6，具有良好的柔韧性和导电性能，成本低廉。为增加碳纤维导电性，在其表面镀银，可以达到较好的使用效果。由于碳纤维较轻，昆虫活动自由，波形更有代表性。相关研究论文《镀银碳纤维在烟粉虱 EPG 记录中的应用》发表在《河南农业大学学报》2015 年第 4 期。

（5）组建无线传感网络：无线传感网络作为新兴的测控网络技术，是能够自主实现数据的采集、融合和信息传输，克服了布线的烦琐，充分体现了智能系统的灵活、方便、高效。无线通信方案的选择和传输模块的设计是组建无线传感网络要解决的主要问题。引入无线传感网络，全面监控和记录昆虫的活动信号，并将 EPG 信号通过无线方式传输回实验室，可将该仪器使用空间拓宽到试验田间。ZigBee 是一组基于 IEEE 批准通过的 802.15.4 无线标准研发的有关组网、安全和应用软件方面的技术标准。它采用跳频技术，使用的频段分别为 2.4GHz（ISM），有效覆盖范围为 75m，可扩展至几百米。所支持的数据传输速率为 10～250kb/s。功耗低，在低耗电待机模式下，两节普通 5 号干电池可以工作 6 个月到 2 年。网络容量大，每个 ZigBee 网络最多可支持 255 个设备。成本低，由于 ZigBee 数据传输速率低，协议简单，所以大大降低了成本。计划开发 EPG 监测系统“互联网+”系统，将 Internet 的远程监控与 ZigBee 短距离控制相结合，实现系统的无线控制和数据传输，可远程检测、控制。

经过以上改进以后，显著提高 EPG 系统的抗干扰能力和分辨率，进而开展昆虫和植物带毒情况检测，从波形的细微变化判断昆虫是否带毒、带毒多少，揭示昆虫传毒过程的新机制和新现象。

（二）软件方面的不足及改进

第一步是常用 EPG 记录和分析软件的汉化，开发专用汉化软件对当前常用 EPG 记录和分析软件进行汉化处理，方便国内初学者使用。

第二步是新软件的开发：DATAQ 公司生产的采集卡为 12～14 位，为了看清 EPG 波形精细结构，需要更多位数的采集卡；美国国家仪器公司（National Instruments）的采集卡为 16～18 位，分辨率更高，而且有功能更为强大的 LabVIEW 软件，下一步将利用 LabVIEW 工作平台研制新型信号记录和分析、处理系统，对接收到的信号进行信号增强处理，主要采用自适应滤波、新型快速傅里叶变换、小波分析、随机共振等算法，有效地提高信噪比。

随机共振的弱信号处理技术与传统方法的区别在于：它是利用噪声而不是消除噪声来达到增强信噪比的目的。随机共振现象是非线性动力系统中一种反直观的现象，它向人们展示了噪声在非线性体系中的积极作用，即输入的弱信号在噪声的帮助下能被放大和优化，并且在合适的噪声强度下，系统的输出信噪比可达到极大值。

研制与新仪器相配套的新型采集软件、分析软件和虚拟仪器软件，全面提高仪器的自动化和数据处理水平。计划在LabVIEW软件环境下对已经采集到的生物电信号进行处理，并调用MATLAB实现信号增强等处理，并在此环境下开发新型的记录分析软件。

LabVIEW软件是一个高效快速的图形化开发环境，编程简单，特别适合于数据采集和控制、数据分析以及数据表达。MATLAB现在是国际公认的最优秀的科学计算与数学应用软件之一，它集数值分析、矩阵运算、信号处理和图形显示于一体，构成了一个方便的、界面友好的用户环境，具有可拓展性。然而，对各种算法的支持方面，LabVIEW软件的工具箱非常有限，限制了大型应用程序的快速开发。MATLAB虽然具有强大的科学计算功能和大量稳定可靠的算法库，但是其界面开发能力较差，并且数据输入、网络通信、硬件控制等方面都比较烦琐。由于两者各有欠缺，因此我们想利用混合编程在LabVIEW软件中调用和操作MATLAB，这样可以相互补充，充分发挥各自的优势。

二、EPG可视化记录系统的原创设计

通过波形来反映昆虫不同阶段的取食行为，已经远远不能满足现代科研的需要，研究者更希望在现有波形的基础上，通过直观的图像来显示昆虫头部和口针内部结构、口针在植物组织间的相对位置，以便更好地研究昆虫的取食行为和传毒机制等。而且，对于从没有进行过EPG记录的昆虫来说，各种波形的生物学意义的阐释，通常要利用口针切断、显微切片和透射电镜等技术判断昆虫口针在植物组织中的位置，从而推测不同波形所代表的昆虫行为。而EPG和动态图像的同步记录和显示，就可以直接判定昆虫的行为，包括唾液分泌、吸取汁液、传播和获取病毒粒子、植物抗性所在组织部位等。

计算机断层成像技术（computed tomography，CT）的快速进步在现代医学诊断和生命科学发展中起到了革命性作用。1972年，英国工程师G. N. Hounsfield首次成功发明了CT技术和设备，并首次完成了对物体内部的精确三维重建。随着成像技术的发展和成像精度要求的日益提高，普通CT已无法满足科学研究对分辨率的苛刻要求，人们开始研发显微CT（micro-computed tomography，Micro-CT）。Micro-CT是一种新型的采用X射线成像原理进行超高分辨率三维成像的设备，可以在不破坏样品的情况下，对各种材质样品或活体小动物进行高分辨三维成像。目前Micro-CT的活体小动物研究对象多为小鼠或兔子，对于刺吸式昆虫的Micro-CT扫描研究尚未见相关报道。

刺吸式昆虫取食与传毒机制可视化记录系统将为刺吸式昆虫研究提供崭新的研究

领域和方法。整个系统的关键是刺吸式昆虫 Micro-CT 精细图像的获取。

Micro-CT 是一种非破坏性的 3D 成像技术，采用微焦点 X 射线成像原理进行超高分辨率三维成像，可以在不破坏样本的情况下清楚了解样本的内部微结构。Micro-CT 不需要样品的制备、染色和切片，只需一次扫描就能得到样品高分辨率的完整内部三维组织结构，实验后保证样品的完整无损。

Micro-CT 系统主要分为两大部分：扫描部分和计算机处理部分。扫描部分包括探测器、X 射线球管、信号转换和与计算机连接的传输系统；计算机部分包括 X 射线和探测器控制台、转台机械控制台、CT 重建的图像显示和存储系统等。

通过 LabVIEW 虚拟仪器开发平台，设计 CT 控制及图像采集系统，与 X 射线源、转台以及探测器等硬件仪器，将一同构建一套完整的高分辨率显微 CT 成像系统，并采用此系统进行图像采集的实验，实现虚拟仪器进行 CT 系统的控制和图像的采集，界面友好，操作容易，简化 CT 系统图像采集的硬件环境。

（潘建斌　吴莉莉　邢玉清　卢少华）

实验篇参考文献

[1] ABUSHAMA F T, KAMBAL M A. The role of sugar in food aelection on the termite *Microtermet sraegardhi* (Sjoat) [J] . *Angew Entomol*, 1977, 84: 250-255.

[2] ALBERT E J. Feeding response of Easton spruce budworm (Choristonewra fumiferana) larvac to sucrose and other carbodydrates [J] . *Chemistry And Ecology*, 1982, 8: 233-40.

[3] BACKUS E A, HOLMES W J, SCHREIBER F, et al. Sharpshooter X wave: correlation of an electrical penetration graph waveform with xylem penetration supports a hypothesized mechanism for *Xylella fastidiosa* inoculation. *Annals of the Entomological Society of America*, 2009, 102 (5): 847-867.

[4] BACKUS E A, HABIBI J, YAN F, et al. Stylet penetration by adult *Homalodisca coagulata* on grape: electrical penetration graph waveform characterization, tissue correlation, and possible implications for transmission of *Xylella fastidiosa*. *Annals of the Entomological Society of America* , 2005, 98 (6): 787-813.

[5] BONANI J P, FERERES A, GARZO E, et al. Characterization of electrical penetration graphs of the Asian citrus psyllid, *Diaphorina citri*, in sweet orange seedlings. *Entomologia Experimentalis et Applicata*, 2010, 134: 35-49.

[6] CALATAYUD P A, RAHBE Y, TJALLINGII W F, et al. Electrically recorded feeding behavior of cassava mealybug on host and non-host plants. *Entomologia Experimentalis et Applicata*, 1994, 72: 219-232.

[7] CID M, FERERES A. Characterization of the probing and feeding behavior of *Planococcus citri* (Hemiptera: Pseudococcidae) on grapevine. *Annals of the Entomological Society of America*, 2010, 103: 404-417.

[8] VAN HELDEN M, TJALLINGII W F . Experimental Design and Analysis in EPG Experiments With Emphasis on Plant reistance Research. //Walker G P, Backus E A. Principles and Applications of Electronic Monitoring and Other Techniques in the Study of Homopteran Feeding Behavior. Lanham, MD: *Thomas Say Publications in Entomology*, *Entomological Society of America*, 2000, 144-171.

[9] JANSSEN JAM, TJALLINGII W F, VAN LENTEREN J C. Electrical recording and ultrastructure of stylet penetration by the greenhouse whitefly. *Entomologia Experimentalis et Applicata*, 1989, 52: 69-81.

[10] JIANG Y X, WALKER G P. Identification of phloem sieve elements as the site of resistance to silver leaf whitefly in resistant alfalfa genotypes [J] . *Entomologia Experimentalis et Applicata*, 2007, 125 (3): 307-320.

[11] JOOST P H, BACKUS E A, MORGAN D, et al. Correlation of stylet activities by the glassy-winged sharpshooter, *Homalodisca coagulata* (Say), with electrical penetration graph (EPG) waveforms. *Journal of Insect Physiology*, 2006, 52: 327-337.

[12] KOYAMA K. Nutritional physiology of the brown rice plant hopper, *Nilaparvate Lugcns* Stal (Hemiptera: Delpbacidae). Effect of sugar on nymphal development [J]. *Applied Entomology and Zoology*, 1985, 20: 292-298.

[13] MATTSON W S. Herbivoryin relation to plant nitrogencontent. *Ann Rcv Ecol Sytt*, 1980, 11: 119-161.

[14] MCLEAN D L, KINSEY M G. A technique for electronically recording aphid feeding and salivation. *Nature*, 1964, 202: 1 358-1 359.

[15] MIRANDA M P, FERERES A, APPEZZATO-DA-GLORIA B, et al. Characterization of electrical penetration graphs of *Bucephalogonia xanthophis*, a vector of *Xylella fastidiosa* in citrus. *Entomologia Experimentalis et Applicata*, 2009, 130: 35-46.
[16] REESE J C, TJALLINGII W F, VAN HELDEN M, et al. Wavefrom comparison among AC and DC electronic monitoring systems for aphid (Homoptera: Aphididae) feeding behavior. In: Principles and applications of electronic monitoring and other techniques in the study of homopteran feeding behavior (eds Walker GP and Backus EA). Thomas Say Publications in Entomology, *Entomological Society of America*, Lanham, MD., 2000: 70-101.
[17] SEO B Y, KWON Y H, JUNG J K, et al. Electrical penetration graphic waveforms in relation to the actual positions of the stylet tips of Nilaparvata lugens in rice tissue. *Journal of Asia-Pacific Entomology*, 2009, 12: 89-95.
[18] TJALLINGII W F. Manual of Giga-4/8. www. epgsystems. eu, 2011.
[19] TJALLINGII W F. Manual of Stylet^{+}. www. epgsystems. eu, 2011.
[20] TJALLINGII W F. Manual of Stylet v01. 27x. www. epgsystems. eu, 2015.
[21] WHITE T C R. The abundance of invetebrate berbivores in relation to the availabilitv of nitrogen in streased foodplants [J]. *Oecologia* (Berl.), 1984, 63: 90-105.
[22] XUE K, DENG S, WANG R J, et al. Leaf surface factors of transgenic Bt cotton associated with the feeding behaviors of cotton aphids: a case study on nontarget effects. *Science in China Series C: Life Science*, 2008, 51 (2): 145-156.
[23] XUE K, WANG X Y, HUANG C H, et al. Stylet penetration behaviors of cotton aphid *Aphis gossypii* on transgenic Bt cotton. *Insect Science*, 2009, 16 (2): 137-146.
[24] EPG昆虫刺吸系统中文使用手册．渠道科技．
[25] 忻介六．昆虫的人工饲料 [J]．昆虫知识，1976，02：62-63.
[26] 傅强，张志涛，胡萃．同翅目昆虫人工饲料饲养技术 [J]．昆虫知识，2001，38 (2)：98-104.
[27] 方杰，朱麟，杨振德，等．昆虫人工饲料配方研究概况及问题研究 [J]．四川林业科技，2003，24 (4)：18-26.
[28] 李晓敏，李静静，汤清波，等．烟碱对B型和Q型烟粉虱取食行为的影响——基于EPG和液体饲囊技术体系 [J]. 中国农业科学，2013，46 (10)：2 041-2 049.
[29] 卢少华，潘建斌，宋熙，等．镀银炭纤维在烟粉虱EPG记录中的应用．河南农业大学学报，2015，49 (2)：229-233.
[30] 汤清波，张大山，姬琨，等．刺吸电位技术应用中的几个问题．应用昆虫学报，2011，48 (5)：1 519-1 527.
[31] 王延年. 昆虫人工饲料的发展、应用和前途 [J]．昆虫知识，1990，27 (5)：310-312.
[32] 忻介六，苏德明. 昆虫、螨类、蜘蛛的人工饲料 [M]．北京：科学出版社，1979.
[33] 王延年，郑忠庆，周永生，等．昆虫人工饲料手册 [M]．上海：上海科学技术出版社，1984.
[34] 闫凤鸣．化学生态学 [M]．2版．科学出版社，2011.
[35] 岳梅，罗晨，徐洪富，等．刺吸电波实验中金丝与烟粉虱的粘连技术．昆虫知识，2005，42 (3)：326-328.
[36] 张闲，陈建群，张鹏飞．EPG即时显示软件Realdisplay的开发和利用．动物学研究，2001，22 (6)：507-510.

第三部分

应用篇

第十二章　棉蚜在转基因抗虫棉上的取食行为

（一）研究背景

棉花 *Gossypium hirsutum* L. 是世界上最重要的经济作物之一，为了控制对棉花产量造成极大损失的棉铃虫而研发的表达苏云金芽孢杆菌 *Bacilus thuringiensis*（Bt）杀虫毒蛋白的转 Bt 基因棉（Schnepf and Whiteley，1981；Schnepf et al.，1998）在我国已经有了大规模的应用（Wang et al.，1997；Wu and Guo，2000），以棉铃虫为代表的鳞翅目害虫得到有效的控制（Wilson et al.，1992；Cui and Xia，2000）。但是，随着转 Bt 基因棉的大面积种植，其生态安全性问题成为公众关注的焦点，其中就包括基因逃逸（Messeguer，2003）和非靶标效应（Andow and Hilbeck，2004）。一些关于转 Bt 基因棉对非靶标生物的影响成为研究热点，如土壤微生物（Liu et al.，2005b）、家蚕（Baines et al.，1997；Li et al.，2005）、蜜蜂（Malone et al.，2001；Malone et al.，1999；Malone and Pham-Delegue，2001；Liu et al.，2005a）、蚯蚓（Zwahlen et al.，2003）和刺吸式口器昆虫（蚜虫、粉虱、叶蝉等）（Wu and Guo，2003；邓曙东等，2003；Jasinski et al.，2003；Sisterson et al.，2004；Liu et al.，2005c；Chen et al.，2006）。

棉蚜 *Aphis gossypii* Glover 作为一种刺吸式昆虫，是转 Bt 基因棉的非靶标昆虫，也是转 Bt 基因棉生态风险评价中一种代表性昆虫。很多研究对不同转基因棉田中刺吸式昆虫种群进行了调查，但得到的结果差异很大，有的甚至相互矛盾。有文献报道，转 Bt 基因棉的种植引发刺吸式昆虫（如烟粉虱和棉蚜）大暴发（Cui and Xia，2000；Reed et al.，2001；邓曙东等，2003），而另外一些研究则发现这些昆虫种群在转 Bt 基因棉田和常规棉田中相似（Wan et al.，2003；Wang et al.，1999；Wu and Guo，2003）。刺吸式昆虫这些异常的表现是否是由于 Bt 基因插入棉花基因组而导致的？如果是，那么这些非预期效应又是如何起作用的？实验室可控条件下的实验应该会为回答这些问题提供一些有用的信息。

Xue et al.（2008，2009）使用直流刺吸电位技术记录棉蚜在转 Bt 基因棉上的刺探行为，通过比较分析，证明了该技术对于以刺吸式昆虫为实验材料的转基因生物安全性评价体系的有效性，并为转 Bt 基因棉对于非靶标的刺吸式昆虫的影响提供实验证据。

（二）材料和方法

1. 植物材料　实验所选两个棉花品系分别为 GK12 和陆地棉（*G. hirsutum*）泗棉 3 号，其中 GK12 是以泗棉 3 号作为亲本，由中国农业科学院生物技术中心研发的，表达

Bt 蛋白 Cry1Ac 的转基因棉品系。这两个品系的棉花种子由中国农业科学院植物保护研究所提供。经处理的棉籽种在塑料花盆（直径 10cm，高 10cm）中，放入光照培养箱培养，培养条件设定为光照黑暗比 16∶8，相对湿度 75%，光照时温度为 25℃，光照度 2 700lx，黑暗温度为 20℃。待棉株有 4 片全展真叶时用于实验。

实验结束后，采用酶联免疫吸附测定方法（enzyme-linked immunosorbant assay, ELISA）（Steven and Berberich，1996）验证实验用植株当中是否有 Bt 蛋白的表达。

2. 蚜虫 实验用棉蚜 *A. gossypii* Glover 采自棉田，在光照培养箱内置罩笼中的棉株上喂养。在两种棉株上保持超过 20 代后，该孤雌蚜品系的无翅成蚜即用于实验，在泗棉 3 号上繁殖的品系称为 CK 蚜虫，在 GK12 上繁殖的品系称为 Bt 蚜虫。

3. EPG 记录 使用 Giga-4 直流放大器（荷兰瓦赫宁根大学）记录棉蚜的刺探行为。实验时用水溶性导电胶将实验棉蚜的背板和金丝（直径 20μm）相连进而接入记录系统。4 只试虫都被安放在棉株第三片全展真叶的叶背。实验记录之前，使试虫饥饿 1h。整个 EPG 记录系统置于法拉第屏蔽笼中以减小噪声。每次使用两株转 Bt 基因棉和两株常规棉，随机安排通道，试虫和植物只使用一次。实验条件为室温，大约 600lx 光照度，记录持续 6h。输出信号使用 WinDaq Lite 采集软件（Dataq instruments，Inc.），通过模拟—数字转化卡（DI-720，Dataq instruments，Inc.）记录在电脑硬盘上。

EPG 实验包括 4 个组合，即 CK 蚜虫-Bt 棉、CK 蚜虫-常规棉、Bt 蚜虫-Bt 棉和 Bt 蚜虫-常规棉。

4. EPG 数据分析 EPG 波形的处理使用 WinDaq Waveform Browser 软件（Dataq instruments，Inc.），依据 Tjallingii（1990）对 EPG 波形进行识别。按照波形发生的情况，EPG 记录的蚜虫行为分为三个阶段：第一阶段从记录开始到第一次韧皮部波形发生，第二阶段从第一次韧皮部波形发生到第一次韧皮部持续取食（持续时间超过 10min 的取食），第三阶段从第一次持续取食到记录结束。EPG 波形代表了蚜虫口针在刺探过程中在植物特定组织层次的活动情况，可以分为两个大类，即路径波（A 波、B 波和 C 波，在本实验中不加区分，均记作 C 波）和韧皮部取食波（E1 波代表分泌唾液，E2 波表示吸食韧皮部汁液）（Tjallingii，1988；1990）。在此基础上，我们选择特定的 EPG 指标来进一步说明这些波形发生的生物学意义。

5. 统计分析 根据蚜虫品系的不同，EPG 数据分为两组，即 CK 蚜虫组和 Bt 蚜虫组。组内数据的统计分析使用 t 检验，百分数的比较使用 χ^2 检验，显著性差异水平为 $P<0.05$，极显著差异水平为 $P<0.01$。同样，根据棉花品系的不同，实验数据也可以分为常规棉组和转 Bt 基因棉组。

（三）结果

1. EPG 记录结果分析 EPG 实验的 4 个组合得到的有效记录数量分别是：CK 蚜虫-转 Bt 基因棉 28 个、CK 蚜虫-常规棉 31 个、Bt 蚜虫-转 Bt 基因棉 27 个、Bt 蚜虫-常规棉 24 个。

在选定的 47 个代表蚜虫在转 Bt 基因棉和常规棉上取食行为的 EPG 指标中，CK 蚜虫有 8 个指标在转 Bt 基因棉和常规棉之间差异显著，10 个指标接近显著差异（0.05<P

<0. 10)，而 Bt 蚜虫的所有指标都不存在显著性差异。所以实验的结果分析集中于 CK 蚜虫组。

2. EPG 通用指标 14 个非持续性的通用指标概括了蚜虫在两个棉花品系上的一般性的行为表现（表 12. 1）。对于 CK 蚜虫来说，在转 Bt 基因棉上的取食记录中，路径波（C 波）的总持续时间和百分比明显大于在常规棉上的情况，而被动吸食波（E2 波）的总持续时间和百分比的趋势和路径波相反，接近显著差异。CK 蚜虫的韧皮部唾液分泌（E1 波）持续时间在两种棉株上没有显著差异。对于偶尔发生的木质部波形 G（Spiller et al.，1990），无论是发生木质部波形的蚜虫比例，还是 G 波的持续时间和百分比，两个品系的蚜虫在两种棉株上都没有显著差异。在比较刺探次数、非刺探次数和电势降落（pd）次数时，我们发现 CK 蚜虫在转 Bt 基因棉上的“电势降落的总次数”显著多于在对照上的次数，这表明 CK 蚜虫更多次地刺入转 Bt 基因棉组织细胞并品尝细胞质。

3. EPG 第一阶段 我们使用 11 个连续和非连续的指标来考察棉蚜从记录开始到第一个韧皮部波形出现的行为表现（表 12. 2），主要反映了棉蚜在植物表皮和叶肉层次搜寻刺吸位点的行为。在这个阶段，CK 蚜虫在两个品系上有 4 个指标在转 Bt 基因棉和常规棉之间具有显著差异。CK 蚜虫在转 Bt 基因棉上的第一次 C 波的长度明显长于常规棉上，这个指标反映了棉蚜在第一次刺探进入植物的行为表现，这个结果说明 CK 蚜虫口针更容易在转 Bt 基因棉的表皮和叶肉当中运动。“第一阶段的刺探占总刺探数的比例”这一指标，代表蚜虫在第一阶段的付出占总记录的比重，结果显示 CK 蚜虫在常规棉上的这一指标明显大于转 Bt 基因棉上，说明常规棉的叶表面作为第一层防御机制，对 CK 蚜虫有更好的防御效果。CK 蚜虫在常规棉上第一阶段的非刺探时间百分比明显长于转 Bt 基因棉，同时在常规棉上第一阶段刺探时间百分比接近显著差异，这两种状态在第一阶段的分布情况说明 CK 蚜虫在常规棉表面上花费更多的时间寻找刺吸位点。CK 蚜虫在常规棉上第一阶段电势降落次数占总次数的百分比明显大于转 Bt 基因棉，但是这个阶段的电势降落数却是基本相等的，这表明这个显著差异是来源于计算，对于行为分析来说意义不大。这个阶段 CK 蚜虫的其他指标在统计上都是相似的。这个阶段中 Bt 蚜虫在两种棉株上的行为表现基本一致，只是第一次刺探即达到韧皮部的蚜虫比例这一指标存在显著差异，转 Bt 基因棉上的第一阶段电势降落数接近显著差异。

4. EPG 第二阶段 第二阶段是指第一次 E 波发生到第一次韧皮部持续取食（E 波持续时间超过 10min）（表 12. 3）。在这个阶段，所有的 E 波都短于 10min，所以在这个阶段不对 E1 波和 E2 波进行区分。这个阶段中，无论是 CK 蚜虫还是 Bt 蚜虫，在两种棉株上的行为指标都不存在显著性差异。其中，对于 CK 蚜虫来说，在转 Bt 基因棉上第一次持续取食之前的刺探持续时间略短于常规棉，转 Bt 基因棉上第二阶段 E 波百分比高于常规棉，差异接近显著。转 Bt 基因棉上 Bt 蚜虫的 E 波百分比高于常规棉，差异同样接近显著。在此阶段，两个蚜虫品系在两种棉株上表现近似，这说明从发现韧皮部到持续取食这个过程当中，棉蚜在棉株上的表现是一致的。

表 12.1　蚜虫在两种棉株上 EPG 记录中通用指标的比较

通用指标	相关的植物组织层次	CK 蚜虫-常规棉 (n=31)	CK 蚜虫-Bt 棉 (n=28)	P 值	Bt 蚜虫-常规棉 (n=27)	Bt 蚜虫-Bt 棉 (n=24)	P 值
刺探总次数	表皮、叶肉和实质	21.65±1.73	24.25±1.75	0.148	27.59±2.58	27.13±2.73	0.451
电势降落总次数	细胞内	150.74±11.41	186.82±12.60	0.019	161.78±10.49	180.42±16.34	0.171
E1 波的总持续时间（s）	韧皮部	1 174.99±128.03	1 457.42±388.05	0.247	1 174.26±204.63	1 248.31±188.22	0.396
E1 波的百分比（%）	韧皮部	5.44±0.59	6.75±1.80	0.247	5.44±0.95	5.78±0.87	0.396
E2 波的总持续时间（s）	韧皮部	8 359.19±900.38	6 626.16±795.62	0.079	5 929.55±758.05	5 642.48±1 013.01	0.410
E2 波的百分比（%）	韧皮部	38.70±4.17	30.68±3.68	0.079	27.45±3.51	26.12±4.69	0.410
非刺探的总持续时间（s）	表面	3 266.29±362.76	3 452.46±369.21	0.360	4 310.37±568.32	3 858.87±616.80	0.296
非刺探的百分比（%）	表面	15.12±1.68	15.98±1.71	0.360	19.96±2.63	17.87±2.86	0.296
非刺探的总次数	表面	17.23±1.49	19.68±1.48	0.125	23.19±2.50	22.04±2.46	0.373
刺探的总持续时间（s）	表面、表皮和叶肉	8 389.33±604.26	9 530.56±583.12	0.091	9 313.30±552.41	9 497.52±826.62	0.427
刺探的百分比（%）	表面、表皮和叶肉	38.84±2.80	44.12±2.70	0.091	43.12±2.56	43.97±3.83	0.427
G 波的总持续时间（s）*	木质部	2 543.28±997.65	2 133.56±404.87	0.339	2 141.64±459.68	3 971.04±1 590.90	0.151
G 波的百分比（%）*	木质部	11.77±4.62	9.88±1.87	0.339	9.92±2.13	18.38±7.37	0.151
发生木质部吸食的蚜虫比率（%）	木质部	16.13	25.00	0.398	40.74	33.33	0.585

* 该指标的重复数分别为 5、7、11 和 8。

5. EPG 第三阶段 本阶段反映了蚜虫在韧皮部阶段的行为表现（表 12.4）。对于 CK 蚜虫来说，在两种棉株上的第三阶段的持续时间相似，但在转 Bt 基因棉上第三阶段的 E2 波百分比极显著地小于常规棉，本阶段刺探次数则明显多于常规棉，发生持续吸食的次数基本相等。这些结果说明 CK 蚜虫吸食更多的常规棉韧皮部汁液，而在转 Bt 基因棉上则更频繁地变换吸食位点。同时，在转 Bt 基因棉上的刺探百分比、非刺探百分比和总 E 波百分比都高于常规棉上的指标，且差异接近显著。这些数据为上面说明的行为提供了进一步的证明。CK 蚜虫的第三阶段的电势降落数存在显著差异，这说明 CK 蚜虫在转 Bt 基因棉上更多次地在组织细胞当中吸食并品尝细胞质。Bt 蚜虫在此阶段的各项指标没有显著差异。

此外，不同蚜虫品系在同样棉株上的行为比较不存在显著差异（表 12.3 中没有标示出来）。

（四）讨论

1. EPG 记录的生物学意义 蚜虫的刺探取食行为可以通过上述 EPG 指标反映出来的生物信息，加上已经详细报道过的关于 EPG 波形和蚜虫在不同植物组织中口针刺探行为的相互关系（Tjalingii，1988；Ellsbury et al.，1994；Walker and Backus，2000）推断出来。更重要的是，因为蚜虫取食行为和植物因素的复杂性，整合足够的 EPG 指标可以得到某些重要结论，所以足够多的严格设计的 EPG 指标是得到足够生物学信息的前提。

根据现在得到的结果，我们可以从 CK 蚜虫的行为推断出一些转 Bt 基因棉和常规棉的差异。从上述的取食行为指标当中，特别是 CK 蚜虫在两种棉株上存在差异的指标，我们可以推断出植物当中肯定存在一些可能影响蚜虫口针刺探行为的因素，这些因素可以定位在组织和细胞层次，如表面、表皮、叶肉和实质、细胞内和韧皮部。

在 EPG 记录前受试蚜虫被饥饿 1 h，所以记录开始时蚜虫对于食物是非常渴望的。然而，第一阶段非刺探百分比表明 CK 蚜虫在转 Bt 基因棉上花费的非刺探时间少于在常规棉上，这说明转 Bt 基因棉表面比常规棉表面更容易被蚜虫口针刺入。可以推定，表面的物理和（或）化学因素在这个层次上影响 CK 蚜虫行为。

在表皮、叶肉和其他实质层次，CK 蚜虫第一阶段的刺探占总刺探的比率，转 Bt 基因棉上的明显低于常规棉上，尽管第一阶段的刺探数基本一致。进入第三阶段，CK 蚜虫在转 Bt 基因棉上刺探次数明显增多（图 12.1）。更多的刺探表明蚜虫不喜欢该植物，不断变换取食位点。综合这些结果，在这个植物层次上，CK 蚜虫更容易在转 Bt 基因棉上进行刺探。

表 12.2　蚜虫在两种棉株上 EPG 记录中第一阶段指标的比较

第一阶段	相关的植物组织层次	CK 蚜虫-常规棉（n=31）	CK 蚜虫-Bt 棉（n=28）	P 值	Bt 蚜虫-常规棉（n=27）	Bt 蚜虫-Bt 棉（n=24）	P 值
第一次 E 波发生的时间（s）	除韧皮部外的其他植物组织	3 562. 36±616. 70	2 881. 83±322. 75	0. 167	4 121. 77±817. 06	4 098. 34±629. 01	0. 491
第一阶段刺探次数	表皮、叶肉和实质	6. 71±0. 86	6. 32±0. 72	0. 366	6. 96±0. 74	7. 17±1017	0. 442
第一阶段刺探次数占总刺探次数的比率（%）	表皮、叶肉和实质	36. 65±4. 45	27. 79±2. 86	0. 050	31. 41±4. 86	31. 94±5. 19	0. 471
第一阶段刺探的百分比（%）	表皮、叶肉和实质	68. 81±3. 50	76. 06±3. 47	0. 074	68. 57±2. 94	72. 40±4. 12	0. 223
第一阶段的电势降落数	细胞内	40. 68±5. 92	37. 36±3. 47	0. 315	35. 52±3. 96	46. 08±5. 52	0. 060
第一阶段电势降落数占总电势降落数的比率（%）	细胞内	32. 30±4. 54	22. 25±2. 21	0. 027	26. 45±4. 72	31. 25±4. 32	0. 230
第一次刺探的发生时间（s）	表面	315. 99±154. 26	150. 36±89. 04	0. 179	208. 95±88. 34	221. 06±111. 86	0. 466
第一阶段非刺探的百分比（%）	表面	30. 52±3. 56	20. 45±3. 12	0. 020	27. 39±2. 52	24. 50±3. 98	0. 271
第一次刺探的持续时间（s）	表面、表皮和叶肉	183. 02±64. 78	394. 38±106. 53	0. 048	340. 90±190. 30	353. 47±104. 41	0. 477
第一次电势降落的发生时间（s）	表面、表皮、叶肉和实质	355. 35±153. 04	192. 14±95. 63	0. 185	282. 24±94. 64	326. 29±115. 48	0. 384
第一次刺探即可到达韧皮部的蚜虫比率（%）	表面、表皮、叶肉和实质	6. 45	7. 14	0. 916	0	16. 67	0. 027

表 12.3　蚜虫在两种棉株上 EPG 记录中第二阶段指标的比较

第二阶段	相关的植物组织层次	CK 蚜虫-常规棉（n=31）	CK 蚜虫-Bt 棉（n=28）	P 值	Bt 蚜虫-常规棉（n=27）	Bt 蚜虫-Bt 棉（n=24）	P 值
第一次持续取食的发生时间（s）	所有植物组织	5 377.01±814.28	4 677.38±733.27	0.265	6 692.92±1 009.35	6 425.22±940.11	0.424
从第一次刺探到第一次持续取食的时间（s）	所有植物组织	5 061.02±822.82	4 527.02±739.17	0.317	6 483.97±1 015.42	6 204.16±945.83	0.421
第一次持续取食前路径波的持续时间（s）	除表面外的所有植物组织	1 142.06±153.31	909.19±72.95	0.089	973.88±128.80	1 030.84±122.09	0.376
第二阶段刺探的百分比(%) *	表皮、叶肉和实质	70.05±2.63	64.98±2.10	0.182	63.91±2.14	69.51±2.93	0.131
第一次持续取食前路径波当中的电势降落数	细胞内	22.87±2.54	22.29±1.99	0.429	21.07±2.23	22.92±2.60	0.295
第二阶段电势降落数 *	细胞内	58.33±7.79	68.09±13.36	0.346	54.00±8.96	58.25±8.30	0.403
第二阶段 E 波发生次数 *	韧皮部	1.83±0.28	1.91±0.16	0.443	1.47±0.18	2.08±0.29	0.094
第二阶段 E 波百分比（%） *	韧皮部	9.25±1.55	15.51±2.33	0.067	14.85±3.42	11.88±1.16	0.275
第二阶段非刺探的百分比（%） *	表面	19.40±2.39	18.09±2.94	0.415	19.42±2.74	16.60±2.85	0.305
第二阶段的持续时间（s） *	所有植物组织	4 687.84±613.84	4 570.49±814.28	0.471	4 628.07±886.00	4 653.76±696.47	0.494

* 该指标的重复数分别为 12、11、15 和 12。

表 12.4　蚜虫在两种棉株上 EPG 记录中第三阶段指标的比较

第三阶段	相关的植物组织层次	CK 蚜虫-常规棉（$n=31$）	CK 蚜虫-Bt 棉（$n=28$）	P 值	Bt 蚜虫-常规棉（$n=27$）	Bt 蚜虫-Bt 棉（$n=24$）	P 值
第三阶段持续时间（s）	所有植物组织	16 222. 99±814. 28	16 922. 62±733. 27	0. 265	14 907. 08±1 009. 35	15 174. 78±940. 11	0. 424
第一次持续取食之后 E 波的次数	所有植物组织	4. 10±0. 47	4. 07±0. 39	0. 484	3. 59±0. 41	3. 71±0. 54	0. 432
第三阶段 C 波百分比（%）	表皮、叶肉和实质	30. 43±2. 75	36. 43±3. 08	0. 075	33. 23±3. 42	33. 43±4. 10	0. 485
第三阶段电势降落数量	细胞内	87. 48±10. 55	122. 71±11. 68	0. 014	96. 26±12. 27	105. 21±16. 03	0. 446
持续取食前的路径波中电势降落的平均数	细胞内	21. 51±1. 65	23. 31±1. 81	0. 232	21. 47±1. 93	24. 27±2. 22	0. 172
第一次持续取食的持续时间（s）	韧皮部	3 914. 26±914. 37	3 998. 36±816. 89	0. 473	3 679. 41±753. 29	3 459. 49±893. 24	0. 425
持续取食的次数	韧皮部	2. 48±0. 23	2. 68±0. 23	0. 275	2. 11±0. 19	2. 50±0. 26	0. 117
第三阶段 E1 波的百分比（%）	韧皮部	7. 29±0. 99	9. 99±3. 17	0. 211	7. 92±1. 64	7. 07±1. 13	0. 335
第三阶段 E2 波的百分比（%）	韧皮部	51. 58±4. 15	37. 69±4. 02	0. 010	41. 87±4. 74	41. 53±6. 14	0. 482
第三阶段 E1 波和 E2 波总和的百分比（%）	韧皮部	57. 20±4. 17	47. 68±4. 61	0. 065	49. 79±4. 92	48. 60±5. 60	0. 437
第三阶段刺探的次数	韧皮部	8. 58±1. 26	11. 57±1. 19	0. 046	11. 74±2. 32	12. 17±2. 01	0. 446
第三阶段非刺探的百分比（%）	表面	10. 53±1. 42	14. 29±1. 83	0. 053	15. 33±3. 07	12. 97±2. 45	0. 279

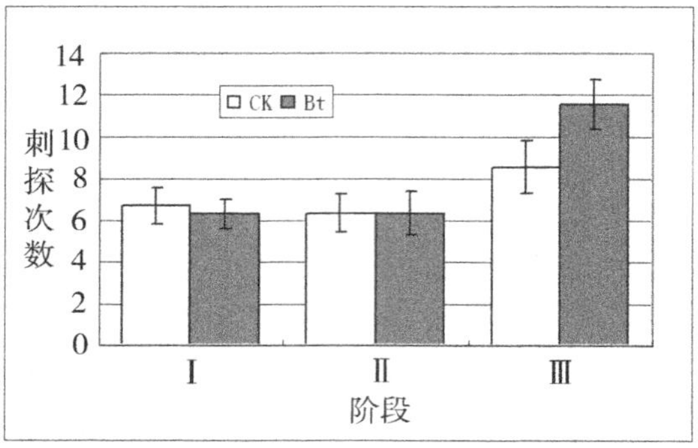

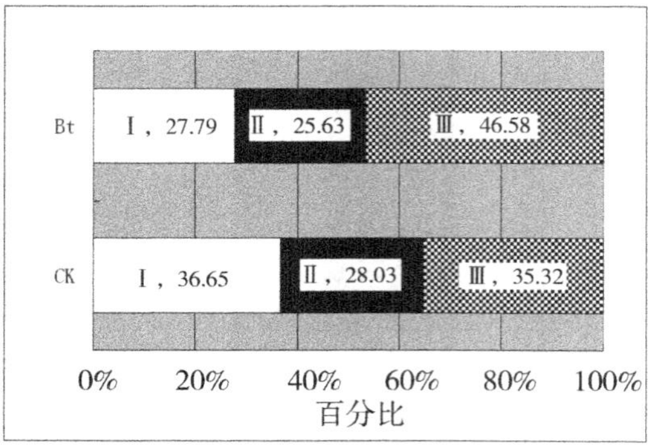

图 12.1 CK 蚜虫在转 Bt 基因棉和常规棉上 EPG 记录当中各个阶段刺探次数（左）和百分比（右）

（Ⅰ、Ⅱ和Ⅲ分别代表 EPG 记录的第一、第二和第三阶段）

电势降落（pd）是路径波的一个特征，是和蚜虫口针刺入叶肉细胞并向其中分泌唾液并取食细胞内容物的过程相联系的（Tjalingii，1988；Walker and Backus，2000）。一般来说，和 pd 相关的指标与取食细胞内容物并判断汁液质量有关。电势降落数量多意味着蚜虫更多地刺入组织细胞并从中抽样取食。通读整个记录，CK 蚜虫在转 Bt 基因棉上刺探了更多次以寻找更好的细胞内容物。在第一阶段，CK 蚜虫在两种棉株上用了几乎相等的次数从植物组织细胞中吸取细胞内容物进而找到第一次吸食位点。持续取食前的 pd 数和第二阶段 pd 数都基本相等。第三阶段中，尽管 CK 蚜虫已经可以在转 Bt 基因棉上持续取食，但是它们还是要更多次地刺探以寻找新的吸位点。

蚜虫是韧皮部吸食昆虫，所以有关韧皮部因子的指标将是评价植物对蚜虫适合程度的重要部分。在上述的结果中，共有 11 个指标反映了韧皮部汁液的质量。E1 波代表蚜虫口针刺入韧皮部筛管并向其中分泌唾液的过程。CK 蚜虫关于 E1 波的指标相等，而代表韧皮部被动吸食的 E2 波的指标则表现出显著的或者接近显著的差异。CK 蚜虫在转 Bt 基因棉的韧皮部取食明显少于在常规棉上的取食。一般来说，韧皮部吸食的减少被认为是汁液不适合的标志。Bt 毒素不能在韧皮部汁液当中表达（Raps et al.，2001；Dutton et al.，2002），蚜虫也不是 Bt 毒素的靶标昆虫，所以 CK 蚜虫在转 Bt 基因棉上韧皮部吸食减少的可能原因是转 Bt 基因棉韧皮部汁液的营养水平低于常规棉。

总而言之，CK 蚜虫更容易刺入转 Bt 基因棉，但是不断地改变吸食位点，并且在转 Bt 基因棉上的吸食明显少于常规棉。但是，Bt 蚜虫在两种棉株上绝大多数指标都无显著差异。基于两个蚜虫品系的行为表现，我们可以得到以下结论：转 Bt 基因棉的表面、叶肉实质和韧皮部汁液存在某些可以影响 CK 蚜虫的因素，但是 CK 蚜虫经过几个世代就可以完全适应转 Bt 基因棉；Bt 蚜虫可以没有障碍地在两种棉株上取食。棉蚜在转 Bt 基因棉上生命表的研究证明棉蚜可以很快适应转基因棉株品系（Liu et al.，2005c；Yang et al.，2006），这一点在 EPG 实验当中也可以得到印证。

2. 棉蚜行为差异的可能机制 CK 蚜虫在两种棉株上的行为有明显差异，这说明转 Bt 基因棉在某些层次同常规棉有不同，但是这些差异的机制还不是很清楚。转 Bt 基因棉对于棉蚜的非靶标效应在本实验中被证实，我们可以对这种非靶标效应的机制进行推测。*bt* 基因的转入是通过土壤农杆菌的 T 质粒，所以 *bt* 基因插入棉花基因组是随机的，这样的遗传操作可能导致植物产生一些表型上的非预期改变。外源毒素的插入可以在多个方面引起植物特征的改变，如随着蛋白的表达，次生代谢物的生物合成有所改变（Aharoni et al.，2003；Yan et al.，2004）。根据上述结果，我们认为一个可能的机制是 *bt* 基因在细胞内表达，使得转 Bt 基因棉对蚜虫刺探的阻碍减少，但转 Bt 基因棉韧皮部汁液的营养水平变低。通过化学分析，排除了转基因棉花和常规棉叶表化学物质的差异，最后找到了造成棉蚜取食差异的原因是转基因棉叶表面毛的不同（Xue et al.，2008）。

3. 棉蚜行为差异的生态学意义　自从转 Bt 基因棉被批准商业化种植以来，不同的转基因棉田中刺吸式昆虫种群的大田观测结果有很大的不同。一些研究报道了刺吸式昆虫，如粉虱、蚜虫和叶蝉，在转 Bt 基因棉田的暴发（Cui and Xia，2000；Reed et al.，2001），而另一些研究中的结果显示转 Bt 基因棉田中这些昆虫的种群动态和常规棉田中的相似（Wan et al.，2003；Wang et al.，1999；Wu and Guo，2003）。根据本实验的结果，棉蚜可以更容易地刺入转 Bt 基因棉表面，在其上定殖并很快适应。考虑到鳞翅目昆虫种群被控制和蚜虫的快速繁殖，蚜虫和其他刺吸式昆虫的暴发是非常可能的。影响这些大田观察结果的因素有很多是未知的，所以结合实验室取食行为实验，设计严谨的大田观察会为刺吸式昆虫在转 Bt 基因棉田的发生这样的争论性研究提供验证。

根据本实验结果，我们可以有另外一个推测。pd 波和刺吸式昆虫传播植物病毒相关（Jiang et al.，2000；Johnson et al.，2002）。在整个刺探过程中，CK 蚜虫将口针刺入转 Bt 基因棉的次数明显多于常规棉上。当蚜虫从感染了植物病毒的常规棉上取食之后转移到健康的转 Bt 基因棉和常规棉上时，转 Bt 基因棉植株比常规棉更容易感染。换句话说，转 Bt 基因棉更容易受到棉蚜携带的植物病毒的感染。以前，病毒造成的棉花损失不是很严重，相关的报道也很少。当鳞翅目昆虫被 Bt 毒素控制后，刺吸式昆虫成为最主要的害虫，刺吸式昆虫传播病毒的机会更高了。所以当转 Bt 基因棉大面积种植时这个问题应该被重视起来。

参考文献

[1] 邓曙东，徐静，张青文，等．转 Bt 基因棉对非靶标害虫及害虫天敌种群动态的影响．昆虫学报，2003，46（1）：1-5.

[2] AHARONI A，GIRI A P，DEUERLEIN S，et al. Terpenoid metabolism in wild-type and transgenic Arabidopsis plants. *Plant Cell*，2003，15：2 866-2 884.

[3] ANDOW D A，HILBECK A. Science - based risk assessment for nontarget effects of transgenic crops. *Bioscience*，2004，54：637-649.

[4] BAINES D，SCHWARTZ J L，SOHI S，et al. Comparison of the response of midgut epithelial cells and cell lines from lepidopteran larvae to CryIA toxins from *Bacillus thuringiensis*. *Journal of Insect Physiology*，1997，43：823-831.

[5] CHEN M, Y E G Y, LIU Z C, et al. Field assessment of the effects of transgenic rice expressing a fused gene of cry1Ab and cry1Ac from *Bacillus thuringiensis* Berliner on nontarget planthopper and leafhopper populations. *Environmental Entomology*, 2006, 35: 127-134.

[6] CUI J J, XIA J Y. Effects of Bt (*Bacillus thuringiensis*) transgenic cotton on the dynamics of pest population and their enemies. *Acta Phytophylacica Sinica*, 2000, 27: 141-145.

[7] DUTTON A, KLEIN H, ROMEIS J, et al. Uptake of Bt-toxin by herbivores feeding on transgenic maize and consequences for predator *Chrysoperla carnea*. *Ecological Entomology*, 2002, 27: 441-447.

[8] ELLSBURY M M, BACKUS E A, ULLMAN, D L. History, Development, and Application of AC Electrical Insect Feeding Monitors. *In*: *Thomas Say Publications in Entomology*, *Entomological Society of America*, Lanham, MD, 1994: 128.

[9] JASINSKI J R, EISLEY J B, YOUNG C E, et al. Select nontarget arthropod abundance in transgenic and nontransgenic field crops in Ohio. *Environmental Entomology*, 2003, 32: 407-413.

[10] JIANG Y, DE BLAS C, BARRIOS L, et al. Correlation between whitefly (Homoptera: Aleyrodidae) feeding behavior and transmission of tomato yellow leaf curl virus. *Annals of the Entomological Society of America*, 2000, 93: 573-579.

[11] JOHNSON D D, WALKER G P, CREAMER R. Stylet penetration behavior resulting in inoculation of a semipersistently transmitted closterovirus by the whitefly *Bemisia argentifolii*. *Entomologia Experimentalis et Applicata*, 2002, 102: 115-123.

[12] LI W, WU K, WANG X, et al. Impact of pollen grains from Bt transgenic corn on the growth and development of Chinese tussah silkworm, *Antheraea pernyi* (Lepidoptera: Saturniidae) . *Environmental Entomology*, 2005, 34: 922-928.

[13] LIU B, XU C, YAN F, et al. The impacts of the pollen of insect-resistant transgenic cotton on honeybees. *Biodiversity and Conservation*, 2005a, 14: 3 487-3 496.

[14] LIU B, ZENG Q, YAN F, et al. Effects of transgenic plants on soil microorganisms. *Plant and Soil*, 2005b, 271: 1-13.

[15] LIU X, ZHAI B, ZHANG X, et al. Impact of transgenic cotton plants on a non-target pest, *Aphis gossypii* Glover. *Ecological Entomology*, 2005c, 30: 307-315.

[16] MALONE L A, BURGESS E P J, GATEHOUSE H S, et al. Effects of ingestion of a *Bacillus thuringiensis* toxin and a trypsin inhibitor on honey bee flight activity and longevity. *Apidologie*, 2001, 32: 57-68.

[17] MALONE L A, BURGESS E P J, STEFANOVIC D. Effects of a *Bacillus thuringiensis* toxin, two *Bacillus thuringiensis* biopesticide formulations, and a soybean trypsin inhibitor on honey bee (*Apis mellifera* L.) survival and food consumption. *Apidologie*, 1999, 30: 465-473.

[18] MALONE L A, PHAM-DELEGUE M H. Effects of transgene products on honey bees (*Apis mellifera*) and bumblebees (*Bombus sp*) . *Apidologie*, 2001, 32: 287-304.

[19] MESSEGUER J. Gene flow assessment in transgenic plants. *Plant Cell*, *Tissue and Organ Culture*, 2003, 73: 201-212.

[20] RAPS A, KEHR J, GUGERLI P, et al. Immunological analysis of phloem sap of *Bacillus thuringiensis* corn and of the nontarget herbivore *Rhopalosiphum padi* (Homoptera: Aphididae) for the presence of Cry1Ab. *Molecular Ecology*, 2001, 10: 525-533.

[21] REED G L, ANDREW S J, JENNIFER R, et al. Transgenic Bt potato and conventional insecticides for Colorado potato beetle management: comparative efficacy and non-target impacts. *Entomologia Experimentalis et Applicata*, 2001, 100: 89-100.

[22] SCHNEPF E, CRICKMORE N, VAN RIE J, et al. *Bacillus thuringiensis* and its pesticidal crystal proteins. *Microbiology and Molecular Biology Reviews*, 1998, 62: 775-806.

[23] SCHNEPF H E, WHITELEY H R. Cloning and expression of *Bacillus thuringiensis* crystal protein gene in *Escherichia coli*. *Proceedings of the National Academy of Science of the United States of America*, 1981, 78: 2 893-2 897.

[24] SISTERSON M S, BIGGS R W, OLSON C, et al. Arthropod abundance and diversity in Bt and non-Bt cotton fields. *Environmental Entomology*, 2004, 33: 921-929.

[25] SPILLER N J, KOENDERS L, TJALLINGII W F. Xylem ingestion by aphids – a strategy for maintaining water balance. *Entomologia Experimentalis et Applicata*, 1990, 55: 101-104.

[26] STEVEN R S, BERBERICH S A. *Bacillus thuringiensis* CryIA protein levels in raw and processed seed of transgenic cotton: determination using insect bioassay and ELISA. *Journal of Economic Entomology*, 1996, 89: 247-251.

[27] TJALLINGII W F. Electrical recording of stylet penetration activities. *In*: *Aphid*, *Their Biology*, *Nature Enemies and Control* (eds. A. K. Minks & P. Harrewijn). Elsevier Science Publishers, Amsterdam, 1988: 95-108.

[28] TJALLINGII W F. Continuous recording of stylet penetration activities by aphids. *In*: *Aphid-Plant Genotype Interactions* (eds. R. K. Campbell & R. D. Eikenbary), Elsevier Science Publishers, Amsterdam, 1990: 89-99.

[29] WALKER G P, BACKUS E A. Principles and Applications of Electronic Monitoring and other Techniques in the Study of Homopteran Feeding Behavior. *In*: *Thomas Say Publications in Entomology*, *Entomological Society of America*, Lanham, MD, 2000: 260.

[30] WAN P, HUANG M, WU K, et al. Effects of transgenic Bt cotton on development and population dynamics of cotton aphid. *Scientia Agricultura Sinica*, 2003, 36: 1 484-1 488.

[31] WANG W, JIANG Y, YANG X, et al. Study on evaluation and utilization of transgenic Bt cotton resistance to cotton bollworm. *Scientia Agricultura Sinica*, 1997, 30: 7-12.

[32] WILSON W D, FLINT H M, DEATON R W, et al. Resistance of cotton lines containing a *Bacillus thuringiensis* toxin to pink bollworm (Lepidoptera: Noctuidae) and other insects. *Journal of Entomological Science*, 1992, 34: 415-425.

[33] WU K, GUO Y. Field resistance evaluations of Bt transgenic cotton GK series to cotton bollworm. *Acta Phytophylacica Sinica*, 2000, 27: 317-321.

[34] WU K, GUO Y. Influences of *Bacillus thuringiensis* Berliner cotton planting on population dynamics of the cotton aphid, *Aphis gossypii* Glover, in northern China. *Environmental Entomology*, 2003, 32: 312-318.

[35] XUE K, WANG X Y, HUANG C H, et al. Stylet penetration behaviors of cotton aphid *Aphis gossypii* on transgenic Bt cotton. *Insect Science*, 2009, 16 (2): 137-146.

[36] XUE K, DENG S, WANG R J, et al. Leaf surface factors of transgenic Bt cotton associated with the feeding behaviors of cotton aphids: a case study on non-target effects. *Science in China C*, 2008, 51 (2): 145-156.

[37] YAN F, BENGTSSON M, ANDERSON P, et al. Antennal response of cotton bollworm (*Heliocoverpa armigera*) to volatiles in transgenic Bt cotton. *Journal of Applied Entomology*, 2004, 128: 354-357.

[38] YANG Y, LU Y, XUE W, et al. Population dynamics of *Aphis gossypii* Glover in transgenic cotton fields and an analysis of the related influencing factors. *Acta Entomologica Sinica*, 2006, 49: 80-85.

[39] ZWAHLEN C, HILBECK A, HOWALD R, et al. Effects of transgenic Bt corn litter on the earthworm *Lumbricus terrestris*. *Molecular Ecology*, 2003, 12: 1 077-1 086.

（薛 堃）

第十三章　烟粉虱的取食行为研究

（一）研究背景

烟粉虱 *Bemisia tabaci*（Gennadius），属半翅目（Hemiptera）粉虱科（Aleyrodidae）小粉虱属（*Bemisia* Quaintance & Baker），又叫棉粉虱、银叶粉虱、甘薯粉虱，是一种具刺吸式口器的植食性昆虫，寄主植物广泛，达600余种（Oleveira，2001）。成虫为淡黄白色，翅展约1.9mm，静止时翅合拢成屋脊形，雌成虫体长约0.95mm，雄成虫体长约0.85mm，成虫喜欢群集并成对取食，喜欢刺吸寄主植物上部嫩叶，随着植物的生长，成虫不断向新叶转移，而卵和伪蛹则留在了下部，故出现由下向上扩散为害的垂直分布。烟粉虱在热带和亚热带地区1年可以发生11~15代，且世代重叠（Husain & Trehan，1993）。另外烟粉虱的形态特征、生理代谢等会随着寄主植物的不同而产生变异，这些变异给种类鉴定和农业防治带来许多困难，如在具毛的叶片上烟粉虱的蛹壳体形较小，边缘深陷，背面乳突或刚毛较长；而在无毛的叶片上蛹壳体形较大，边缘整齐，背面乳突或刚毛短（闫凤鸣，2001）。

烟粉虱是由许多生物型组成的复合种。在形态上，烟粉虱的分类主要依靠观察蛹壳，而蛹壳有很大的变异性（闫凤鸣，1991），从而给烟粉虱生物性的划分带来许多困难，褚栋等（2006）在世界各国对烟粉虱核糖体ITS1（rDNA ITS1）、线粒体COI（mtDNA COI）基因大量测序的基础上，进一步分析了烟粉虱不同地理种群的遗传分化，结果显示烟粉虱不同地理种群可分为5组：即亚洲组、美洲组、非洲组、澳洲组、B型/地中海/中东/北非/Ms型组，此外还包括3个没有特定组的种群，即乌干达种群、科特迪瓦种群和台湾种群。目前研究表明，烟粉虱至少可分为36个生物型（Perring，2001）。

烟粉虱广泛分布于除南极洲外的世界各地，给农业生产造成了严重的经济损失，已成为一种世界性的重要农业害虫，数据显示美国仅在烟粉虱发生的10年间所造成的经济损失平均每年就超过10亿美元（White，1998）。烟粉虱在中国为害棉花的报道始见于1953年的台湾省（李凤荪，1953），之后在大陆各地陆续发现，在我国许多棉区因烟粉虱为害而造成的损失严重时可高达50%（吴孔明等，2001）。烟粉虱对寄主植物的为害，不仅仅是刺吸植物汁液造成植物干枯，而且它分泌的蜜露还可以诱发煤污病影响植物光合作用和生长，在烟粉虱取食过程中还会传播病毒造成病害，如烟粉虱传播的棉花皱缩病毒（CdTV）可使棉铃减少15%~87%，得病的棉花产量损失80%（徐文华，2006）。B型烟粉虱是中国普遍发生的生物型，相关研究也较多（林克剑等，

2008；周隆等，2011），但自2003年在云南发现Q型烟粉虱后，又相继在其他地区发现，最近几年Q型烟粉虱在中国各地发生日趋严重，并逐渐取代B型烟粉虱成为优势生物型（褚栋等，2005a；徐婿等，2006；姬琨等，2010），因而备受人们关注。与B型烟粉虱相比，Q型烟粉虱具有更强的生物学优势、更强的传毒能力和更高的抗药性（纠敏，2008）。

烟粉虱体型较小，生物型分化较多，取食行为复杂，因此常规的昆虫学研究法在烟粉虱的研究上有较多困难，而刺吸电位技术则在烟粉虱的行为学研究中具有得天独厚的条件。本章综述近年来本实验室利用EPG技术在研究烟粉虱的取食行为等方面取得的进展。

（二）B型烟粉虱在4种不同寄主植物上的DC-EPG取食行为研究

张大山（2010）运用DC-EPG技术研究B型烟粉虱在烟草（K326）、黄瓜（绿剑一号）、茄子（优罐一号）、西葫芦（新玉二号）4种寄主上的取食行为。结果发现，在一般性指标如总的刺探时间、刺探次数、总的细胞内刺探次数等指标上烟粉虱在这4种寄主上差异不显著。而在到达韧皮部（Phloem）有关的指标中（表13.1），有4个指标达到显著差异（$P<0.05$）：从B型烟粉虱口针到达韧皮部的次数上（N）来看烟草的次数最多，茄子的最少（6.00±4.05）；B型烟粉虱第一次到达韧皮部所用的时间（1st E），烟草最长（15 356.02±1 741.97），西葫芦最短（6 796.32±3 659.93）；第一次刺探就达到韧皮部所花费的时间，黄瓜最长（3 062.83±1 959.04），西葫芦最短（113.32±21.49）；第一次刺探中达到韧皮部所经历的细胞内刺探数，黄瓜最多（5.00±0.00），烟草次之（3.00±0.00），西葫芦最少（2.00±0.00）。

表13.1　B型烟粉虱在4种寄主上与韧皮部行为有关的EPG指标（张大山，2010）

指标	烟草	黄瓜	茄子	西葫芦
韧皮部时间总和E（E1+E2）（s）	6 061.58±1 672.68	4 077.76±708.07	4 103.86±1 115.67	3 292.03±1 814.77
韧皮部分泌唾液时间总（E1）（s）	3 797.25±1 581.32	1 247.42±448.89	2 724.84±841.27	2 160.70±1 557.45
韧皮部吸食时间总和（E2）（s）	3 113.45±1 313.16	4 055.91±1 086.75	3 033.86±1 019.41	1 131.33±374.98
第一次韧皮部吸食开始前分泌唾液时间 E_2（s）	432.67±126.89	219.79±140.14	711.17±322.32	76.12±24.68
第一次韧皮部持续吸食开始时间 E_2>10min（s）	14 472.89±1 585.26	10 791.36±1 304.21	11 073.52±3 454.16	12 121.31±4 643.77
开始刺探至开始持续吸食时间 E_2>10min（s）	12 364.85±1 966.97a	8 310.08±1 843.00d	12 288.76±0.00b	7 733.08±1 571.03c
刺探内到达持续吸食时间 E_2>10min（s）	466.19±7.16	3 668.00±3 265.13	0.00±0.00	67.94±3.59
刺探内到达持续吸食时间历经pd数（N）	0.00±0.00	5.00±0.00	2.00±0.00	4.00±0.00

与韧皮部取食行为有关的指标，一个指标差异显著（表13.2）。B型烟粉虱在开始刺探至开始取食所花费的时间上（12 364.85±1 966.97）显著多于其他作物，而在黄瓜

上的最短（8 310.08±1 843.00）；B 型烟粉虱在韧皮部花费的时间，在烟草上较长（6 061.58±1 672.68），黄瓜次之（4 077.76±708.07），西葫芦最短（3 292.03±1 814.77）；韧皮部分泌唾液的时间，黄瓜的最短（1 247.42±448.89），烟草的最长（3 797.25±1 581.32）；B 型烟粉虱在韧皮部吸食植物汁液的时间，黄瓜的时间最长（4 055.91±1 086.75），烟草的次之（3 113.45±1 313.16），西葫芦最短（1 131.33±374.98）。

表 13.2　B 型烟粉虱在 4 种寄主上与韧皮部行为有关的指标（张大山，2010）

指标	烟草	黄瓜	茄子	西葫芦
韧皮部时间总和 E（E1+E2）(s)	6 061.58±1 672.68	4 077.76±708.07	4 103.86±1 115.67	3 292.03±1 814.77
韧皮部分泌唾液时间总和（E1）(s)	3 797.25±1 581.32	1 247.42±448.89	2 724.84±841.27	2 160.70±1 557.45
韧皮部吸食时间总和（E2）(s)	3 113.45±1 313.16	4 055.91±1 086.75	3 033.86±1 019.41	1 131.33±374.98
第一次韧皮部吸食开始前分泌唾液时间 E_2 (s)	432.67±126.89	219.79±140.14	711.17±322.32	76.12±24.68
第一次韧皮部持续吸食开始时间 $E_2>10min$	14 472.89±1 585.26	10 791.36±1 304.21	11 073.52±3 454.16	12 121.31±4 643.77
开始刺探至开始持续吸食时间 $E_2>10min$ (s)	12 364.85±1 966.97a	8 310.08±1 843.00d	12 288.76±0.00b	7 733.08±1 571.03c
刺探内到达持续吸食时间 (s) $E_2>10min$	466.19±7.16	3 668.00±3 265.13	0.00±0.00	67.94±3.59
刺探内到达持续吸食时间历经 pd 数 $E_2>10min$	0.00±0.00	5.00±0.00	2.00±0.00	4.00±0.00

以上数据可以看出，B 型烟粉虱在烟草上的非刺探时间较其他几种寄主植物短，刺探次数也少，第一次刺探开始的时间也较其他几种寄主植物较早。从取食偏好性上来说，B 型烟粉虱更容易接受烟草作为取食对象，这可能与烟草表面的颜色、叶表面毛以及叶表面化学物质有关，使得烟草在 B 型烟粉虱寻找食物的过程中更容易被发现，并被接受为食物对象。

总体来讲，B 型烟粉虱在烟草上的取食行为特点为刺探时间长、细胞内刺探次数多、韧皮部分泌唾液时间长和取食时间相对较短等，表明了 B 型烟粉虱在烟草韧皮部中的取食过程存在一定的抗性；B 型烟粉虱在西葫芦上的取食行为特点为刺探时间短、刺探次数少、韧皮部取食尝试时间短等，表明 B 型烟粉虱在取食西葫芦叶片表皮至韧皮部阶段没有遇到较大阻力，而在韧皮部取食的过程遇到了一定程度上的阻力，导致了韧皮部取食时间短；B 型烟粉虱在黄瓜上时开始刺探至开始取食时间最短，韧皮部取食时间最长，这表明了 B 型烟粉虱在刺探尝试中遇到的阻力较小，在进入了取食植物汁液的阶段之后，就基本上能够持续取食；而在其他的 3 种寄主植物上，即使到达了韧皮部，它也可能只分泌唾液不进行取食，或者短时间的取食之后就放弃。也就是说，虽然 3 种寄主植物都是 B 型烟粉虱可以接受的取食寄主，但在接受程度上，黄瓜更易于作为其取食寄主。

（三）烟粉虱在棉花上的DC-EPG取食行为研究

高佳（2012）利用DC-EPG技术研究Q型烟粉虱在转基因棉花上取食行为，以便为转基因棉的生物安全性评价积累数据和提供理论依据。棉花为转WSA（野生荠菜凝集素）基因棉、转PTA（半夏凝集素）基因棉和转GAN（雪花莲凝集素）基因棉。

结果显示（表13.3），Q型烟粉虱在不同品系的棉花上取食6h内，第一次刺探时间、反映未取食的np波在转基因棉花和亲本间差异都不显著，但反映刺探的C波平均出现时间以及反映被动取食的E波平均出现时间在抗刺吸式害虫转WSA基因棉、转PTA基因棉和其对应受体亲本棉之间差异显著，表现为Q型烟粉虱在抗刺吸式害虫转WSA基因、转PTA基因棉上刺探的平均时间较长而取食的平均时间较短。根据每次E波持续的时间长于8min，蚜虫口针才能真正吸取汁液（Kimmins et al.，1985），可初步判断转WSA基因棉和转PTA基因棉相对于亲本表现出对靶标害虫烟粉虱的取食有显著的抑制作用。

表13.3 Q型烟粉虱在不同品系转基因棉的部分EPG行为比较（高佳，2012）

参数	第一次刺探时间		各波平均出现时间（min）					
			np波		C（A+B+C）波		E（E1+E2）波	
品系	秒（s）	Sig.	时间	Sig.	时间	Sig.	时间	Sig.
转WSA基因棉	638.26±149.23	0.813	15.67±4.13	0.584	4.24±0.50	0.036	7.87±0.98	0.005
转WSA棉亲本	692.81±163.44		13.41±1.53		2.91±0.34		15.54±1.95	
转PTA基因棉	642.94±166.72	0.588	13.90±2.00	0.931	3.75±0.28	0.033	9.84±2.21	0.041
转PTA亲本	541.60±122.17		13.63±2.24		2.97±0.15		16.65±2.06	
转GAN基因棉	684.0±134.96	0.067	10.28±0.65	0.642	3.96±0.46	0.308	14.27±2.75	0.915
转GAN棉亲本	338.33±100.39		10.94±1.34		3.36±0.27		13.92±1.25	

进一步比较Q型烟粉虱在不同品系转基因棉与受体亲本上取食6h内各波出现时间的百分比（%）（表13.4），结果np波出现的时间在整个记录过程中所占的比例、在抗刺吸式害虫转基因棉与其对应受体亲本棉之间没有显著差异；Q型烟粉虱在转WSA基因、转PTA基因棉上C波出现时间所占的百分比显著大于其对应的受体亲本，而在转GAN基因棉和其受体亲本之间没有显著差异；Q型烟粉虱在转PTA基因棉、转GAN基因棉上E1波出现时间所占的百分比与其受体亲本间没有差异，而在转WSA基因和其受体亲本之间存在显著差异；E2波和E波出现时间在整个记录过程中所占的比例均是

转基因品系显著小于其对应受体亲本，表明抗刺吸害虫转基因棉对 Q 型烟粉虱的取食有一定的抑制作用，表现出一定的抗性。

表 13.4　Q 型烟粉虱在不同品系转基因棉与受体亲本上各波出现时间百分比（%）比较（高佳，2012）

参数	各波出现时间百分比（%）									
	np 波		C（A+B+C）波		E1 波		E2 波		E（E1+E2）波	
品系	百分比	Sig.	百分比	Sig.	百分比	Sig.	百分比	Sig.	百分比	Sig.
转 WSA 基因棉	50. 96±3. 56	0. 647	30. 33±2. 02	0. 013	3. 52±0. 40	0. 016	6. 50±1. 13	0. 002	10. 03±1. 26	0. 001
转 WSA 棉亲本	48. 87±2. 79		22. 85±1. 76		5. 48±0. 56		14. 8±1. 80		20. 28±1. 84	
转 PTA 基因棉	53. 58±3. 01	0. 169	29. 42±1. 35	0. 008	3. 51±0. 57	0. 166	6. 25±1. 64	0	9. 77±2. 08	0. 001
转 PTA 亲本	47. 84±2. 49		22. 89±1. 57		4. 82±0. 70		17. 89±1. 97		22. 71±2. 33	
转 GAN 基因棉	7. 55±1. 87	0. 476	27. 32±2. 05	0. 185	4. 43±0. 54	0. 672	11. 31±1. 16	0. 019	15. 75±1. 37	0. 031
转 GAN 棉亲本	49. 75±2. 42		23. 62±1. 51		4. 76±0. 49		16. 85±1. 88		21. 61±2. 19	

（四）B 型和 Q 型烟粉虱的烟草上取食行为比较研究

姬琨（2011）以中烟 100 作为材料，利用 EPG 比较烟粉虱 B 型、Q 型的取食行为差异。在通用指标上，两个生物型在刺探总次数和刺探总持续时间上差异极显著，B 型高于 Q 型烟粉虱，这表明整个刺探取食过程中 B 型比 Q 型烟粉虱更频繁地变换吸食位点。电势降落次数差异极显著，B 型小于 Q 型烟粉虱，这表明整个刺探取食过程中 Q 型烟粉虱更多次地刺入中烟 100 组织细胞并品尝细胞质。E 波、E2 波持续时间差异极显著，Q 型长于 B 型烟粉虱，这表明 Q 型更适宜取食中烟 100。E1 波持续时间差异极显著，B 型长于 Q 型烟粉虱，表明在取食前 B 型烟粉虱唾液分泌时间较长，传播效率与 E1 波总的历期高度相关（Jiang et al.，1999）。此结果与一贯报道的 B 型烟粉虱的高传毒效率相吻合。这说明 Q 型烟粉虱相比于 B 型烟粉虱具有更高的寄主适应性，同时它也为我们日后探索新的烟粉虱生物型鉴定方法提供了理论基础。

而在第一阶段的指标（从刺吸开始到第一次到达韧皮部）中，B 型第一次到达韧皮部所用时间显著长于 Q 型烟粉虱，这表明 Q 型烟粉虱更适宜取食中烟 100。以上结果显示，从表皮-叶肉层开始，Q 型烟粉虱就显示出比 B 型烟粉虱对寄主高的适合度；在第二阶段（第一次到达韧皮部到第一次韧皮部持续取食）的指标中，B 型烟粉虱第一次持续取食韧皮部前唾液分泌总时间差异极显著长于 Q 型烟粉虱，这说明在取食前 B 型烟粉虱唾液分泌时间较长，传播效率与 E1 波的总的历期高度相关（Prado et al.，1994），此结果与一贯报道的 B 型烟粉虱的高传毒效率相吻合。B 型烟粉虱在第二阶段刺探次数差异显著多于 Q 型烟粉虱，这表明在第二阶段中 B 型比 Q 型烟粉虱更频繁地变换吸食位点；而 B 型烟粉虱第二阶段电势降落数差异显著少于 Q 型烟粉虱，这表明第二阶段中 Q 型烟粉虱更多次地刺入中烟 100 组织细胞并品尝细胞质。

对于第三阶段（从第一次韧皮部持续取食至记录结束）的 EPG 行为指标（表 13.5）：B 型烟粉虱第一次持续取食后的刺探次数、第三阶段刺探的次数差异极显著多于 Q 型烟粉虱，这表明此阶段 B 型比 Q 型烟粉虱更频繁地寻找、变换吸食位点。第一次韧皮部持续吸食后，韧皮部相所占时间的百分比、第三阶段 E2 波的百分比差异极显著，B 型<Q 型烟粉虱，这表明 Q 型烟粉虱更适宜取食中烟 100。B 型烟粉虱在第三阶段 E1 波的百分比显著高于 Q 型烟粉虱，这说明在取食前 B 型烟粉虱唾液分泌时间较长，传毒效率与 E1 波的总的历期高度相关（Prado et al.，1994），结果与一贯报道的 B 型烟粉虱的高传毒效率相吻合。

表 13.5　B 型、Q 型烟粉虱第三阶段指标的比较（姬琨，2011）

指标	B 型烟粉虱	Q 型烟粉虱
第一次持续取食的持续时间（h）	1.11±0.15	1.49±0.16
第一次到达韧皮部及能持续吸食的粉虱比例（%）	69.26±1.95	66.73±1.37
第一次韧皮部持续吸食后，韧皮部相所占时间的百分比（%）	60.35±1.14	66.54±1.84 * *
第一次持续取食后的刺探次数	13.8±1.49	4.7±0.22 * *
持续取食的次数	2.24±0.17	2.33±0.18
第三阶段 E1 波的百分比（%）	9.93±0.35	7.03±0.29 * *
第三阶段 E2 波的百分比（%）	39.32±2.09	52.21±1.75 * *
第三阶段刺探的次数	10.1±0.87	7.76±0.61 * *
第三阶段电势降落数	87.48±4.02	122.71±3.9 * *
第三阶段持续时间（h）	3.23±0.12	3.22±0.14

（表中数据为平均数+标准误统计分析使用 T-test 测验法。* * 表示 1%水平上显著、* 表示 5%水平上显著、空格表示不显著）。

总之，B 型和 Q 型烟粉虱取食波形差异不大，35 个指标中，有 16 个指标差异极显著（1%显著水平）、3 个指标差异显著（5%显著水平），主要表现为 B 型烟粉虱刺探次数较多、Q 型烟粉虱电势降落数较大且 E 波持续时间较长，这表明 Q 型更适宜取食中烟 100，这同时说明在取食前 B 型烟粉虱唾液分泌时间较长，传毒效率与 E1 波的总的历期高度相关。

（五）利用 EPG 技术进行烟粉虱生物型的鉴定

不同的烟粉虱生物型在地理分布、寄主范围、抗药性及传毒等方面存在显著差异，在形态学上极为相似，虽然烟粉虱的 B 型和其他生物型的 4 龄蛹壳在第四前亚缘毛和气门外的蜡缨等性状存在一定差异（Bellows et al.，1994），但这些差异很难作为鉴定烟粉虱生物型的标准（Rosell et al.，1997）。对于烟粉虱生物型的鉴定主要根据它们在生物学、酶位点和核酸的多态性等差异进行分析。近年来 mtDNA CO 分子标记在鉴定烟粉虱生物型上应用比较广泛，如 Legg 等发现乌干达的 Ug1、Ug2 烟粉虱种群和 B 型烟粉虱有 8%的遗传距离，因此推断它们不是 B 型烟粉虱（Legg et al.，2002）。罗晨等利用 mtDNA COI 分子标记，证实了我国大部分地区发生为害的烟粉虱主要是外来的 B 型烟粉虱（罗晨等，2002）。利用此方法还证实了在河南省郑州市、云南省昆明市、北京市海淀等地区存在着另一种外来的 Q 型烟粉虱，而且浙江地区还存在一种非 B/Q 型烟粉虱（褚栋等，2005b）。

虽然 mtDNA COI 序列分析广泛应用于烟粉虱生物型的鉴定，是目前为止烟粉虱生物型鉴定及遗传分化研究中最有效的一种方法（Brown et al.，1995b；De Barro et al.，1998），但此法要将虫体匀浆处理，比较耗时，成本也较高。姬琨（2011）比较 B 型和 Q 型烟粉虱的 EPG 行为发现，这两种生物型的烟粉虱在刺探次数、刺探持续时间、pd 波次数、E1、E2 波持续时间上存在稳定的差异，这就为鉴别 B 型、Q 型烟粉虱提供了新思路、新途径。

利用 EPG 鉴别烟粉虱的生物型，有其独特的优势：第一，时间短，EPG 鉴定烟粉虱的生物型一次记录仅需 6h，而生物学实验需要很长的时间才能进行初步鉴定；第二，成本低，EPG 仪器相较于 PCR 等分子设备成本低廉，且 EPG 实验分析数据时仅仅使用相应的免费软件即可，而分子实验需要测序，耗费大量的人力财力；第三，昆虫可以存活再利用，EPG 实验中使用的烟粉虱成虫，在 6h 实验完成后，从仪器上取下来仍可正常生活，而分子实验时需要匀浆虫体，昆虫数量少的时候无法进行验证。

要利用 EPG 进行生物型鉴定，需要制定一套标准体系。根据我们的经验，主要需要考虑以下几点：第一，寄主植物的选择，即根据不同生物型的烟粉虱对不同寄主的喜好程度，来判断烟粉虱的生物型；第二，评价指标的选择，即根据不同生物型烟粉虱在同一种寄主植物上取食时 EPG 波形的差异，来判断烟粉虱的生物型，如 B 型、Q 型烟粉虱在烟草上取食时刺探次数、电势降落数以及 E 波的时间存在差异；第三，数量差异范围的判定，即根据不同生物型烟粉虱在同一种寄主植物上取食时 EPG 波的某一具体波形的差异范围，来判断烟粉虱的生物型，如本实验证明 B 型、Q 型烟粉虱取食波形的 35 个指标中有 16 个指标差异极显著（1%显著水平）、3 个指标差异显著（5%显著水平）。

当然，利用 EPG 进行烟粉虱生物型的鉴定或鉴别属于新的探索，需要在实践中不断完善，使之成为一个烟粉虱生物型鉴定可靠的方法和现有方法的有益补充。

（六）烟粉虱 B 型和 Q 型竞争能力的室内比较分析

卢少华等（2015）利用刺吸电位技术测定烟粉虱 B 型和 Q 型雌、雄成虫在黄瓜叶片上的取食行为，并在室内黄瓜上混合饲养两种烟粉虱生物型，多代监测其种群动态和雌雄比例，以比较其在室内无农药压力条件下的竞争力。EPG 结果显示：①总体上，烟粉虱 B 型的总刺探次数（$P=0.020$）和刺探的总持续时间（$P=0.048$）显著高于 Q 型；在第二阶段 B 型 E 波时间百分比显著低于 Q 型（$P=0.001$），刺探时间百分比差异不显著；B 型 pd 波总次数（$P=0.012$）和 pd 波总持续时间（$P=0.016$）均显著高于 Q 型。②烟粉虱 B 型雌性非刺探波 np 的时间百分比显著低于雄性（$P = 0.035$）；而第二阶段刺探时间的百分比显著大于雄性（$P=0.011$），与唾液分泌相关的第二阶段 E 波百分比却显著短于雄性（$P = 0.001$）；但 B 型雌、雄性成虫在第三阶段吸食韧皮部汁液的时间则差异不显著（$P = 0.05$）。Q 型雌性不仅在第二阶段刺探的百分比显著高于雄性（$P = 0.031$），而且在整个记录时间内的总持续时间也显著高于雄性（$P = 0.039$），而雌性的非刺探时间 np 波却显著短于雄性（$P = 0.038$）。③对于雌性，烟粉虱 B 型被动吸食植物汁液的 E2 10min 的个数显著多于 Q 型（$P = 0.036$）；但是，烟粉虱 Q 型雌性第二阶段 E 波百分比显著高于 B 型雌性（$P = 0.001$）。这些结果表明，烟粉虱 B 型在黄瓜上的取食能力和种群竞争能力均高于 Q 型，两种生物型雌性的取食和竞争能力均高于雄性。

（七）烟粉虱的取食行为总结

行为观察发现，当有多种寄主同时存在时，烟粉虱会频繁地在不同寄主间移动，并不断地尝试取食，但是在每一个寄主上尝试取食的时间短于只有一种寄主存在时的时间，这种不同寄主存在时的不同行为可能是烟粉虱对不同生境的适应行为的表现。

扫描电镜观察发现，在到达韧皮部之前，粉虱口针经过的可能路径为：外侧表皮细胞壁→原生质体→内侧细胞壁→叶肉细胞间隙（Freeman et al.，2001）。而多个研究证明粉虱类昆虫在口针到达韧皮部之前就已经决定是否接受该植物为取食寄主（Walker，1985），这意味着决定粉虱取食的植物因子可能存在于表皮、叶肉细胞间隙或叶肉细胞内。

张大山（2010）的研究表明，在烟草、黄瓜、茄子和西葫芦 4 种寄主植物上，烟粉虱在到达韧皮部之前都要经历几十次甚至几百次的刺探，这些刺探持续时间较短（< 2min），说明烟粉虱在口针接触了叶表皮以及表皮以下较浅范围内的一些信息之后就决定不向韧皮部继续刺吸，而是继续寻找合适的表皮/叶肉细胞间隙。另外，Freeman 等（2001）利用扫描电镜观察发现，烟粉虱的口针也可能会部分地拔出，在刺探路径中间改变方向继续刺探。由此可以推测，因为口针没有脱离植物表面，在 EPG 记录到的波形当中，这样的情况可能表现为一次不间断的刺探，而在对 4 种植物的 EPG 记录过程中确实也包含了很多这样长时间的刺探，有时一次刺探就有 1~2h 之久。这些都说明了粉虱在刺探中不仅仅是为了到达韧皮部，同时也在感知口针沿途所接触到的一些物理和化学信息，以便对是否接受该植物做出选择。

参考文献

[1] BELLOWS T S，PERRING T M，GILL R J，et al. Description of a species of *Bemisia*（Homoptera：Aleyrodidae）. *Annals of the Entomological Society of America*，1994，87（2）：195-206.

[2] BROWN J K，FROHLICH D R，ROSELL R C. The sweet potato or silver leaf whiteflies：biotypes of *Bemisia tabaci* or a species complex. *Annual Review of Entomology*，1995，40：511-534.

[3] DE BARRO PJ, LIEBRECTS W, CARVER M. The distribution and identity of biotypes of *Bemisia tabaci* (Gennadius) (Hemiptera: Aleyrodidae) in member countries of the Secretariat of the Pacific Community? *Australian Journal of Entomology*, 1998, 37: 214-218.

[4] FREEMAN T P, BUCKNER J S, NELSON D R, et al. Stylet penetration by *Bemisia argentifolii* (Homoptera: Aleyrodidae) into host leaf tissue. *Annals of the Entomological Society of America*, 2001, 94 (5): 761-768.

[5] HUSAIN M A, TREHAN K N. The life history, bionomics and control of the whitefly of cotton (*Bemisia gossypipera* M. & L.). *Indian Journal of Agricultural Science*, 1993, 30 (3): 701-705.

[6] JIANG Y X, LEI H, COLLAR J L, et al. Probing and feeding behavior of two distinct biotypes of *Bemisia tabaci* (Homoptera: Aleyrodidae) on tomato plants. *Journal of Economic Entomology*, 1999, 92 (2): 357-366.

[7] KIMMINS F A, TJALLINGII W F. Ultrastructure of sieve element penetyation by aphid stylets during electrical recording. *Entomologia Experimentalis et Applicata*, 1985, 39: 135-141.

[8] LEGG J, FRENCH R, ROGAN D, et al. A distinct *Bemisia tabaci* (Gennadius) (Hemiptera: Sternorrhyncha: Aleyrodidae) genotype cluster is associated with the epidemic of severe cassava mosaic virus disease in Uganda. *Molecular Ecology*, 2002, 11 (7): 1219-29.

[9] OLEVEIRA M R V, HENNEBERRY T J, AANERSON P. History current status and collaborative research projects for *Bemisia tabaci*. *Crop Protection*, 2001, 20 (9): 709-723.

[10] PERRING T M. The *Bemisia tabaci* species complex. Crop Prot, 2001, 20: 725-737.

[11] PRADO E, TJALLINGII W F. Aphid activities during sieve element punctures. *Entomologia Experimentalist Applicata*, 1994, 72 (2): 157-165.

[12] ROSELL R C, BEDFORD I D, FROHLICH D R, et al. Analysis of morphological variation in distinct populations of *Bemisia tabaci* (Homoptera: Aleyrodidae). *Annals of the Entomological Society of America*, 1997, 90 (5): 575-589.

[13] WALKER G P, BACKUS E A. Principles and applications of electronic monitoring and other techniques in the study of homopteran feeding behavior. Entomological Society of America, Lanham, MD: Thomas Say Publications in Entomology, 2000.

[14] WHITE J. Silverleaf whitefly extends range. California Agriculture, 1998, 11 (52): 6-7.

[15] 褚栋，张友军，毕玉平，等．警惕Q型烟粉虱在我国进一步扩散．植物检疫，2005a，19 (3)：171-174.

[16] 褚栋，张友军，丛斌，等．烟粉虱不同地理种群的mtDNA COI基因序列分析及其系统发育．中国农业科学，2005b，38 (1)：76-85.

[17] 褚栋，刘国霞，范仲学，等．烟粉虱复合种不同地理种群的遗传分化．中国农业科学，2006，39 (8)：1 571-1 580.

[18] 高佳．不同转基因抗虫棉对Q型烟粉虱、棉蚜的靶标和非靶标效应研究．河南农业大学硕士研究生论文，2012.

[19] 姬琨，白润娥，汤清波，等．河南省非露地烟粉虱越冬生物型的鉴定及系统进化．河南农业大学学报，2010，44 (5)：560-565.

[20] 姬琨．B型和Q型烟粉虱河南种群的遗传背景分析及取食行为比较．河南农业大学硕士研究生论文，2011.

[21] 纠敏，刘树生．抗同翅目昆虫的相关基因及其转基因抗虫植物研究进展．中国生物防治，2004，20 (4)：228-233.

[22] 李凤荪. 中国经济昆虫（中册）. 湖南农学院丛刊第 1 号，1953：666.
[23] 林克剑，吴孔明，张永军，等 . B 型烟粉虱成虫对五种寄主植物的取食和产卵行为 . 植物保护学报，2008，35（3）：199-204.
[24] 卢少华，李静静，刘明杨，等 . 烟粉虱 B 型和 Q 型竞争能力的室内比较分析，中国农业科学，2017，48（7）：1339-1347.
[25] 罗晨，姚远，王戎疆，等 . 利用 mtDNA COI 基因序列鉴定我国烟粉虱的生物型 . 昆虫学报，2002，45（6）：759-763.
[26] 吴孔明，徐广，郭予元 . 华北北部地区棉田烟粉虱成虫季节性动态 . 植物保护，2001，27（2）：14-15.
[27] 徐文华，王瑞明，丁志宽. 盐城农区棉田烟粉虱的危害特征与治理对策. 江苏农业科学，2006（6）：172-174.
[28] 徐婿，王文丽，刘树生 . Q 型烟粉虱在浙江局部地区大量发生危害 . 植物保护，2006，32（4）：121.
[29] 闫凤鸣 . 非形态特征在粉虱分类中的应用 . 昆虫分类学报，2001，23（2）：107-113.
[30] 闫凤鸣 . 烟粉虱的形态变异 . 北京农学院学报，1991，6（1）：68-71.
[31] 张大山 . 不同寄主植物对 B 型烟粉虱生长发育和取食行为的影响 . 河南农业大学硕士研究生论文，2010.
[32] 周隆，文礼章，王少丽，等 . 不同寄主种群 B 型烟粉虱体内营养和成虫体长差异比较 . 植物保护，2011，37（4）：87-90.

（汤清波）

第十四章　扶桑绵粉蚧取食行为的刺吸电位（EPG）研究

（一）研究背景

扶桑绵粉蚧 *Phenacoccus solenopsis* Tinsley 属半翅目 Hemiptera 粉蚧科 Pseudococcidae，绵粉蚧属 *Phenacoccus*，源于北美洲，最早记载于 1897 年（Tinsley，1898）。1990 年在美国得克萨斯州被确定为棉花害虫；1992 年始，开始向中美洲、加勒比海、厄瓜多尔等地区扩散；之后，在其他地区如阿根廷、智利、巴西、巴基斯坦、加纳、尼日利亚、印度也相继发生；特别是在 2006~2008 年，大量报道反映该虫在巴基斯坦和印度的转基因棉区暴发，造成严重的经济损失（Hodgson et al.，2008）。扶桑绵粉蚧主要为害植物的幼嫩部位，以雌成虫和若虫吸食植物汁液为害，受害植物生长势衰弱，分泌的蜜露诱发的煤污病可导致叶片脱落，严重时可造成植株成片死亡（Aheer et al.，2009）。扶桑绵粉蚧繁殖扩散能力极强（黄芳等，2011），对我国农业生产具有巨大威胁（王艳平等，2009；Wang et al.，2010），已引起国内专家的高度关注（张润志，2010），于 2010 年被我国农业部确定为入侵害虫。

扶桑绵粉蚧食性杂，寄主范围广，其寄主植物多达上百种，仅在巴基斯坦 Punjab 地区一处半径 20km 的棉田农业生态系统中，记录的寄主植物就有 53 科 154 种（Arif et al.，2009）；在原产地美国寄主植物则已多达 183 种（Ben-Dov et al.，2009）；在我国，以浙江省 2009 年和 2010 年的调查数据表明，在该虫传入浙江后不久便已侵染 19 科 29 种植物（周湾等，2010；李月红等，2011）。

取食行为及效率是考察昆虫对宿主植物适应性的重要指标之一。EPG 是专门针对植食性刺吸式昆虫用于研究其在植物上刺探、取食行为的一种技术手段。因此，对扶桑绵粉蚧的 EPG 图谱进行研究具有重要意义。

（二）材料与方法

1. 扶桑绵粉蚧的来源及饲养　扶桑绵粉蚧采自浙江省金华市婺城区的观赏植物木槿上，采回后在人工气候室温度：(27±1)℃，相对湿度：(70±5)%，光照 L/D＝12h/12h，用棉花（*Gossypium hirsutum* Linn，品种名：浙凤棉 1 号）进行饲养。

2. EPG 记录　采用四通道直流电型（DC-EPG 型）昆虫刺吸电位仪进行取食观察。先用软毛刷轻柔地刷去扶桑绵粉蚧中胸背部的蜡粉，重复多次直至该部位的蜡粉完全被清理干净，若稍有残余将影响信号的收集；用导电银胶将一根细金线（直径为

10μm）的一端与中胸背部粘连，金线另一端相连至昆虫电极。将植物电极插入栽有 1 株 3~4 叶龄的棉花苗的盆土中，接通电流，使悬挂的试虫接触棉叶表面，开始记录。每个重复连续记录至少 12h。所有记录均在室内进行，温度为（27±1）℃，相对湿度为（70±5）%。采用随机附带的 ACQ 和 ANA 软件分别用于昆虫取食行为的记录和数据的分析统计。

（三）结果

1. 扶桑绵粉蚧取食棉花的 EPG 波形 扶桑绵粉蚧取食棉花的 EPG 波形有路径波和取食波（表 14.1）。

表 14.1 扶桑绵粉蚧的 EPG 波形及其与其他粉蚧 EPG 波形的比较

波形			扶桑绵粉蚧的波形特性				其他粉蚧对应波形	
扶桑绵粉蚧			电压水平	相对振幅（%）	电势波动	频率（Hz）	木薯绵粉蚧	橘臂纹粉蚧
A			胞外	100	电阻	8-11	A	A
B			胞外	78.8±6.9	电阻	0.05-0.15		
C			胞外	49.4±4.0	电阻	混合	C（CI/CII/CIII）	C
pd			胞内		电动势		pd/E1	pd/E1
E1e		波峰	胞外		电动势	1-3		
		波	胞外		电阻	4-8		
E1	E11	波峰	胞内		电动势/电阻	混合	E2I	E21
		波	胞内		电动势	6-8		
	E12	波	胞内		电动势	4-9	E2II	E22
E2		波峰	胞内		电阻	2-4	E23	
		波	胞内		电动势/电阻	混合		
F			胞外		电阻	13-19		
G			胞外		电阻	3-6		
								H

路径波含 A 波、B 波和 C 波（图 14.1）。A 波表示穿刺进入叶面组织，其波形不规则，相对振幅可为记录波形中的最大值（图 14.1A）。B 波表示唾液鞘形成，发生时紧随 A 波，平均 5~10s 发生一次，其相对振幅由变化剧烈逐渐趋于平缓，同时慢慢过渡到 C 波，在记录数据中，多为 B 波和 C 波叠加后的复合波形（图 14.1A-B）。C 波表示口针在胞间游走，每隔（10.9±0.9）s 发生一次重复，重复内小波的频率由低到高，振幅由大变小（图 14.1C）。当口针在细胞间钻探时，会对周围细胞进行尝试性的刺探，一旦口针刺入细胞，由于细胞对外的一个压力，在波形上会出现一个电动势突然下降（potential drop，pd）的现象，这种尝试性的刺探时间为 10s 左右，非常规则；pd 波包含两个亚波为 pd1 和 pd2，持续时间分别为（5.78±1.20）s 和（5.42±1.62）s，pd1 的频率（12~16Hz）和相对振幅较 pd2（9~12Hz）高，两个亚波之间会有一段为时 2~4s 的过渡期，

期间为两种亚波形的叠加（图 14.1C）。有时，pd 波出现时会发生其他波的叠加（图 14.1D）。通常，单个 pd 波内只发生一次 pd1 和 pd2，但有时两个亚波交替出现的次数会增加，因此整个 pd 波持续的时间会很长，可长达数小时（图 14.1E），此种 pd 波被定义为 P-pd（Prolonged potential drop）。P-pd 结束后通常回复到 C 波，个别情况下可直接发生至取食波。取食波是口针长时间处于韧皮部营养细胞内进行持续取食时的波形，包括 E1e 波、E1 波和 E2 波。E1e 波的波形包含两种亚波：一种发生频率为 1~4Hz，属 EMF 成分；另一种发生频率为 8~12Hz，属 R 成分（图 14.1D）。

E1e 波发生时间通常<1min，发生于胞外水平，所以常叠加于 C 波，有时也会叠加于 pd。E1 波发生于胞内水平，持续时间可长达数小时（图 14.2A），包含 E11 和 E12 两种亚波。E1 发生时，E11 和 E12 交替出现（图 14.2B）。E11 波属 EMF 和 R 的混合成分，EMF 成分波峰的发生频率类似于 E1e 的 EMF 成分，但有时会出现规律性的变化，根据波峰出现的频率可分为四个阶段，阶段 Ⅰ（<0.5Hz），阶段 Ⅱ（0.5~1Hz），阶段 Ⅲ（1~2Hz）和阶段 Ⅳ（> 2Hz），而 R 成分亚波不发生变化（图 14.2D）。当四个阶段完成时，EMF 成分波峰会忽然消失，而 R 成分波形的相对振幅会忽然升高后慢慢下降至原来的水平，随后 EMF 成分波形再次规律性的发生（图 14.2E）。E12 亚波则为 R 成分，发生时较稳定。E2 波表示在取食的同时，唾液通过口针进入植物组织内；E2 包含两个亚波，EMF 成分波发生频率为 2~4Hz，R 成分波形的发生频率则为 6~8Hz（图 14.2C）。其他波形有 G 波和 F 波。F 波的发生频率为 13~19Hz（图 14.3A），G 波的发生频率为 3~6Hz（图 14.3B）。

2. 扶桑绵粉蚧取食棉花的 EPG 各波形的发生情况　扶桑绵粉蚧取食棉花的 EPG 各参数如表 14.2 所示。在记录扶桑绵粉蚧取食棉花的 8h 内，粉蚧取食的暂停间歇平均发生 3 次左右，平均约为 10min，个别记录中不发生取食暂停；pd 发生的次数平均为 100 次左右；到第一次持续性取食（>10min）的时间为 1h 左右；取食波发生的频率约 1 次 · h^{-1}；含有 E 波和不含 E 波的刺探个数均发生 1~2 次，持续时间含 E 波的刺探时间较长为 4.4h 左右，不含 E 波的刺探时间约为 0.8h。扶桑绵粉蚧在棉花上取食的 4 个主要行为，即不取食（np）、口针胞间游走（C 波）、试探性取食（pd 波）、取食（E 波），分别占总时间的 7%、41%、4%和 48%（图 14.4）。

表 14.2　扶桑绵粉蚧取食棉花的 EPG 各参数（总记录时间：8h）

	平均数 ± SE	单位	参数含义
n-np	3.33 ± 1.37	#	np 的个数
n-pd	98.72 ± 6.25	#	pd 的个数
t-Ei	1.34 ± 0.67	H	到第一个 E 波（>10min）的时间
n-E	7.83 ± 3.60	#	E 波的个数
n-pen（E）	1.93 ± 0.73	#	含有 E 波的刺探个数
n-pen（xE）	1.50 ± 0.76	#	不含有 E 波的刺探个数
d-pen（E）	4.43 ± 1.62	H	含有 E 波的刺探持续时间
d-pen（xE）	0.81 ± 0.43	H	不含有 E 波的刺探持续时间

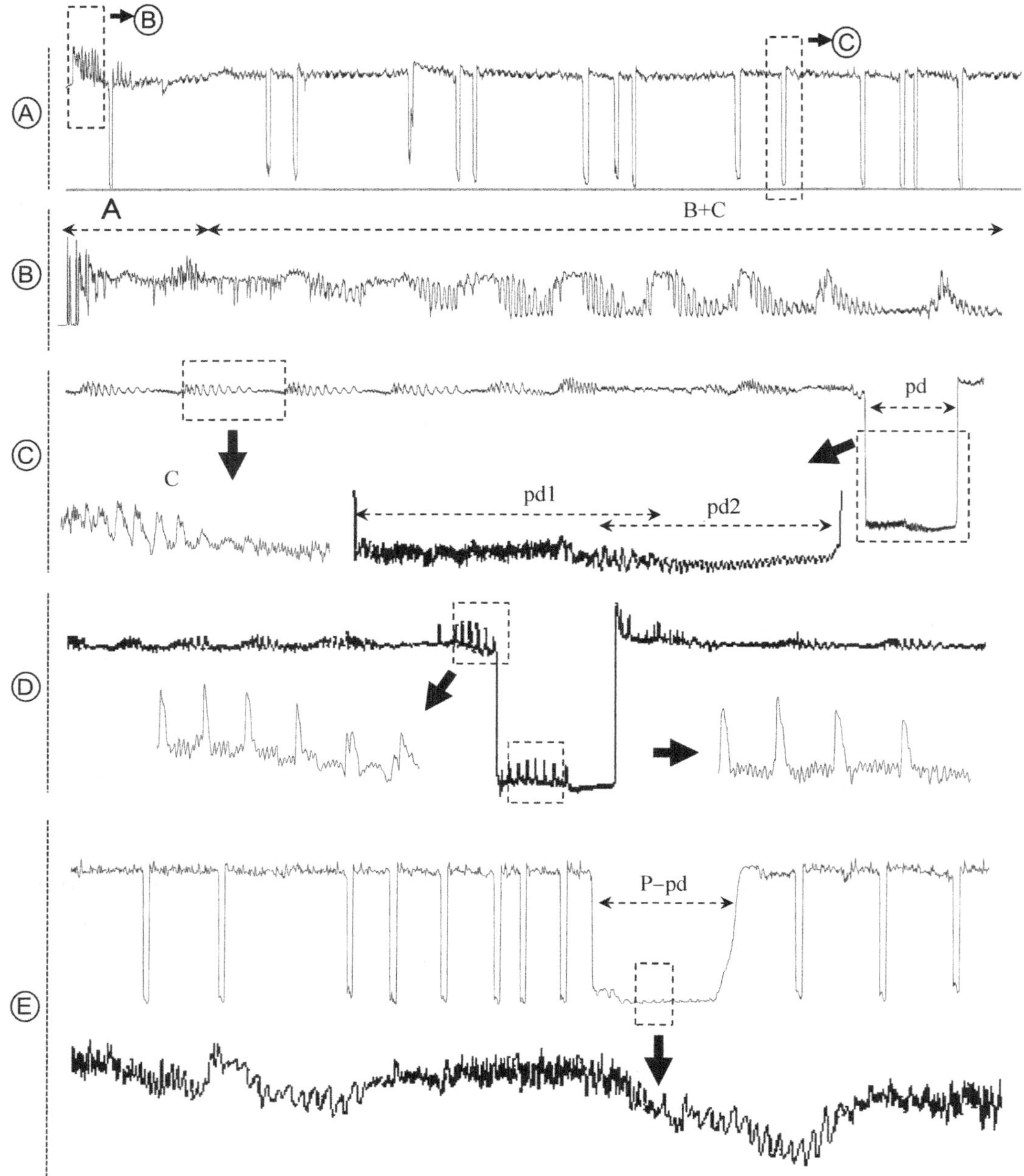

图 14.1　扶桑绵粉蚧在棉花上的 EPG 的路径波

扶桑绵粉蚧取食棉花的 EPG 各波形的发生转换情况如图 14.5 所示。扶桑绵粉蚧取食行为发生时首先产生 A 波，B 波紧随其后，但有时 B 波并不明显，C 波也可能直接跟随 A 波之后发生；有时，扶桑绵粉蚧暂停取食后，口针仍滞留于植物细胞间，处于胞外水平，一旦发生继续取食时，EPG 记录表现为直接出现 C 波。C 波开始后，其后可发生 E 波、P-pd 波、G 波和 F 波等。跟随 C 波发生的 E 波主要为 E1 波，E1 波可转

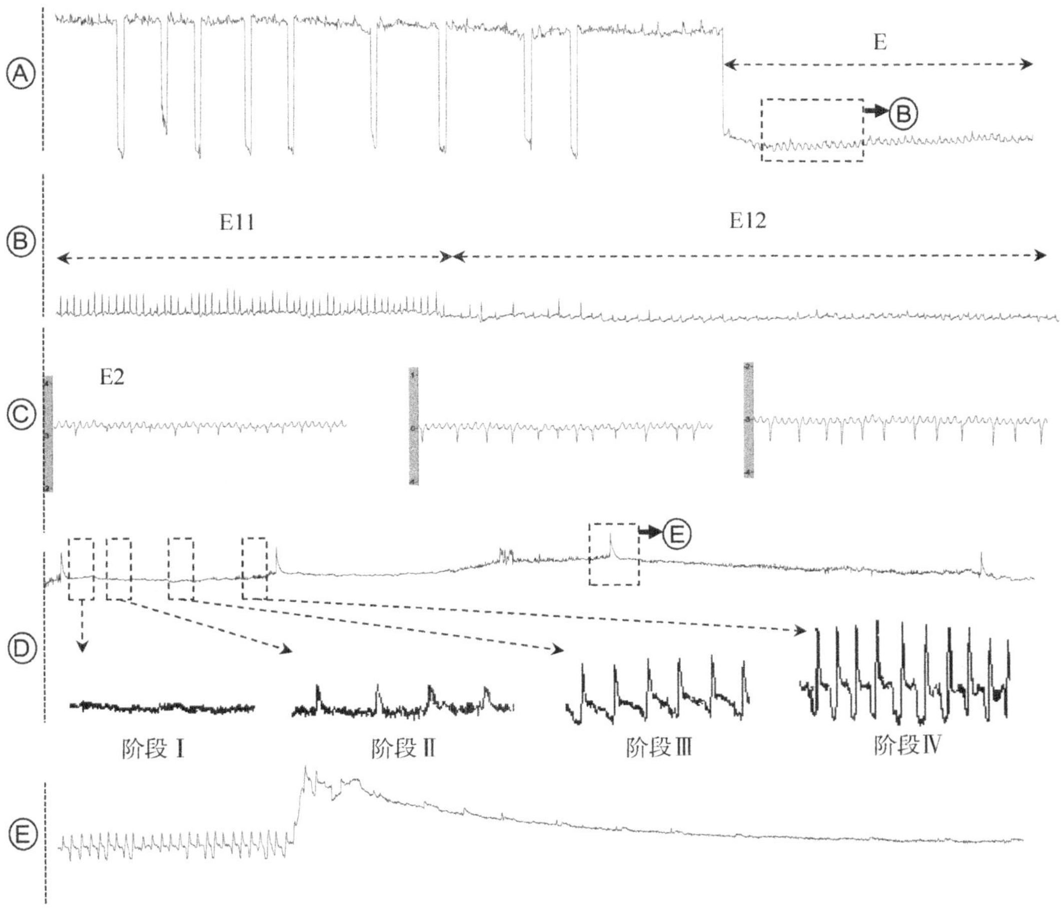

图 14.2 扶桑绵粉蚧在棉花上的 EPG 取食波

换为 E2 波，E2 波发生之后一般不会再恢复到 E1 波，转而直接回到 C 波。P-pd 波发生时，通常是回到 C 波，但个别情况下也可直接发展至 E1 波。值得注意的是，在 C 波转为 E1 波时，通常在最初电动势突然下降的 10~20s 会出现一段 pd 波，为单个 pd1 和 pd2 的相继发生或单个 pd1 的发生。当口针进入木质部，且扶桑绵粉蚧也可能会发生取食，即 G 波，G 波一旦发生，其持续时间较长，为数十分钟到数小时均有可能，G 波结束后主要恢复到 C 波，不能进行其他波形的转换。F 波的发生频率最低，结束后恢复到 C 波。

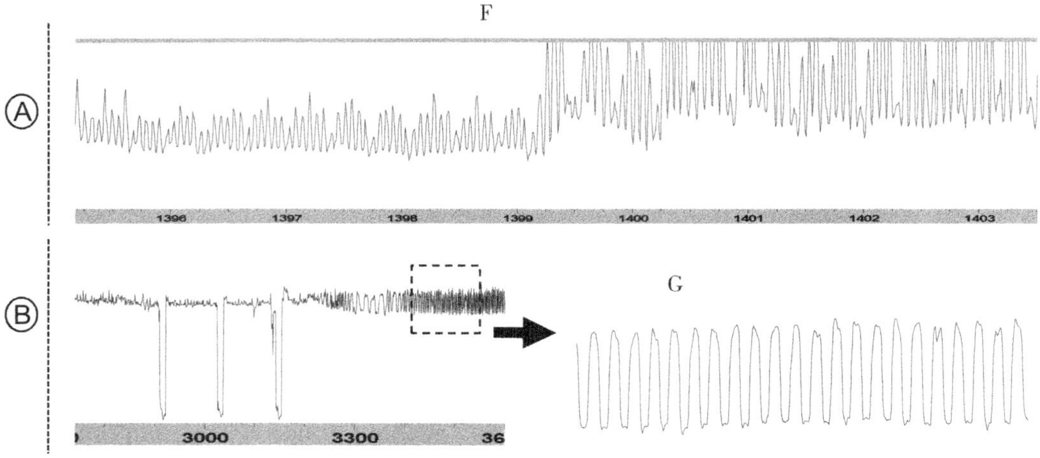

图 14.3　扶桑绵粉蚧在棉花上的 EPG 的其他波（除路径波和取食波）

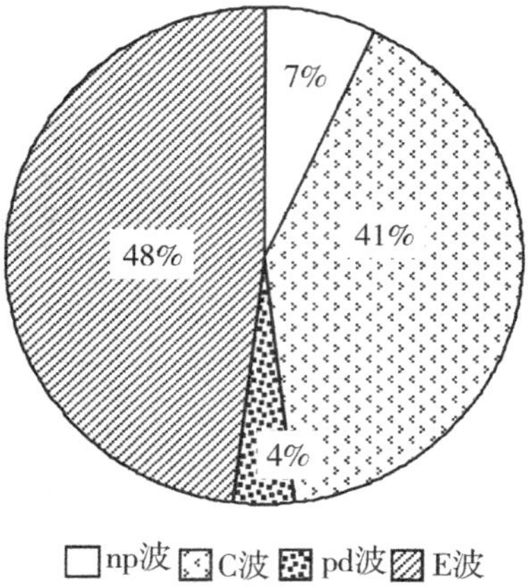

图 14.4　扶桑绵粉蚧在棉花上取食的 4 个主要行为

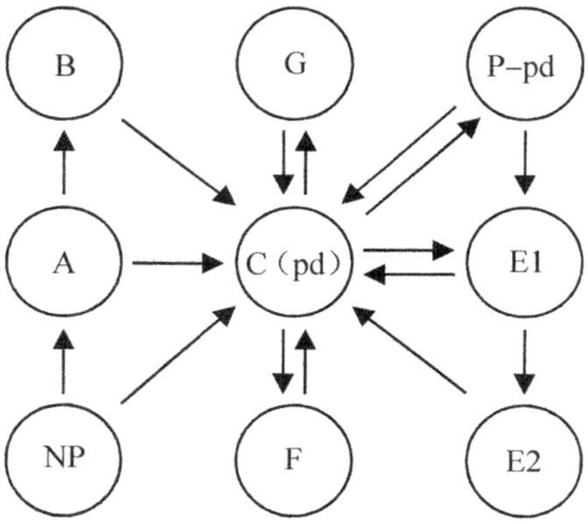

图 14.5　扶桑绵粉蚧取食棉花的 EPG 各波形的发生转换

（三）讨论

在 EPG 技术出现的半个多世纪内时，大部分的研究起始并集中于蚜虫。因此，我们将蚜虫的取食波形，称为经典波形。以蚜虫取食过程的 EPG 研究为例（Tjallingii, 1988；1990），首先是代表穿刺进入叶表皮的 A 波，接着是表示唾液鞘形成的 B 波，事实上，B 波的出现通常都伴随着 C 波，即在 EPG 图谱上是两种波形的叠加，然后就是表示口针在细胞间钻探的 C 波，那么在 C 波发生的过程中通常会出现一系列尝试性的细胞穿刺动作，就形成了 pd 波；当尝试性结果得到肯定时，即口针已经达到韧皮部时，E1 波开始出现，即水溶性唾液分泌，然后是持续吸食的 E2 波。在尝试性取食的过程中可能会发生口针配合出错的情况，此时 F 波出现；而当口针到达木质部的时候，

G 波出现。以上经典波形均在扶桑绵粉蚧的 EPG 研究中被发现。以蚜虫的取食过程作为参考，扶桑绵粉蚧在取食过程中会出现新的波形，就是 P-pd 波。P-pd 波波形与 pd 波类似，频率也相近，但是其持续时间可长达数小时，且有时该波形可逐渐发展到 E1 波形，这就说明此时口针已经到韧皮部，但又不呈现典型的取食波，这个新波形所针对的具体行为目前还不清楚。此外，扶桑绵粉蚧 EPG 相对于蚜虫 EPG 的不同点还在于：①B 波不是 A 波过渡到 C 波的必然发生波形；②pd 波为 12s 左右，要比蚜虫的 20s 左右明显短；③E1 波的波峰出现可呈现规律性的变化，这与蚜虫稳定的取食波形是不一样的。

针对粉蚧的 EPG 研究较少，目前已有 EPG 研究的粉蚧为两种，木薯绵粉蚧 *Phenacoccus manihoti*（Calatayud et al.，1994）和橘臀纹粉蚧 *Planococcus citri*（Cid and Fereres，2010）。对比分析以上两种粉蚧的 EPG 和本文的研究结果，笔者发现不同作者对粉蚧同一行为的 EPG 波形定义并不十分一致，对各粉蚧取食行为的 EPG 波形命名进行总结，结果如表 14.1 所示。Calatayud 等将木薯绵粉蚧的 C 波分为 3 个亚波，但并没有针对各个亚波所指向的具体行为进行研究。经典 C 波的主要特征是一小段行为的不断重复，在这个行为片段内，频率和相对振幅均会发生一系列相对规则的变化，如图 14.1C 中第一个虚线框所放大的片段；在木薯绵粉蚧中，其 C 波相当于单个 CⅠ和 CⅡ两个亚波的交替出现。在橘臀纹粉蚧中也发现类似的 EPG 记录，但相关研究中没有提到 C 波的亚波形；这与蚜虫、烟粉虱等 EPG 中的 C 波定义是相一致的。在扶桑绵粉蚧的 C 波发生过程中，并没有出现与木薯绵粉蚧的 CⅢ亚波相似的波形。为了方便数据分析，我们认为对扶桑绵粉蚧 C 波进行亚波分类的意义并不大。此外，借鉴于其他昆虫的 EPG 研究，扶桑绵粉蚧 EPG 的 A 波和 B 波也被整合入 C 波，以便于数据的分析。对比三种粉蚧和蚜虫的 E1 波形，我们发现木薯绵粉蚧和橘臀纹粉蚧的 E1 波与其自身的 pd 波是相似的，而这段 pd 波（E1 波）和扶桑绵粉蚧的 pd 波也是类似的。经典 pd 波被定义为是胞内水平的一个尝试性取食行为，而在尝试性取食行为发生后紧随持续性取食（经典 E 波）的发生在行为学上是合理的，即 pd 波后可跟随 E 波的发生（Huang et al.，2012）。鉴于此，我们认为木薯绵粉蚧和橘臀纹粉蚧的 E1 波应属 pd 波；依此类推，木薯绵粉蚧和橘臀纹粉蚧中的 E2 波应为 E1 波，两者 E2 波中的两个亚波形命名相应改为 E11 和 E12。木薯绵粉蚧的 E23 波形与扶桑绵粉蚧 E2 波形相同，据此 E23 波形命名应为 E2 波。经过更正的粉蚧 E1 和 E2 波，不管在波形还是具体行为含义上都和经典 EPG 呈一致性。

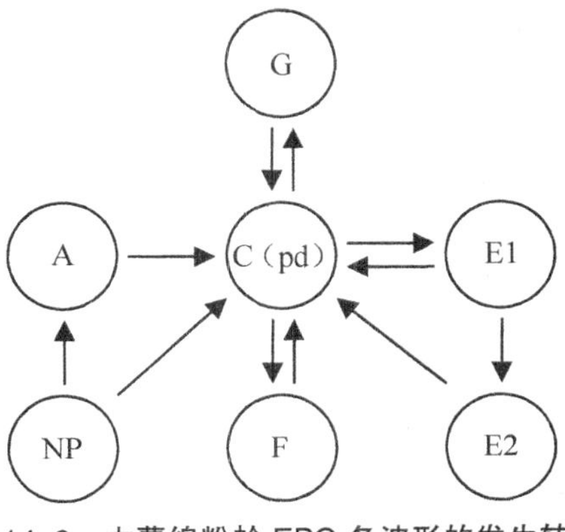

图 14.6　木薯绵粉蚧 EPG 各波形的发生转换

在对三种粉蚧取食行为的 EPG 波形的定义进行统一后，笔者对比了三种粉蚧取食波形的发生转换情况（图 14.5～14.7），结果发现除叶表穿刺时存在一些差异外，其他的动作路径都相同，这可能与三者相同的刺吸取食习性相关。而叶表穿刺时产生的差异则可能与三者喜好的宿主植物相关。

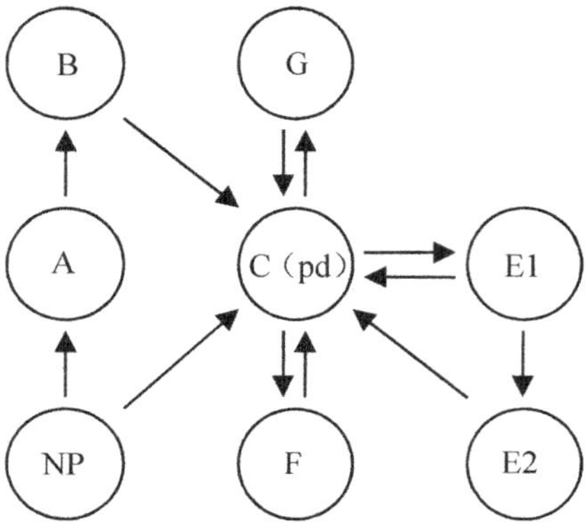

图 14.7　橘臀纹粉蚧 EPG 各波形的发生转换

本文研究了扶桑绵粉蚧取食行为 EPG 图谱，依据经典的 EPG 定名系统，类比已知的两种粉蚧的 EPG，对各波形进行了定义，并对粉蚧类昆虫的 EPG 定名系统进行了修正。在对三种粉蚧的 EPG 各波形命名进行梳理后，其命名系统与经典 EPG 定名系统（蚜虫）已具有较高的统一性，有效提高了类比刺吸式口器昆虫 EPG 数据的可行性。此外，本文结果还为评价扶桑绵粉蚧在各宿主植物上适应性评价提供了理论基础。

参考文献

[1] AHEER G M，SHAH Z，SAEED M. Seasonal history and biology of cotton mealy bug，*Phenacoccus solenopsis* Tinsley. *Journal of Agricultural Research*，2009，47（4）：423-431.

[2] ARIF M I，RAFIQ M，GHAFFAR A. Host plant of cotton mealybug（*Phenacoccus solenopsis*）：a new menace to cotton agroecosystem of Punjab. *International Journal of Agriculture and Biology*，2009，11（2）：163-167.

[3] BEN-DOV Y，MILLER D R，GIBSON GAP. ScaleNet：A searchable information system on scale insects［J］. 2009.

[4] CALATAYUD PA，RAHBE Y，TJALLINGII WF，et al. Electrically recorded feeding behavior of cassava mealybug on host and non-host plants［J］. *Entomologia Experimentalis et Applicata*，1994，72：219-232.

[5] CID M，FERERES A. Characterization of the probing and feeding behavior of *Planococcus citri*（Hemiptera：Pseudococcidae）on grapevine［J］. *Annals of the Entomological Society of America*，2010，103：404-417.

[6] KARAR H. *Phenacoccus solenopsis* Tinsley（Sternorrhyncha：Coccoidea：Pseudococcidae），an invasive mealybug damaging cotton in Pakistan and India，with a discussion on seasonal morphological variation［J］. *Zootaxa*，2008，1913：1-35.

[7] HUANG F，TJALLINGII W F，ZHANG P J，et al. EPG waveform characteristics of solenopsis mealybug stylet penetration on cotton［J］. *Entomologia Experimentalis et Applicata*，2012，143：47-54.

[8] TINSLEY JD. An ant's nest coccid from New Mexico［J］. *Canadian Entomologist*，1898，30（2）：47

-48.
[9] TJALLINGII W F. Electrical recording of stylet penetration activities. Aphids, Their Biology, Natural Enemies and Control, Vol. 2B (ed. by AK Minks & P Harrewijn), 1988, 95-108. Elsevier, Amsterdam, the Netherlands.
[10] TJALLINGII W F. Continuous recording of stylet penetration activities by aphids. Aphid-Plant Genotype Interaction (ed. By RK Campbell & D Eikenbary), 1990. 89-99. Elsevier, Amsterdam, the Netherlands.
[11] WANG Y P, WATSON G W, ZHANG R Z. The potential distribution of an invasive mealybug Phenacoccus solenopsis and its threat to cotton in Asia [J]. *Agricultural and Forest Entomology*, 2010, 12: 403-416.
[12] 黄芳，张蓬军，章金明，等. 三种寄主植物对扶桑绵粉蚧发育和繁殖的影响. 植物保护，2011，37 (4): 58-62.
[13] 李月红，阮永明，颜宗德，等. 金华市扶桑绵粉蚧的发生与防治 [J]. 长江蔬菜，2011 (5): 33-34.
[14] 王艳平，武三安，张润志. 入侵害虫扶桑绵粉蚧在中国的风险分析 [J]. 昆虫知识，2009，46 (1): 101-106.
[15] 周湾，林云彪，许凤仙，等. 浙江省扶桑绵粉蚧分布危害调查 [J]. 昆虫知识，2010，47 (6): 1 231-1 235.
[16] 张润志. 扶桑绵粉蚧（棉花粉蚧）. 北京：中国农业出版社，2010.

（黄　芳）

第十五章　刺吸电位技术中固定昆虫的三种方法比较研究及其应用

（一）研究背景

EPG实验属于行为学和电生理学研究，受影响因素多，实验中很多细节因素都可能成为影响实验成功与否的关键（汤清波等，2011）。其中，连接昆虫电极时供试昆虫的固定处理是一个至关重要的实验环节（岳梅等，2005）。如果固定不好，昆虫还处于可活动状态，就会增加昆虫电极连接的难度；再者，若固定过程中对昆虫本身造成伤害，也会直接影响到昆虫适合度及其取食，进而对实验的可靠性和准确性构成影响。例如，烟粉虱 *Bemisia tabaci*（Gennadius）在冷冻室中经过20~30s即不能活动，取出置于操作台上即可进行电极的连接；而褐飞虱 *Nilaparvata lugens*（Stål）的活动受温度影响较大，经过冷冻处理的褐飞虱恢复活动后其取食能力较差，对实验结果有较大影响（汤清波等，2011）。

在EPG实验中通常会用到三种昆虫固定方法，即负压法、冷冻法和麻醉法。Seo（2009）和荆裴（2013）等先后做过稻飞虱的EPG研究，但均没对稻飞虱的固定方法做详细描述与介绍。稻飞虱体形小，活跃，成虫有翅善飞翔。稻飞虱EPG研究中，在不损伤稻飞虱情况下固定起来比较困难。本文以灰飞虱为例，通过对三种昆虫固定技术优化后做了较为详细的阐述，并进行了系统比较，为EPG研究中昆虫固定这一关键技术提供参考。

（二）材料与方法

1. 供试水稻　本研究选用水稻品种为感稻飞虱的TN1。实验中，将TN1种子分期分批地播种在恒温（26℃±1℃）、恒湿（65%±5%RH）、光周期为L/D=14h/10h的温室内。实验时，随机选取生长10d左右长势一致的三叶期幼苗，将稻苗洗净，去掉枯萎的叶片备用。

2. 供试虫源　本研究以灰飞虱为供试昆虫，选用在TN1水稻上连续饲养10代以上的灰飞虱种群，饲养在恒温恒湿的光照培养箱（型号：RXZ智能型；厂家：宁波江南仪器厂），环境温度设置为（26℃±1℃）、环境相对（RH）湿度设置为65%±5%、光周期设置为L/D=14h/10h。灰飞虱饲养期间均使用生长到10d左右的三叶期幼苗，适时更换。EPG实验开始时，随机选取当天羽化的短翅雌成虫为供试虫源。

3. EPG实验方法　EPG信号放大器采用荷兰瓦赫宁根大学昆虫实验室研制的

Giga-8型直流放大器，实验用银胶、铜钉、铜丝、金丝（直径 20μm）均为购买仪器附带。实验中，将灰飞虱饲水饥饿处理 1h 后（Diaz-Montano et al.，2007；荆裴等，2013），再迅速进行负压法、冷冻法和麻醉法三种昆虫固定方法的 EPG 实验。实验在恒温（26.5℃±1℃）、灯光常亮的温室内记录 4h，整个记录过程在法拉第电磁笼中进行（Jiang and Walker，2007）。

（1）负压法：把供试灰飞虱置于干燥的 1.5mL 离心管中。由于该虫趋光而喜欢向上爬行，所以将离心管底部斜向上放置，使之自动留在离心管底部，且保持头部朝上状态，从而方便使用吸取定位装置吸附其腹部。本研究利用自主研发的吸虫设备“一种针对飞虱等小型昆虫 EPG 实验的吸取定位装置”（图 15.1；专利授权号：ZL201420133901.3）来固定灰飞虱。首先把枪头的大小修剪成与所选取灰飞虱腹部大小相近程度，然后将负压调节器调整到最佳吸力来吸附灰飞虱；之后，把该设备昆虫吸附头伸入离心管底部吸住灰飞虱腹部，然后将吸附头固定在装置托架上，而后迅速用导电银胶将灰飞虱中胸背板与一根直径为 20μm、长 2~3cm 金丝黏合（Seo et al.，2009；Civolani et al.，2011），再通过铜丝连接于 8 通道昆虫刺吸信号电子记录仪上（型号：Giga-8 DC-EPG，荷兰瓦赫宁根农业大学昆虫系研制）；待昆虫电极连接好后，再将灰飞虱放置在供试稻苗茎秆（叶鞘）上，并适当调整金丝，避免金丝碰到植株或仪器，调整完毕后单击采集按钮开始 EPG 波谱记录。

（2）冷冻法：将供试灰飞虱放入干燥的 1.5mL 离心管中并盖紧盖子，之后将其放入（-20℃冰箱内并开始计时，冷冻处理 60~80s 后取出灰飞虱（此时灰飞虱在低温作用下可瞬时麻痹 40s 左右），在其麻痹期间用柔软的毛笔将其挑起并使其中胸背板朝上，迅速用导电银胶将其与金丝黏合（此过程与负压法相同），等待灰飞虱复苏后再将其连接于同一刺吸信号电子记录仪上并进行 EPG 数据记录。

（3）麻醉法：麻醉法是利用气体进行麻醉，包括乙醚麻醉法和 CO_2 麻醉法两种，在此选用乙醚麻醉法。具体做法如下：首先，将灰飞虱置于干燥的 1.5mL 离心管中并盖好盖子备用；其次，把脱脂棉塞到 2mL 离心管底部，使其刚好塞满底部约 5mm 深度；然后，用移液器（量程：100mL）吸取 80mL 100%乙醚均匀地洒在脱脂棉上并盖好盖子，挥发 30s 后将装灰飞虱的 1.5mL 离心管与装乙醚气体的 2mL 离心管两口对接持续 10~20s（此时灰飞虱可以保持麻醉状态 40s 左右），之后，迅速用柔软的毛笔将灰飞虱挑起，并使其中胸背板朝上，再用导电银胶将灰飞虱与金丝黏合（此过程与负压法相同）；再次，待灰飞虱复苏后再将其连接于同一型号刺吸信号电子记录仪上；最后，将连接好昆虫的电极的灰飞虱放置在供试植株茎秆上，并调整金丝以避免金丝碰到植株或仪器，调整完毕后开始记录 EPG 波谱数据。

4. 灰飞虱取食行为的波形鉴定及统计分析 灰飞虱取食行为原始数据使用 EPG Stylet⁺a software（Wageningen University，2012）进行波形识别和数据转化的初步处理。鉴定的依据参照已有的关于稻飞虱或者其他刺吸式口器昆虫取食行为的报道（Ghaffar et al.，2011；He et al.，2011；Seo et al.，2009；Kimmins，1989）。未知波形的可参照取食行为研究较多的蚜虫进行鉴定（Tjallingii，1988）。

此类数据均使用 SPSS 20.0 进行统计分析。对每组数据统计分析之前，首先使用

Levene's test（$P>0.10$）和 Shapiro-Wilk test 分别检验数据的方差齐性和正态分布性。以上检测结果如果不符合正态分布，则根据数据的特点对数据进行正态化变换，以符合方差分析要求（本部分对取食次数数据进行了平方根转换；对取食时长数据进行了对数转换）。实验的统计分析步骤如下：以 RSV 感染情况为处理因子，以所鉴定到的波形为单位，分别对转换后的取食次数和取食时长数据进行单因素方差分析（$P<0.05$）。

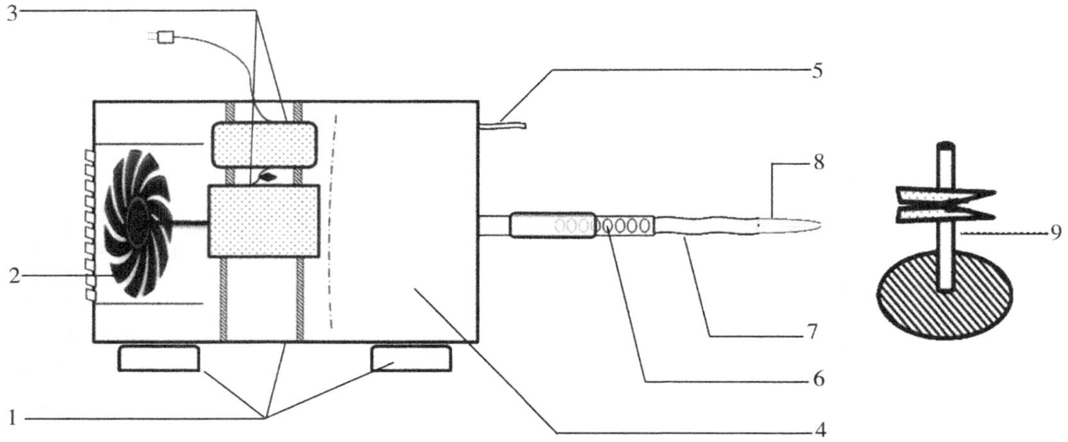

图 15.1　刺吸电位技术实验中昆虫负压固定装置示意图

1. 外壳和减震系统　2. 叶轮　3. 供电与驱动系统　4. 气室　5. 安全阀　6. 控压系统　7. 导气管　8. 可更换式吸头　9. 支架

（三）结果与分析

1. 负压法、冷冻法和麻醉法的便捷性与可操作性比较　实验中，负压法所需辅助工具仅为一种针对小型昆虫的吸取定位装置（专利授权号：201420133901.3）。该设备体积小巧、方便携带，控制简单、运行稳定，并且可根据试虫大小更换相应吸附头的枪头即可。该设备适应性更广，可适应每一种昆虫的固定。冷冻法则需要一台冰箱或者其他制冷设备，这些设备一般较大且不利于携带，在冷冻处理时间控制上有严格要求，如冷冻时间短了，昆虫麻痹时间太短，很快恢复活跃状态而完不成昆虫与电极连接；如冷冻时间长了，昆虫容易死亡，既浪费人力、物力，又因为不同昆虫或同种昆虫不同个体都对低温的承受能力有差异，有的甚至受温度影响太大而不适宜用冷冻法，所以冷冻法局限性大、不易控制。麻醉法相对也较简易，但是在昆虫麻痹时间上也是有一定局限的。熏的时间越长则昆虫麻痹时间越久而苏醒时间延长，耽误实验进行；熏的时间短则昆虫苏醒太快而完不成昆虫与电极的连接，所以相对也不太好控制。后两种方法还都需要用毛笔将昆虫挑起或扶正来连接昆虫和导电金丝，增加了实验操作难度。综上所述，昆虫固定方法的便捷性和可操作性比较为依次为：负压法>麻醉法>冷冻法。

2. 负压法、冷冻法和麻醉法对昆虫本身的影响比较　负压法是物理方法，昆虫在负压的作用下被吸附在塑料枪头上，减轻了对昆虫的影响，将外界条件对昆虫本身的影响降到了最低。冷冻法也是物理方法，但在低温作用下，昆虫肢体可能被冻伤，并且人工用毛笔挑起或扶正昆虫过程中如力度把握不好也可能伤及昆虫肢体，实验中昆虫可能由于身体损伤而直接影响实验结果。麻醉法是化学方法，运用气体使昆虫麻醉，如乙醚就是作用于神经系统来使昆虫麻醉的，而昆虫个体一般较小，使用乙醚的量不好把握，用量过大会导致昆虫死亡，而用量不足起不到麻痹作用。此外，实验中发现，使用乙醚法固定的灰飞虱活跃度明显降低，且此方法也需要用毛笔挑起或扶正，也可能伤及昆虫肢体。三种昆虫固定方法对昆虫本身的影响程度依次为：负压法<麻醉法<冷冻法；其中，负压法的负面作用最小。

3. 负压法、冷冻法和麻醉法对 EPG 波谱质量的影响 从 EPG 实验记录的波谱图形看出，使用三种方法固定的灰飞虱 EPG 波形均有 8 种波（Seo et al.，2009；Kimmins，1989)，但波形质量上有差异。使用负压法（图 15.2a）固定的灰飞虱 EPG 波形清晰、易于辨别，各个波的过渡波也可明显区分且干扰波较少。使用麻醉法和冷冻法固定的灰飞虱波形干扰波相对较多，并且部分过渡波的干扰波较多，增加了波形辨别难度（图 15.2 中 A、C、E、G、H、I 图)；但麻醉法的 EPG 波形优于冷冻法。综合来看，三种固定方法对 EPG 波形质量的影响大小依次为：负压法<麻醉法<冷冻法。

4. EPG 技术在水稻条纹叶枯病毒（RSV）传播中的应用 Wan 等（2015）运用负压法将灰飞虱进行固定，针对灰飞虱传播 RSV 病毒进行取食行为的 EPG 实验研究，实验结果从感染 RSV 病毒对灰飞虱雌、雄成虫取食过程各特征波形出现的总频次和总持续时间的影响进行分析。

（1）RSV 感染对灰飞虱成虫取食过程各特征波形出现总频次的影响：灰飞虱初羽化成虫取食过程各特征波形出现总次数的单因素方差分析结果显示，RSV 感染显著减少了灰飞虱初羽化雌虫取食过程中的 *P* 形波（-47.08%；$P<0.05$)，而雄虫中的 *P* 形波出现总次数未受到 RSV 感染的显著影响（$P>0.05$)。另外，带毒（RSV）的雌、雄虫的 NX 波形总次数均明显低于未带毒雄虫（$P>0.05$)；雄虫中，感染 RSV 的灰飞虱的 np 波形总次数明显高于未感染 RSV 灰飞虱，但差异不显著（$P>0.05$)；同时雌虫感染 RSV 后，N7 波形出现总次数明显低于未感染灰飞虱，但差异不显著（$P>0.05$)。在灰飞虱取食波形出现总次数的单因素方差分析中，感染或者未感染 RSV 的初羽化灰飞虱成虫取食过程中各个波形出现总次数在雌、雄虫之间均无显著差异（$P>0.05$)（图 15.3)。

（2）RSV 感染对灰飞虱成虫取食过程各特征波形出现总持续时间的影响：灰飞虱初羽化成虫取食过程各特征波形出现总持续时间的单因素方差分析结果进而显示，RSV 感染显著缩短了灰飞虱初羽化雌虫取食过程中的 np 形波总时长（-71.85%；$P<0.05$)，并显著延长了 N4b 波形总时长（$+67.20\%$；$P<0.05$)。而在雄虫中，RSV 感染后，np（$+1176.83\%$；$P<0.05$）和 N4b（-56.44%；$P<0.05$）波形出现总时长的变化趋势与雌虫的相反，并且无论带毒与否，雌、雄之间均存在显著差异（$P<0.05$)。另外，与取食过程中唾液分泌有关的 N4a 波形，感染 RSV 的雌、雄初羽化成虫中的出现总时长均长于未感染 RSV 的，且在雌虫中该差异达到显著水平（$+32.25\%$；$P<0.05$）（图 15.4)。

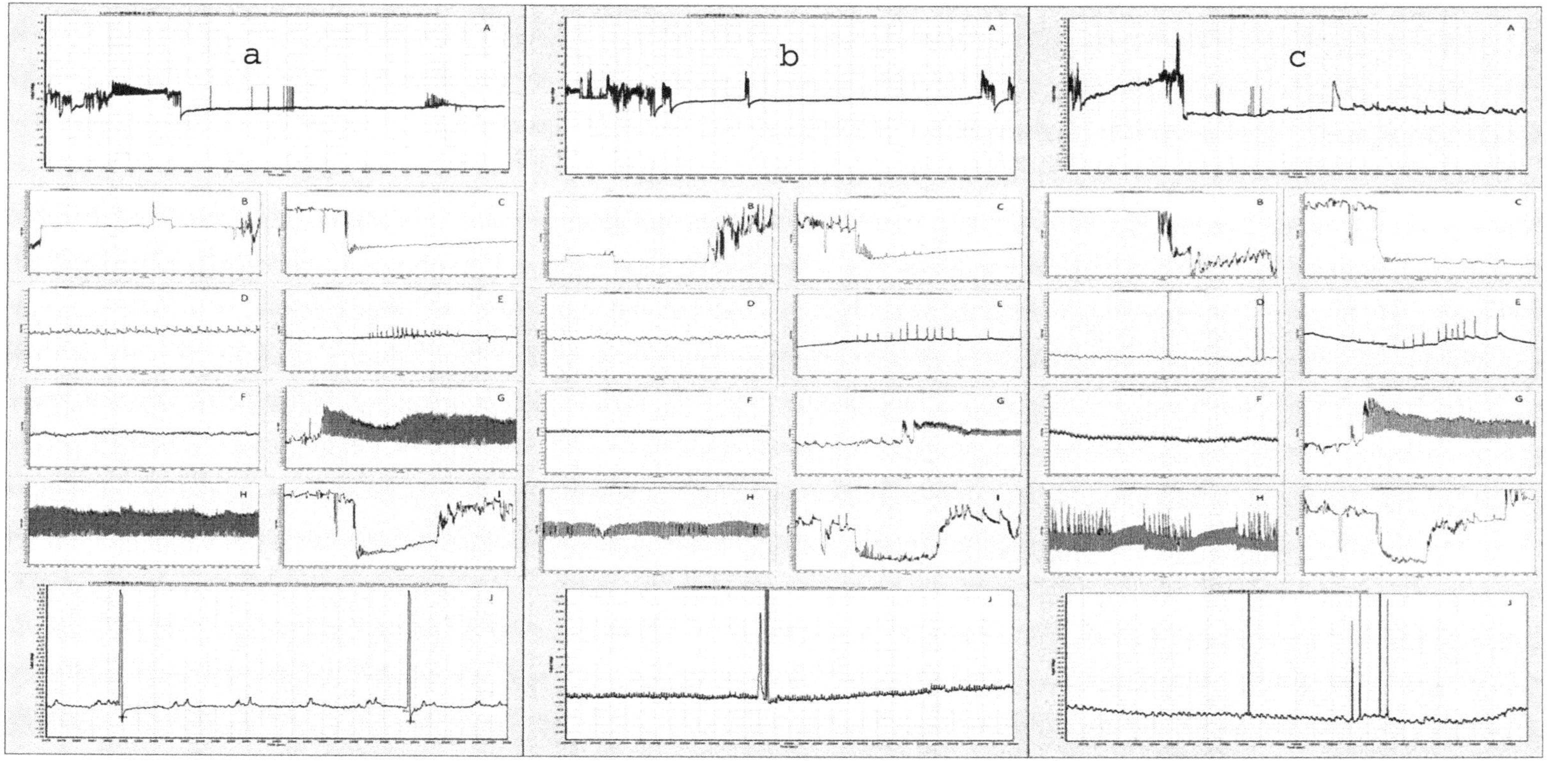

图15.2　EPG实验中应用负压法(a)、乙醚法(b)和冷冻法(c)固定灰飞虱后的刺吸取食波形图

A. 全波形；B.np波和p波；C. p波至N4-a波的过渡波；D. N4-a波；E. N4-a波至N4-b的过渡波，N4-ab波；F. N4-b波；G. p波至N5波的过渡波；H. N5波；I. N7波；J. Nx波

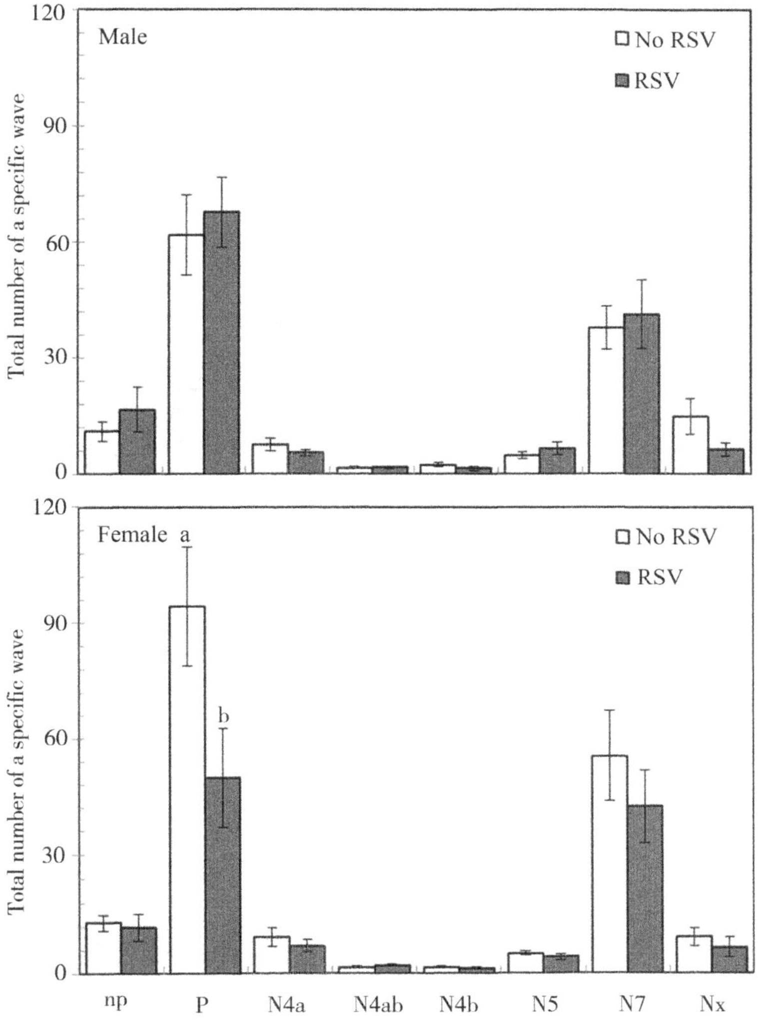

图 15.3 感染和未感染水稻条纹病毒（RSV）的灰飞虱成虫刺吸取食水稻幼苗的总刺吸次数

注：图示柱形图为平均值±标准误差。仅当差异显著时才标字母。不同的小写字母表示雌、雄灰飞虱成虫的某个取食波形出现的总次数在是否感染 RSV 的处理间差异显著。显著性通过单因子方差分析进行检验（$P<0.05$）。(Wan et al.，2015)

研究发现，感染 RSV 后可促进雌虫韧皮部营养取食的 N4b 波形，表明感染 RSV 的雌虫可以获得更多有利于自身生长的养分。对 np 波和 N4b 波的分析发现，感染 RSV 后灰飞虱的此两种波形的出现总次数和持续总时长表现出雌雄二型现象，类似病毒对介体昆虫造成二型现象的报道也为数不多。此外研究还证实，RSV 感染后增加了传毒介体灰飞虱取食的 N4a 波形，该波形被确定为在口针吸取韧皮部或细胞内营养物质前的唾液分泌过程（Stafford et al.，2011；Alvarez et al.，2007；Seo et al.，2009；Jin et al.，2012），唾液分泌与病毒通过介体昆虫取食植物宿主接种密切相关（Prado and Tjallingii，1994），因此 RSV 感染对此波形持续时间的延长效应有利于 RSV 自身的水平

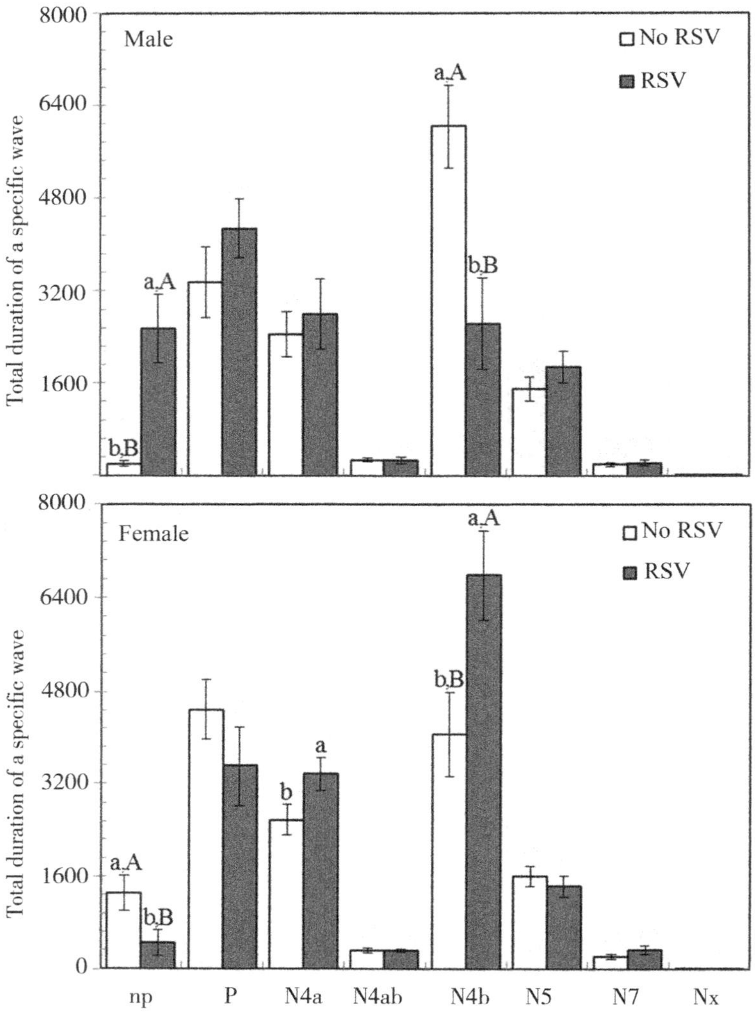

图 15.4 感染和未感染水稻条纹病毒（RSV）的灰飞虱成虫刺吸取食水稻幼苗的总刺吸次数

注：图示柱形图为平均值±标准误差。仅当差异显著时才标字母。不同的小写字母表示雌、雄灰飞虱成虫的某个取食波形出现的总次数在是否感染 RSV 的处理间差异显著。显著性通过单因子方差分析进行检验（$P<0.05$）。（Wan et al.，2015）

传播。而且，研究发现，RSV 感染后 N4a 波的持续时长虽然在雌、雄虫均有所增加，但在雌虫中的增加达到显著水平，这也首次从取食行为的角度为之前研究发现的雌虫较雄虫有更高的传毒效率做出了解释（Gingery，1988）。类似的雌雄传毒能力的差异也在番茄斑萎病毒（Tomato spotted wilt virus，TSWV）和西花蓟马互作对介体昆虫取食行为的改变研究中发现（Stafford et al.，2011；Wilkinson and Ishikawa，2001）。

（四）讨论

与其他生物研究一样，利用 EPG 技术进行昆虫学试验研究要进行精心设计、减少误差，以达到揭示自然规律之目的。与生物化学和分子生物学等可以严格控制条件的实验不同，EPG 实验属于行为学和电生理学交叉学科研究，且受多种因素影响，实验误差大。可见，在 EPG 实验的每个环节都要尽量降低产生误差的可能性。

对 EPG 实验研究而言，在选择昆虫固定方法的时候，负压法是比较有优势的，可尽可能地避免人为因素对实验结果造成的不利影响。负压法中，通过使用吸取定位装置不仅可根据不同供试昆虫设定适宜压力，还可依据试虫的体形大小来更换合适的枪头，从而使负压法的适用范围更广。负压法的装置简单，方便操作且稳定性好，可为实验的顺利开展提供良好的基础条件；此外，在吸附时间上没有限制，对昆虫本身也不会造成危害。另外，因为 EPG 就是记录昆虫的行为变化时所对应的电信号变化的技术，所以在昆虫受到外力作用而导致其肢体损伤时，极有可能会影响到昆虫的各种行为，包括取食行为，EPG 所反映出来的图谱也就会产生干扰波。目前对波形分类还是人工操作，若干扰波太多，势必会影响波形判断，进而在一定程度上影响实验结果的准确性。而负压法的干扰波少，有利于后期对波形的处理。麻醉法是次于负压法的昆虫固定技术之一。由于麻醉法使用的气体是通过影响神经系统使昆虫麻醉的，实验开展也会对操作人员的健康造成不利影响。所以，在使用麻醉法的时候要选择在通风处操作。以往文献中也经常会用到 CO_2 麻醉法（He et al.，2011），其原理是通过使用 CO_2 来隔绝氧气，从而使昆虫缺氧而晕厥。此方法也需要控制好处理时间，但要想固定昆虫并使其背部朝上，这依然需借助毛笔挑起或扶正来连接昆虫和导电金丝，从而会对昆虫造成伤害。冷冻法在操作上较为烦琐，可选择在冰箱或冰块上处理，对于像稻飞虱这类活跃昆虫，使用这种方法就略显烦琐。综上所述，在 EPG 实验研究中，固定昆虫最好的方法是负压法，其次是麻醉法，最后再考虑冷冻法。再者，应考虑到昆虫自身特点以选取最适方法，比如蚜虫活动性很弱，可以直接在叶片上进行昆虫电极的连接（Prado and Tjallingii，1997）。

目前，EPG 技术还处于发展与完善阶段，除了昆虫固定方法需要优化以外，仍在其他一些方面存在一定的不足。由于昆虫连接导电金丝的缘故，其活动受到了限制，这有可能会导致 EPG 波形不能准确反映自然状态下昆虫与寄主植物的相互活动（Caillaud，1999），因此，实验中应该根据应试昆虫的大小考虑金丝的直径和长度，因为这些因素可以影响到应试昆虫被束缚的程度，即昆虫是不是在更接近自由状态下取食（Kindt et al.，2003；Jacobson and Kennedy，2014）。此外，EPG 实验结果需要有自由活动昆虫的发生为害与取食数据加以补充才能更加准确（Prado and Tjallingii，1999）。因此，EPG 研究必须与其他研究方法相结合，才能更准确有效地揭示刺探电位波形与昆虫刺探和取食行为等之间的对应关系，并借助组织学技术、蜜露化学成分分析技术、放射性同位素技术，同时结合生物染色液处理植物根系、观察植物表面残留口针中有无汁液溢出、直接观察人工食物中口针的行踪以及直接观察昆虫口针在植物组织中的行踪等相关技术手段（陈建明等，2002）。目前，EPG 技术已在多种昆虫实验中得到广泛应用。相信，随着 EPG 技术的不断完善并辅以更多的其他研究方法，有望在刺吸式口器昆虫的寄主专化型、植物的抗虫性（Lei et al.，1999）、昆虫的获毒及传毒（Wetering et al.，1998）等方面取得更为积极有效的成果。

参考文献

[1] ALVAREZ A E, GARZO E, VERBEEK M, et al. Infection of potato plants with potato leafroll virus changes attraction and feeding behaviour of *Myzus persicae* [J]. *Entomologia Experimentalis et Applicata*, 2007, 125 (2): 135-144.

[2] CAILLAUD M C. Behavioural correlates of genetic divergence due to host specialization in the pea aphid, Acyrthosiphon pisum [J] . *Entomol. Exp. Appl.* , 1999, 91: 227-232.

[3] CIVOLANI S, LEIS M, GRANDI G, et al. Stylet penetration of *Cacopsylla pyri*; an electrical penetration graph (EPG) study [J] . *Journal of Insect Physiology*, 2011, 57 (10): 1 407-1 419.

[4] DIAZ-MONTANO J, REESE JC, LOUIS J, et al. Feeding behavior by the soybean aphid (Hemiptera: Aphididae) on resistant and susceptible soybean genotypes [J]. *J. Econ. Entomol.* , 2007, 100: 984-989.

[5] GHAFFAR M B, PRICHARD J, FORD-LLOYD B. Brown planthopper (*N. lugens* Stal) feeding behaviour on rice germplasm as an indicator of resistance [J] . *Plos ONE*, 2011, 6 (7): e22137.

[6] GINGERY R E. The rice stripe virus group [M] . The plant viruses. Springer US, 1988: 297-329.

[7] HE Y P, ZHANG J F, CHEN J M, et al. Influence of pymetrozine on feeding behaviors of three rice planthoppers and a rice leafhopper using electrical penetration graphs [J]. *Journal of Economic Entomology*, 2011, 104 (6): 1877-1884.

[8] JACOBSON A L, KENNEDY G G. Electrical penetration graph studies to investigate the effects of cyantraniliprole on feeding behavior of *Myzus persicae* (Hemiptera: Aphididae) on *Capsicum annuum* [J] . *Pest Management Science*, 2014, 70 (5): 836-840.

[9] JIANG Y X, WALKER G P. Identification of phloem sieve elements as the site of resistance to silverleaf whitefly in resistant alfalfa genotypes [J] . *Entomol. Exp. Appl.* , 2007, 125 (3): 307-320.

[10] JIN S, CHEN Z M, BACKUS E A, et al. Characterization of EPG waveforms for the tea green leafhopper, *Empoasca vitis* Gothe (Hemiptera: Cicadellidae), on tea plants and their correlation with stylet activities [J] . *Journal of Insect Physiology*, 2012, 58 (9): 1 235-1 244.

[11] KIMMINS F M. Electrical penetration graphs from *Nilaparvata lugens* on resistant and susceptible rice varieties [J] . *Entomologia Experimentalis et Applicata*, 1989, 50 (1): 69-79.

[12] KINDT F, JOOSTEN N N, PETERS D, et al. Characterisation of the feeding behaviour of western flower thrips in terms of electrical penetration graph (EPG) waveforms [J]. *Journal of Insect Physiology*, 2003, 49 (3): 183-191.

[13] LEI H, LENTEREN J C VAN, TJALLINGII W F. Analysis of resistance in tomato and sweet pepper against the green-house whitefly using electrically monitored and visually observed probing and feeding behavior [J] . *Entomologia Experimentalis et Applicata*, 1999, 92 (2): 299-309.

[14] MCLEAN D L, KINSEY M G. A technique for electronically recording of aphid feeding and salivation [J] . *Nature*, 1964, 202: 1 358-1 359.

[15] PRADO E, TJALLINGII W F. Aphid activities during sieve element punctures [J]. *Entomologia Experimentalis et Applicata*, 1994, 72 (2): 157-165.

[16] PRADO E, TJALLINGII W F. Effects of experimental stress factors on probing behavior by aphid [J] . *Entomologia Experimentalis et Applicata*, 1999, 90: 289-300.

[17] PRADO E, TJALLINGII W F. Effects of previous plant infestation on sieve element acceptance by two aphids [J] . *Entomologia Experimentalis et Applicata*, 1997, 82: 189-200.

[18] SEO B Y, KWON Y H, JUNG J K, et al. Electrical penetration graphic waveforms in relation to the actual positions of the stylet tips of *Nilaparvata lugens* in rice tissue [J]. *Journal of Asia-Pacific Entomology*, 2009, 12 (2): 89-95.

[19] STAFFORD C A, WALKER G P, ULLMAN D E. Infection with a plant virus modifies vector feeding behavior [J] . *Proceedings of the National Academy of Sciences of the United States of America*, 2011, 108 (23): 9 350-9 355.

[20] TJALLINGII W F. Electrical recording of stylet penetration activities [J] . 1988.

[21] TJALLINGII W F. Electrical nature of record signals during styled penetration by aphids [J] . *Entomol. Exp. Appl.* , 1985a, 38: 177-186.

[22] TJALLINGII W F. Electrical recording of stylet penetration activities. MINKS A K, HARREWJJN P. Aphids, their biology, natural enemies and control [J]. Amsterdam: Elsevier, 1987, 95-108.

[23] TJALLINGII W F. Electronic recording of penetration behaviour by aphids [J]. *Entomol. Exp. Appl.* , 1978, 24: 721-730.

[24] TJALLINGII W F. Membrance potentials as an indication for plant cell penetration by aphids stylets [J] . *Entomol. Exp. Appl.* , 1985b, 38: 187-193.

[25] TJALLINGII W F. Wire effects on aphids during electrical recording of stylet penetration [J] . *Entomol. Exp. Appl.* , 1986, 40: 89-98.

[26] WAN G J, JIANG S L, WANG W J, et al. Rice stripe virus counters reduced fecundity in its insect vector by modifying insect physiology, primary endosymbionts and feeding behavior [J] . *Scientific Reports*, 2015, 5: 12527.

[27] WETERING F. VAN D E, HULSHOF J, POSTHUMA K, et al. Distinct feeding behavior between sexes of *Frankliniella occidentalis* results in higher scar production and lower tospovirus transmission by females [J] . *Entomol. Exp. Appl.* , 1998, 88 (1): 9-15.

[28] WIKINSON T L, ISHIKAWA H. On the functional significance of symbiotic microorganisms in the Homoptera: a comparative study of *Acyrthosiphon pisum* and *Nilaparvata lugens* [J] . *Physiological Entomology*, 2001, 26 (1): 86-93.

[29] 陈建明，俞晓平，程家安，等．定量研究刺吸式昆虫取食行为的有效方法——电子取食监测仪的原理和应用技术 [J] . 浙江农业学报，2002，14 (4)：237-243.

[30] 胡想顺，刘小凤，赵惠燕．刺探电位图谱（EPG）技术的原理与发展 [J] . 植物保护，2006，32 (3)：1-4.

[31] 荆裴，白素芬，刘芳．灰飞虱取食行为刺吸电位波形的初步研究 [J] . 应用昆虫学报，2013，50 (3)：758-763.

[32] 罗晨，岳梅，徐洪富，等．EPG 技术在昆虫学研究中的应用及进展 [J] . 昆虫学报，2005，48 (3)：437-433.

[33] 汤清波，张大山，姬琨，等．刺吸电位技术应用中的几个问题 [J] . 应用昆虫学报，2011，48 (5)：1519-1527.

[34] 岳梅，罗晨，徐洪富，等．刺吸电波图实验中金丝与烟粉虱的粘连技术 [J] . 昆虫知识，2005，42 (3)：326-328.

(陈法军　江守林　万贵钧)

第十六章　西花蓟马取食行为的刺吸电位（EPG）研究

（一）研究背景

西花蓟马 *Frankliniella occidentalis*（Pergande）作为一种对大多数蔬菜和花卉等园艺作物具有毁灭性为害的世界性入侵害虫，其为害情况一直备受关注（Kirk and Terry，2003）。西花蓟马最早记载于 1895 年，曾是美国和加拿大西部山区最常见的害虫（张友军等，2004）。此后不久，西花蓟马对不同环境适应性增强，对杀虫剂抗药性增强，因此逐渐向外扩展，成为许多国家和地区蔬菜和花卉上的重要害虫（Kirk and Terry，2003；Andrew et al.，2013；吕要斌等，2011）。我国大陆则于 2003 年首次报道了该害虫在北京郊区大棚内辣椒上暴发并为害严重，随后西花蓟马在我国迅速扩散，目前我国十几个地区有分布（程俊峰等，2006；吕要斌等，2011；刘若思等，2015），并有进一步扩散的趋势。

西花蓟马以其锉吸式口器吸取植物叶片、花、果实等组织的汁液，造成叶片产生斑点、畸形、扭曲，果实表面出现疤痕斑点等，严重时导致植物生长不良，甚至死亡。不仅如此，西花蓟马还能通过取食花粉、蜜露、粉虱、木虱和螨类的卵，提高其自身的繁殖力（Agrawal and Karban，1997；Agrawal and Klein，2001；Faraji et al.，2002；Maanen et al.，2012；Martini et al.，2013；Hall，2014），使其种群数量增长很快，因此西花蓟马的为害非常严重。除此之外，西花蓟马还能传播番茄斑萎病毒（tomato spotted wilt virus，TSWV）、凤仙花坏死斑病毒（impatiens necrotic spot virus，INSV）和玉米枯黄斑点病毒（maize chlorotic mottle virus，MCMV）等多种病毒（Zhao et al.，2014；程俊峰等，2005）。传毒造成的损失巨大，如番茄斑萎病毒是全世界 10 种为害性最大的植物病毒之一（Scholthof et al.，2011），侵染烟草、大豆、番茄、花生、辣椒、莴苣、菊花等作物及多种杂草，为害严重时造成绝收。国内外对西花蓟马生物学特性、发生规律和防治方法、抗药性以及西花蓟马与其他昆虫间的互作等方面都已有较深入的研究（Gao et al.，2014；Hall et al.，2014；Herron et al.，2014；Kos et al.，2014；盖海涛等，2010；王健立等，2011；徐学农等，2011；莫利锋等，2013；谢志娟，2013；马鹤娟，2014；王圣印等，2014）。

目前已知的西花蓟马寄主植物有 50 科 500 种之多，寄主植物包括各种重要的经济作物、蔬菜、花卉，如花生、番茄、生菜、芹菜、胡椒、洋葱、甜椒、黄瓜、菊花、玫瑰、凤仙花等（Yudin，1986；Robb，1989；Parker et al.，1995；Daughtrey，1996），并且随着研究的深入，新的寄主植物还在不断地被发现。有研究表明，番茄对西花蓟马有明显的忌避作用（裴昌莹和郑长英，2011），是西花蓟马的非适宜寄主（袁成明等，2011）。

昆虫的取食行为与其寄主密切相关，通过 EPG 技术可掌握昆虫口针在植物体内的位置以及发生的各种行为过程，因此，研究西花蓟马在不嗜食寄主上的 EPG 图谱具有重要意义，有望在植物的抗虫性、西花蓟马的获毒及传毒、各种防治措施等方面取得更大的研究成果。

（二）材料与方法

1. 西花蓟马的来源及饲养 西花蓟马采自贵州省贵阳市花溪区蔬菜基地的菜豆（*Phaseolus vulgaris* Linn）上，在人工气候箱［L/D = 14h/10h；温度（25±1）℃；湿度（70±5）%］中用菜豆豆荚饲养 3 代以上供试。

2. EPG 记录 采用（DC-EPG）Giga-8 刺吸电位图谱记录仪，将西花蓟马成虫与供试植物分别连接生物电流放大器的昆虫电极和植物电极。昆虫电极是一段长 1.5～2cm，直径 10μm 的金丝，末端用水溶性导电银胶粘在西花蓟马雌成虫前胸背板上，粘连后对其进行 1h 饥饿处理，植物电极插在植物生长的土壤中。随后将西花蓟马放置在待试植物叶背面进行记录，记录时间为 6h。所有记录均在室内进行，温度为（27±1）℃，相对湿度为（70±5）%。采用 ANA 软件用于昆虫取食行为的记录和数据的分析统计。

（三）结果分析

西花蓟马取食番茄的 EPG 各波形的发生情况：西花蓟马在植物上刺吸时 EPG 系统会记录电压的变化，不同的电压波动会产生不同的波形。西花蓟马刺吸产生的波形分为 6 种：np、P、S、Q、R 和 X。这些特定的波形见图 16.1。

当西花蓟马口针没有刺入植物时，EPG 记录电压为零，把此时的波形定义为基线（波形 np）；当西花蓟马的口针刺入植物细胞时，电压会急剧上升（波形 P）。P 波的持续时间一般不超过 1s，其电压也往往高于其他任何波形，所有的刺吸行为都是以波形 P 开始，在 P 波之后会出现一个小的波峰，为 S 波。此后有时会出现一个下降的电压，为 Q 波。一般情况下，电压又会回升并出现一系列的方形波，该波即为 R 波，R 波是西花蓟马的主要刺吸波。一般情况下，在 R 波结束时，波形渐渐消失，但有时在 R 波结束时会出现 X 波。

西花蓟马在番茄上取食的 6 个主要行为：非刺探（np）、上颚口针刺入（P 波）、上颚口针重复刺吸（S 波）、下颚口针刺入（Q 波）、细胞内取食（R 波）和 X 波。在记录西花蓟马取食番茄的 6h 内，西花蓟马大部分时间都处于非刺吸阶段，而刺吸番茄的 P 波平均发生约 49 次 · h^{-1}，其发生频率仅次于不取食阶段；取食波 R 发生的频率次于上颚口针刺入波 P，持续取食时间平均大于 10s；西花蓟马取食行为发生时首先产生 P 波，S 波紧随其后，在 S 波之后有时会出现下降的 Q 波，持续时间 1～2s。X 波偶尔出现在 R 波之后，其持续时间较短。

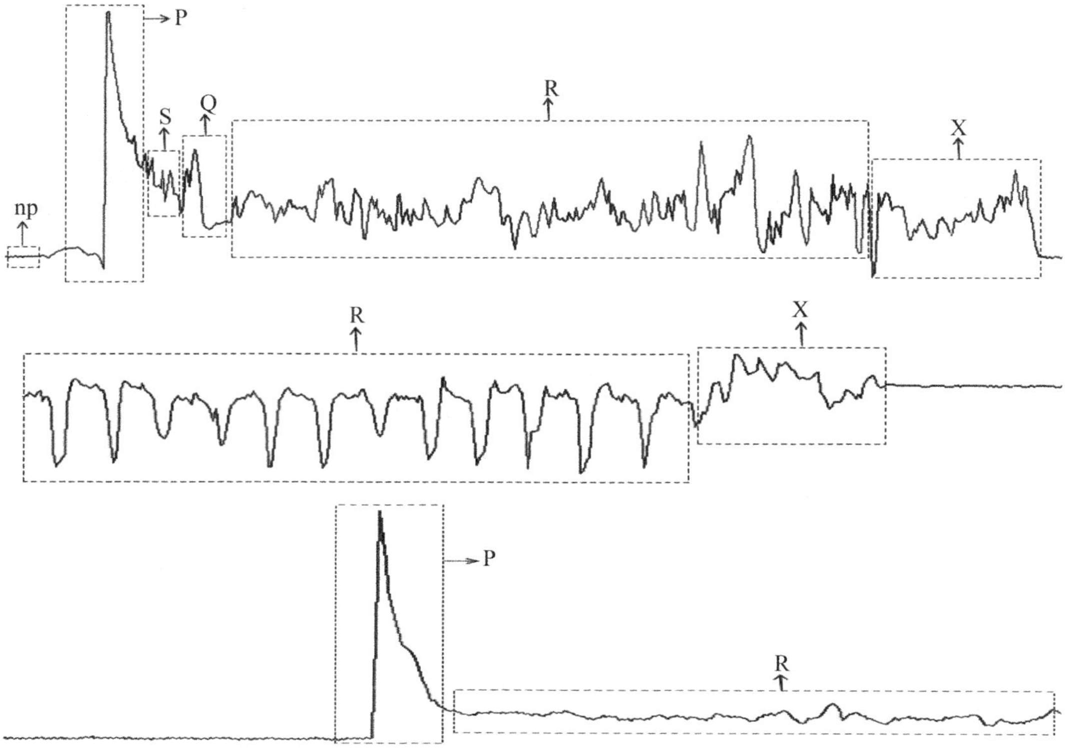

图 16.1　西花蓟马取食番茄的 6 种典型 EPG 波形

np：不取食；P：上颚口针刺入；S：上颚口针重复刺吸；Q：下颚口针刺入；
R：细胞内取食；X：不常规波形

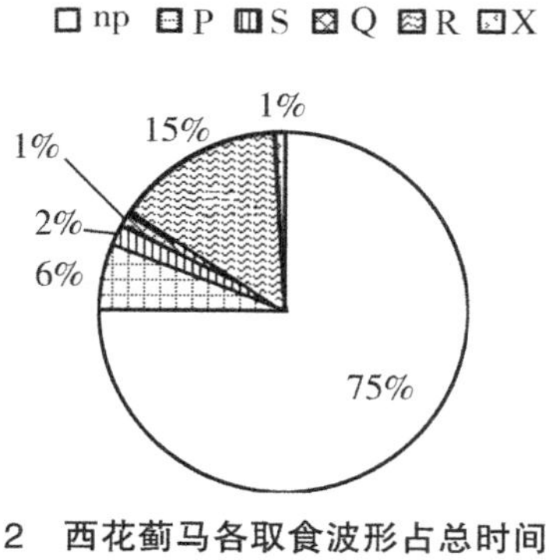

图 16.2　西花蓟马各取食波形占总时间比例

（四）讨论

刺吸式口器昆虫取食植物时，产生的 EPG 波形跟昆虫和植物的种类有关，正确识别 EPG 波形是研究此类昆虫取食行为的关键。蚜虫是最早应用 EPG 技术对其刺吸行为进行研究的，也是研究最深入的。西花蓟马的取食波形较其他刺吸式昆虫不同。Kindt 等（2003）在用 EPG 对西花蓟马刺吸辣椒和 10%糖溶液的过程中，记录了口针刺入细胞、分泌唾液和取食过程，将西花蓟马的取食波形分为非刺探（np）、上颚口针刺入（P 波）、上颚口针重复刺吸（S 波）、下颚口针刺入（Q 波）、细胞内取食（R 波）和 X 波等 6 种波形，并且认为 X 波形为不常规波形。在早期，Harrewijin 等（1996a）曾将西花蓟马在黄瓜上的取食波形记录为 P、Q 和 R 波。

许多昆虫在嗜食寄主上的生长发育要优于非嗜食寄主，一方面是由于嗜食寄主中的营养物质更有利于促进其生长发育，另一方面是由于其在嗜食寄主上的取食量要高于非嗜食寄主，在嗜食寄主上的高取食量同时也导致了对嗜食寄主为害程度的加深，而在非嗜食寄主上的取食量可能会减少。裴昌莹和郑长英（2011）的研究表明，番茄对西花蓟马有明显的忌避作用，据此推测，西花蓟马在番茄上的取食量比嗜食寄主少。Harrewijin 等（1996b）对比了西花蓟马在抗性黄瓜和非抗性黄瓜上的取食行为，与非抗性黄瓜相比，西花蓟马在抗性黄瓜上的非取食阶段增加。本研究表明西花蓟马在非嗜食寄主番茄上的非取食阶段明显高于取食阶段，与该结果一致。而张玉秀等（2014）用 EPG 技术分析了西花蓟马在 4 种不同嗜食寄主植物上的取食行为，发现西花蓟马的取食行为与寄主适宜性之间无显著的相关性。不同学者对西花蓟马对不同寄主的取食行为结果不完全相同，具体原因还需要进一步分析。

参考文献

[1] AGRAWAL A A, KARBAN R. Domatia mediate plant - arthropod mutualism. *Nature*, 1997, 387 (6633): 562-563.

[2] AGRAWAL A A, KLEIN C N. What omnivores eat: direct effects of induced plant resistance on herbivores and indirect consequences for diet selection by omnivores. *Journal of Animal Ecology*, 2001, 69: 525-535.

[3] ANDREW G M, BROCK R D, MADALINE A H. Biological control of western flower thrips, *Frankliniella occidentalis* (Pergande) (Thysanoptera: Thripidae), in gerberas, chrysanthemums and roses. *Australian Journal of Entomology*, 2013, 52: 246-258.

[4] DAUGHTREY M L. Detection and identification of tospoviruses in greenhouses. *Acta Horticulturae*, 1996, 431: 90-98.

[5] FARAJI F, JANSSEN A, SABELIS M W. The benefits of clustering eggs: the role of egg predation and larval cannibalism in a predatory mite. *Oecologia*, 2002, 131 (1): 20-26.

[6] GAO C F, MA S Z, SHAN C H, et al. Thiamethoxam resistance selected in the western flower thrips *Frankliniella occidentalis* (Thysanoptera: Thripidae): Cross - resistance patterns, possible biochemical mechanisms and fitness costs analysis. *Pesticide Biochemistry and Physiology*, 2014, 114: 90-96.

[7] HARREWIJIN P, TJALLINGII W F, MOLLEMA C. Electrical recording of plant penetration by western flower thrips. *Entomologia Experimentalias et Applicata*, 1996a, 79: 345-353.

[8] HARREWIJIN P, PIRON P G M, MOLLEMA C. Electrically recorded probing behavior of thrips species on optimal and suboptimal hosts. *Entomologia Experimentalias et Applicata*, 1996b, 80: 43-45.

[9] HALL DG. Interference by Western Flower Thrips in rearing Asian citrus psyllid: Damage to host plants and facultative predation. *Crop Protection*, 2014, 60: 66-69.

[10] HERRON G A, GUNNING R V, COTTAGE E L A. Spinosad resistance, esterase isoenzymes and temporal synergism in *Frankliniella occidentalis* (Pergande) in Australia. *Pesticide Biochemistry and Physiology*, 2014, 114: 32-37.

[11] KINDT F, JOOSTEN N N, PETERS D, et al. Characterisation of the feeding behavior of western flower thrips in terms of electrical penetration graph (EPG) waveforms. *Journal of Insect Physiology*, 2003, 49: 183-191.

[12] KIRK W D J, TERRY L I. The spread of the western flower thrips *Frankliniella occidentalis* (Pergande) . *Agricultural and Forest Entomology*, 2003, 5: 301-310.

[13] KOS S P, KLINKHAMER P G L, LEISS K A. Cross-resistance of chrysanthemum to western flower thrips, celery leafminer, and two-spotted spider mite. *Entomologia Experimentalis et Applicata*, 2014, 151: 198-208.

[14] MAANEN R V, BROUFAS G, OVEJA M F, et al. Intragulid predation among plant pests: western flower thrips larvae feed on whitefly crawlers. *BioControl*, 2012, 57: 533-539.

[15] MARTINI X, KINCY N, VAUGHN K, et al. Positive association between thrips and spider mites in seedling cotton. *Agricultural and Forest Entomology*, 2013, 15: 197-203.

[16] PARKER B L, SKINNER M, LEWIS T. Thrips Biology and Management. NATO ASI Series. Series A: Life Sciences. New York: Plenum press, 1995.

[17] ROBB K L. Analysis of *Frankliniella occidentalis* (Pergande) as a pest of floricultural crops in California greenhouses. Ph. D. Dissertation, University of California, Riverside, 1989.

[18] SCHOLTHOF K B, ADKINS S, CZOSNEK H, et al. Top 10 plant viruses in molecular plant pathology. *Molecular Plant Pathology*, 2011, 12: 938-954.

[19] YUDIN L S, CHO J J, MICHELL W C. Host range of western flower thrips, *Frankliniella occidentalis* (Thysanoptera: Thripidae), with special reference to Leucaena glauce. *Environ Entomol*, 1986, 15: 129-1295.

[20] ZHAO M, HO H, WU Y, et al. Western Flower Thrips (*Frankliniella occidentalis*) transmits *Maize Chlrotic Mottle* virus. *Journal of Phytopathology*, 2014, 162: 532-536.

[21] 程俊峰，万方浩，郭建英，等．外来有害入侵生物——西花蓟马．中国生物防治，2005 ，21 (2)：74-79.

[22] 程俊峰，万方浩，郭建英．西花蓟马在中国适生区的基于 CLIMEX 的 GIS 预测．中国农业科学，2006，39 (3)：525-529.

[23] 盖海涛，郅军锐，李肇星，等．西花蓟马和花蓟马在温度逆境下的存活率比较．生态学杂志，2010 ，29 (8)：1 533-1 537.

[24] 莫利锋，郅军锐，陈祥叶．温度对南方小花蝽捕食西花蓟马功能反应的影响．中国生物防治学报，2013，29 (2)：187-193.

[25] 刘若思，刘燕，王军，等．重要外来入侵害虫西花蓟马在吉林省部分地区的首次发现．北京农学院学报，2015，30 (2)：2-5.

[26] 吕要斌，张治军，吴青君，等．外来入侵害虫西花蓟马防控技术研究与示范．应用昆虫学报，2011，48 (3)：488-496.

[27] 马鹤娟．加州新小绥螨对西花蓟马的控制作用初探．华中农业大学，2014.

[28] 裴昌莹，郑长英．西花蓟马对不同蔬菜寄主的选择性研究．中国生态农业学报，2011，19 (2)：383-387.

[29] 王健立，王俊平，郑长英．西花蓟马与烟蓟马在紫甘蓝上的实验种群生命表．植物保护学报，

2011，38（5）：390-394.
[30] 王圣印，张安盛，李丽莉，等．西花蓟马田间种群对常用杀虫剂的抗性现状及防治对策．昆虫学报，2014，57（5）：621-630.
[31] 徐学农，Borgemeister Christian，Poehling Hans-Michael. 西花蓟马、二斑叶螨与黄瓜新小绥螨的相互关系研究．应用昆虫学报，2011，48（3）：579-587.
[32] 谢志娟．西花蓟马对阿维菌素抗性的分子机制．扬州大学，2013.
[33] 袁成明，郅军锐，曹宇，等．西花蓟马对蔬菜寄主的选择性．生态学报，2011，31（6）：1 720-1 726.
[34] 张友军，张治军，徐宝云，等．外来入侵害虫——西花蓟马的发生、为害与防治．中国蔬菜，2004（5）：51-52.
[35] 张玉秀，张治军，吕要斌．利用 EPG 技术分析西花蓟马在 4 种寄主植物上的取食行为．浙江农业学报，2014，26（3）：714-721.

（郅军锐　温　娟）

第十七章 Armet 蛋白影响豌豆蚜取食行为的研究

（一）研究背景

豌豆蚜 *Acyrthosiphon pisum* 属于半翅目（Hemiptera）胸喙亚目（Stemorhyncha）蚜科（Aphidaidaea）。豌豆蚜体形较大，多为扁椭圆形，身体较柔软，体色丰富多样。豌豆蚜的生活周期分为若虫期和成虫期，在食物充足和环境良好的情况下生殖方式是孤雌生殖，在环境恶化和食物来源不足的情况下，部分豌豆蚜会进行有性生殖。蚜虫繁殖能力惊人，平均生活周期为 1 周左右（Karley et al.，2004）。作为一种多寄主的刺吸式昆虫，豌豆蚜在农业植物的生长季迅速繁殖，常以若虫和胎生雌虫群居形态对植物的叶片、嫩茎、花和幼果等进行大规模的取食，为害香豌豆、豌豆、蚕豆、苜蓿、草木樨、大豆等草本豆科植物的生长，严重影响农业生产。此外，蚜虫大量取食还造成植物营养恶化、生长停滞，组织提前老化、早衰。蚜虫唾液含有的氨基酸或植物生长激素，注入寄主植物组织内，引起植株出现斑点、缩叶、卷叶、虫瘿和肿瘤等畸形现象。蚜虫还是植物病毒的传播介体，可携带 30 多种植物病毒（Ponsen，1977），这些病毒都是通过取食这一途径完成病原传播的。

唾液分泌是蚜虫取食过程中的最重要一环。蚜虫唾液腺由一个主唾液腺和两个副唾液腺组成，主唾液腺中的 8 个分泌细胞受神经细胞调控，分泌管道从主唾液腺伸出并分别连接左右副唾液腺（Ponsen，1972）。一般认为蚜虫唾液腺蛋白通过唾液传递到植物中（Bellafiore et al.，2008；Davis et al.，2004），病原性效应因子能抑制和扰乱植物非特异免疫系统，甚至能操纵免疫响应（Jones et al.，2006；Chisholm et al.，2006）。很多证据表明，蚜虫唾液蛋白的某些成分与植物病原体引起的抗性反应相关，特别是与线虫的效应因子具有高度序列同源性和功能相似性，因此很多唾液蛋白被预测为潜在的效应因子。Carolan 等人利用豌豆蚜唾液腺转录组和蛋白组数据，构建豌豆蚜唾液腺蛋白组，预测了包含 Armet 在内的 42 个可能的蚜虫效应因子蛋白（Carolan et al.，2011）。

Armet 预测为一种钙结合蛋白，而钙结合蛋白在植物防御反应中发挥重要作用（Will et al.，2007）；在组织表达上，*Armet* 在唾液腺的表达量要远高于其他组织；而在不同的取食条件下，豌豆蚜中取食蚕豆植物与取食人工饲料相比，*Armet* 基因的转录水平升高了约 5 倍（Wang et al.，2015），表明豌豆蚜在寄主植物上取食比在人工饲料中对 *Armet* 蛋白有了更高的需求。由此推测在豌豆蚜取食植物行为中，特别是在蚜虫与植

物互作方面，*Armet* 蛋白发挥重要作用。

RNA 干扰（RNA interference，RNAi）是动植物中广泛存在的一种转录后基因沉默（PTGS）机制，它是由一段基因特异性的长链 dsRNA（double-stranded RNA，dsRNA）被切割成 20~50 nt 长短的片段，即小干扰 RNA（small interference RNA siRNA ）所引起的细胞内 mRNA 表达水平降低，导致了目标基因的转录后表达沉默的现象（Fire et al. 1998；Zamore et al. 2000）。Mutti 等将 RNAi 技术应用于蚜虫，利用显微注射技术将 siRNA 注入蚜虫，对 c002 干扰结果表明豌豆蚜取食行为有了显著的改变（Mutti et al.，2008）。

Armet 作为豌豆蚜唾液腺高表达的分泌蛋白，可能是具有调控作用的效应分子，它影响昆虫与植物的相互作用，对于促进豌豆蚜取食、促进蚜虫适应寄主植物有重要作用，甚至能够调控植物防御免疫机制来提高种群的生存度。通过研究干扰 *Armet* 基因的豌豆蚜生存表型和取食行为的变化，为研究该蛋白参与蚜虫-植物相互作用的机制提供有力证据和基础。

（二）材料与方法

1. 豌豆蚜的饲养及龄期控制　豌豆蚜 *Acyrthosiphon pisum* 采自云南玉溪农田的豌豆苗（*Pisum sativum*）上，挑选健康蚜虫放置到温室智能人工气候箱［温度 =（21±1）℃，相对湿度 =（60±5）%，光周期：暗周期 = 16：8］的养虫笼（40cm×40cm×40cm）内，以蚕豆为寄主植物。取一头健康的孤雌生殖成蚜作为母体，单独饲养，使其建立一个遗传背景一致种群，即单雌系。

在 23℃的培养条件下，将 30~50 只健康的豌豆蚜成虫放置在新鲜的蚕豆苗上 12h 后，移走成虫，留下 1 龄幼虫，3d 后即可生长成 3 龄幼虫待用。

2. 干扰豌豆蚜 *Armet* 基因　以已成功鉴定的重组质粒 pGEM-T-EASY：*Armet* 为模板，扩增包含 *Armet* 基因的 dsRNA 区域的 PCR 产物（*Armet*-RNAi-T7-F：TAATACGACTCACTATAGGATTAGAAGATTCTGCAACTGG；Armet-RNAi-T7-R：TAATACGACTCACTATAGGCTCTTCCTTGACGTAAGAGG），连接转化和测序鉴定，并提取重组质粒 pGEM-T-EASY：*dsArmet*。并以其为模板，扩增得到两端分别带有 T7 RNA 聚合酶的 dsRNA 的 PCR 产物，并纯化（Wizard SV Gel and PCR Clean-Up System，Promega）。按照说明书合成 *Armet* 双链 RNA（T7 Ribo MAX Express RNAi System，Promega）。测定 RNA 的浓度及纯度（Nanodrop ND1000 核酸定量仪），稀释分装后-80℃保存待用。同样合成绿色荧光蛋白（GFP）的 dsRNA，即 dsGFP 作为对照。

用小毛笔轻轻将豌豆蚜 3 龄幼虫从蚕豆苗上取下，放置在解剖镜下，使用微量注射仪（Nanoject II nanoliter injector，USA，Warner）将合成的 dsRNA（23nL，6μg/μL）从腹部注射到蚜虫体内血淋巴中。

将注射过 dsRNA 的豌豆蚜放置在蚕豆叶片上培养待用。每 5 只蚜虫为一个重复，共 6 个重复，用 Trizol 试剂提取其 RNA，反转录成 cDNA（M-MLV Reverse Transcriptase，Promega），cDNA 作为模板，进行实时荧光定量 PCR。（Armet-qPCR-F：GTTACTATTTGGCGGATTA；Armet-qPCR-R：TATCGCAGACTTGAGCAT）。

3. 豌豆蚜生存曲线的观察和统计

（1）观察注射双链 RNA 后豌豆蚜在模拟蚕豆植株上的生存状况：剪取 2 片新鲜的蚕豆叶片，将叶柄立即插入盛有清水的 45mm 塑料培养皿的底壳中，用保鲜膜封好，并用橡皮筋固定。再将叶片和塑料培养皿置于 120mm 玻璃培养皿中，用打有细小气孔的保鲜膜盖住玻璃培养皿，并用橡皮筋固定。将注射过 dsRNA 的豌豆蚜放到蚕豆叶片上，每个培养皿中放入 15 只蚜虫作为一个生物学重复，并将培养皿放入恒温培养箱中，从注射后 24h 后开始对存活的豌豆蚜计数，每隔 24h 计数一次。

（2）观察注射双链 RNA 后豌豆蚜在人工饲料上的生存状况：将一只注射过 dsRNA 的蚜虫放入直径约 1.5cm 的塑料白色小盖，用 Parafilm 膜拉制成半透明状薄膜盖在塑料小盖上，使之密闭封好。将 20μL 蚜虫人工饲料（Febvay，1987）滴加在小盖上的半透明封口膜上后，再用另一张半透明封口膜封好，使蚜虫饲料能够密闭在两层封口膜之间。每隔 2d 按照上述的方法换一次新鲜的蚜虫饲料。将 15 只注射过 dsRNA 的蚜虫作为一个生物学重复，放在恒温培养箱中培养。注射 24h 后开始对存活的豌豆蚜计数，每隔 24h 计数一次。

以上两种方法连续记录 9d 后，使用 SPSS17.0 软件、Kaplan-Meier Method 分析统计生存曲线的显著性。

4. EPG 记录 用八通道昆虫刺吸电位测量系统（Giga-8 型，EPG system，Wageningen，Netherlands）对注射 dsRNA 3d 的豌豆蚜进行连续 8h 取食观察。将 EPG 系统的植物电极插入栽有 2~3 株发芽 1~2 周的蚕豆苗的花盆土（湿润）中。将昆虫电极的铜钉上缠绕 3~4cm 的细金线（直径为 18μm），在解剖镜下用导电银胶将金丝的另一端粘在豌豆蚜的背部。当 8 只豌豆蚜都粘好金丝后，尽量同时固定并调整昆虫电极，使悬挂的豌豆蚜接触到蚕豆苗的嫩芽上，并迅速开始记录。每个重复连续记录至少 9h，第 9 小时开始部分仍可以记录到取食波形的数据为有效数据，将其中 0~8h 用于数据统计。所有记录均在室内进行，温度为 25℃，光周期与培养箱的光周期吻合。采用 Giga-8 附带的 Stylet$^+$daq 和 ana 软件分别用于昆虫取食行为的记录和数据的分析。两组之间的差异通过独立样本 t 检验，使用 SPSS 17.0 软件统计分析。

（三）结果

1. 干扰 *Armet* 基因对豌豆蚜生存的影响 在对豌豆蚜 3 龄幼虫注射 dsRNA 后，分别在 1d 和 3d 两个时间点取样，使用荧光定量 PCR 对 *Armet* 基因表达水平进行检测，如图 17.1a 所示，*Armet* 在 24h 和 72h 后转录水平分别下降了 38.7%和 34.4%，表明瞬时干扰的效果至少能维持 3d。由于 *Armet* 基因在唾液腺高表达，因此为了验证 *Armet* 基因转录水平的下降主要是由唾液腺基因转录水平下降所导致，因此我们收取干扰后的豌豆蚜包含唾液腺的头部组织和肠道组织做进一步的定量验证，结果表明，头部的 *Armet* 基因转录水平的降低是导致蚜虫全身转录水平下降的原因（图 17.1b）。

如图 17.2 所示，干扰 *Armet* 的豌豆蚜比对照在模拟蚕豆植株装置上的生存率明显下降（$P<0.001$），而它们的生存率在人工饲料上却基本没有差异（$P=0.08$）。

2. 注射 dsRNA 的豌豆蚜在蚕豆上取食行为 与野生型蚜虫取食的 EPG 经典波形

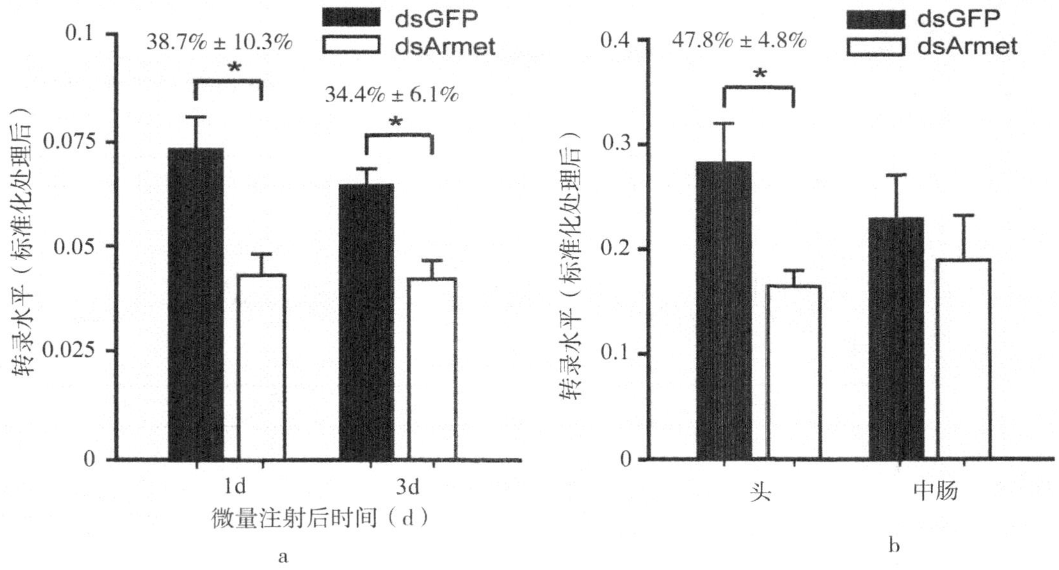

图 17.1　豌豆蚜在不同时间和组织的 *Armet* 基因干扰效率

每组处理至少有 6 个生物学重复，误差线表示标准误（SE）；“ * ”代表处理与对照之间存在显著性差异（$P<0.05$）

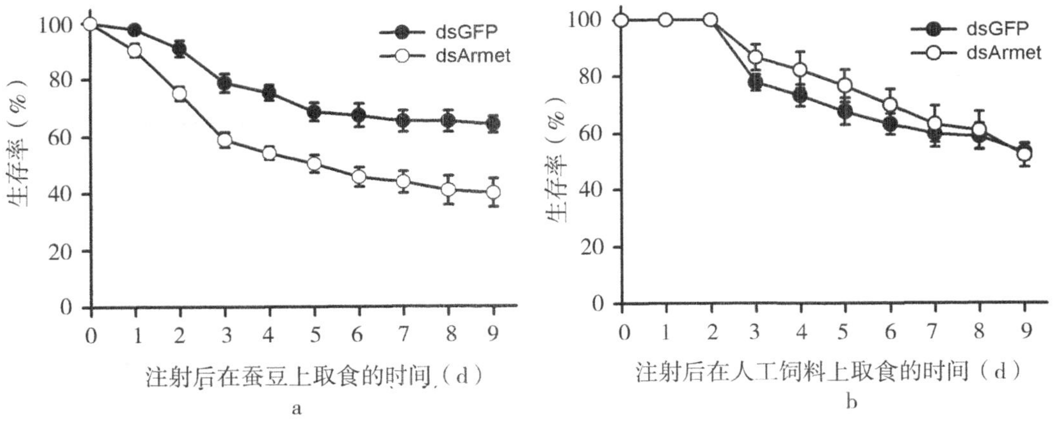

图 17.2　干扰 *Armet* 基因对蚜虫在不同取食环境下生存率的影响

每组处理至少有 6 个生物学重复，误差线表示标准误（SE）；“ * ”代表处理与对照之间存在显著性差异（$P<0.05$）

类似［Tjallingii，1988；1990；Prado（wang et al.，2015），1999］，注射 dsRNA 的豌豆蚜在蚕豆上取食波形也表现出了 7 种典型的基本波形：np 波，包含 A 波、B 波和 C 波的路径波，pd 波，E1 波，E2 波，F 波，G 波。

无论是注射了 dsGFP 还是 dsArmet 的豌豆蚜，大部分取食行为记录的数据显示如下的 EPG 波形（图 17.3a）。

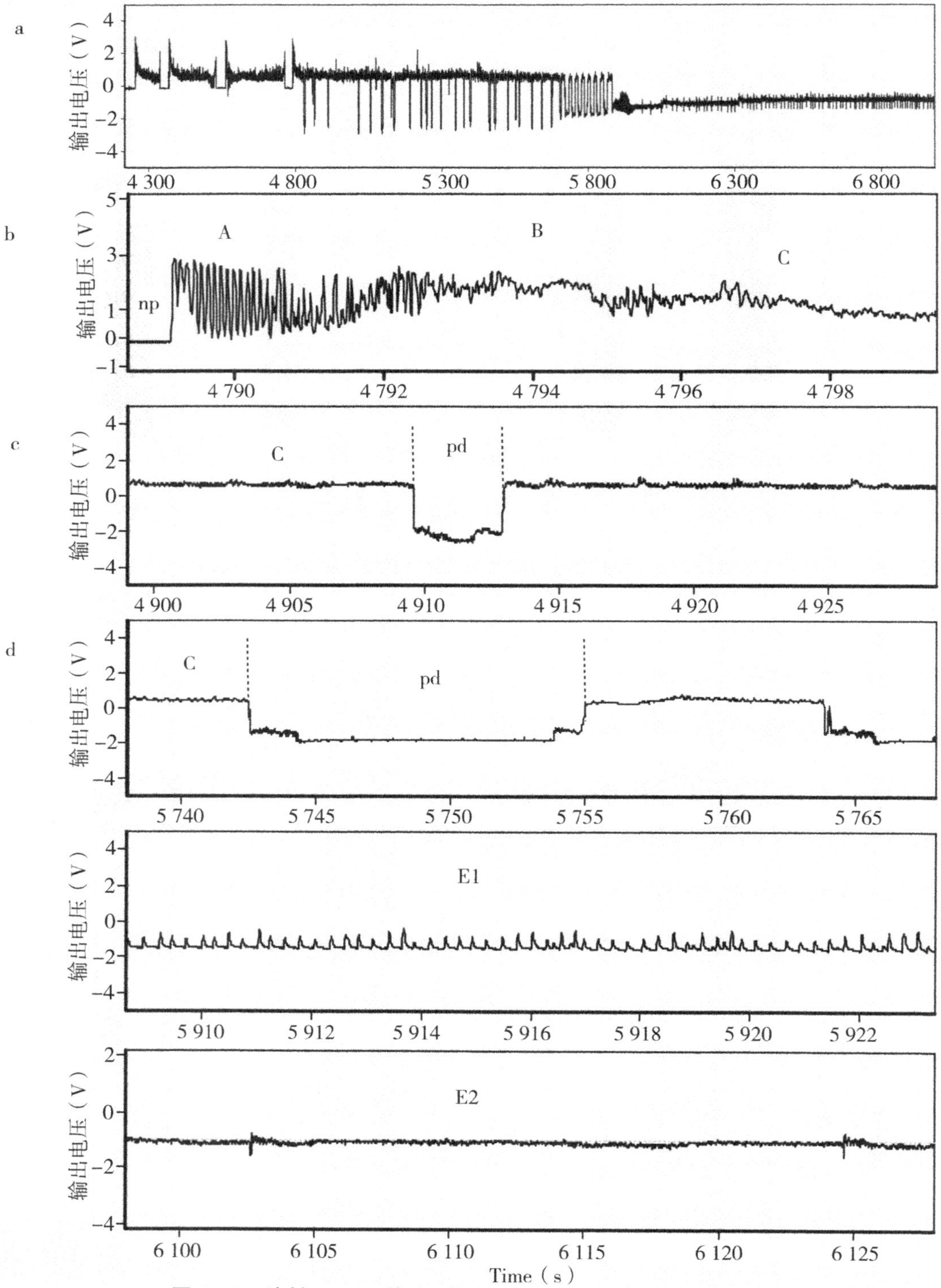

图 17.3　注射 dsRNA 的豌豆蚜下蚕豆植株上取食的 EPG 波形

a 图选自某只注射 dsGFP 豌豆蚜记录的 EPG 波形，开始后 70～116min 的片段。b～f 图是 a 图局部放大后的展示

首先发生的是 np 波（图 17.3b），即非刺探波波形，EPG 取食行为的电路没有接通，豌豆蚜在蚕豆植株上爬行，寻找合适的取食部位。

当豌豆蚜到达蚕豆嫩芽合适的位置后，首先口针与表皮层初步接触，并伴随着少量水溶性唾液的分泌，引发 A 波（图 17.3b），A 波幅度变化较大，波形不规则。伴随着胶状唾液分泌，形成了保护口针的唾液鞘，口针到达蚕豆的表皮和叶肉细胞之间，引发 B 波。之后发生的 C 波与 B 波之间没有明显的界限，是口针尝试在蚕豆的各个组织细胞之间游走形成的波形，波形复杂变化。在 C 波间夹杂了大量的 pd 波，这是口针尖端在细胞内穿刺引起的电势下降（potential drop）现象，其间口针向细胞内分泌水溶性唾液并品尝细胞内容物，每次持续 5～15s（图 17.3c、图 17.3d）。在 EPG 数据一般统计中，将 A 波、B 波、C 波和一些不能明确区分的波形都归入广义的 C 波，也简称 C 波（文中之后提到的 C 波都是指这种广义 C 波）。当豌豆蚜在蚕豆组织中刺探（C 波）结束后，如果蚜虫发现在该位置无法刺探到适合取食的植物组织，会拔出口针重新在蚕豆植株上寻找合适刺探的位置，波形会转变成 np 波，接着开始下一轮的刺探（C 波）。但是有部分 C 波结束后，蚜虫成功发现韧皮部细胞，并开始发生取食波。

表 17.1　干扰 *Armet* 基因后豌豆蚜取食行为的 EPG 参数（Wang et al. 2015）

指数	均值±标准误		*P* 值
	dsGFP（*n*=23）	dsArmet（*n*=29）	
韧皮部取食			
口针第一次抵达韧皮部的时间（min）[1]	116.38±30.49[a]	254.09±34.62[b]	0.005＊＊
水溶性唾液分泌（E1）次数	5.83±1.04	3.93±1.06	0.214
E1 波持续总时间（min）	7.68±1.55	5.23±1.55	0.274
E1 波持续时间占总记录时间的百分比	1.60±0.32	1.02±0.30	0.192
被动取食（E2）次数	2.87±0.48	1.90±0.67	0.265
E2 波持续总时间（min）	188.23±28.03	49.93±15.21	0.000＊＊
有效韧皮部取食次数（E2>10min）	2.26±0.38	0.97±0.32	0.011＊
E2 波持续时间占总记录时间的百分比	39.24±5.84	9.25±3.08	0.000＊＊
E1+E2 波持续时间占总记录时间的百分比	40.85±5.87	10.27±3.24	0.000＊＊
刺探植物细胞			
路径波（C 波）持续总时间（min）	169.68±18.05	250.24±10.96	0.000＊＊
C 波持续时间占总记录时间的百分比	35.38±3.76	52.90±2.22	0.000＊＊
pd 波次数	196.74±23.30	233.17±14.63	0.174
pd 波持续总时间（min）	20.48±2.73	20.15±1.39	0.915
路径波中 pd 波发生频率（no./min）	1.09±0.10	0.88±0.05	0.069

续表

指数	均值±标准误		P 值
	dsGFP（$n=23$）	dsArmet（$n=29$）	
pd 波持续时间占总记录时间的百分比	4.27±0.57	4.02±0.28	0.695
未刺探			
未刺探波（np）发生次数	27.83±4.93	42.62±5.10	0.046 *
np 波持续总时间（min）	64.05±13.90	138.76±12.05	0.000 * *
np 波持续时间占总记录时间的百分比	13.35±2.90	29.62±2.50	0.000 * *

1 如果在记录时间内，实验昆虫一直未到达韧皮部取食，第一次抵达韧皮部的时间按 8h 计算。
* 差异显著 $\alpha=0.05$. * * 差异极显著 $\alpha=0.01$.

取食波（E 波）的电势落差相当于 pd 波水平，反映了口针在韧皮部筛管取食的过程。当豌豆蚜寻找到筛管后，主动分泌水溶性唾液到筛管中，发生 E1 波（图 17.3e），接着由于植物内压强大于外界，筛管中的汁液会冲入蚜虫的口器中，蚜虫进入被动吸收阶段，发生 E2 波（图 17.3f）。一般来说，豌豆蚜在蚕豆植株上取食的 E 波会持续几分钟到几小时不等。

在极少数情况下，蚜虫刺探（C 波）后，蚜虫口针在细胞膜外胞间或细胞壁内的机械穿刺会受到阻碍（F 波）或者口针在木质部主动吸食水分（G 波），一般十几分钟或一两小时后蚜虫会重新开始刺探。

注射 dsGFP 和 dsArmet 的豌豆蚜取食蚕豆植株的 EPG 各波形发生类型和发生的先后顺序大体一致，但注射不同 dsRNA 的豌豆蚜的 EPG 波形所占比例和发生次数却有显著差异。

3. 干扰 *Armet* 基因后豌豆蚜取食蚕豆 EPG 波形发生的影响 干扰 *Armet* 基因豌豆蚜取食蚕豆植株的 EPG 各波形情况如表 17.1 所示。在记录取食的 8h 内，*Armet* 转录水平的干扰会导致豌豆蚜在蚕豆上取食波发生重大变化：处理组豌豆蚜的被动取食（E2）时间只有对照组蚜虫 1/3 不到，由对照的 188.23min 下降到 49.93min，但是与对照组相比被动取食的次数却没有显著下降，这说明 *Armet* 基因干扰后豌豆蚜每次被动取食的平均时间显著下降。而且有效取食次数，即被动取食大于 10min 的次数明显下降，由对照豌豆蚜 8h 内每只 2.26 次下降到干扰 *Armet* 的豌豆蚜每只不到 1 次。口针第一次到达韧皮部的时间也有明显的延迟，由对照豌豆蚜每只 116.38min 增加到干扰 *Armet* 的豌豆蚜每只 254.09min，时间基本多出一倍。

与之相对的口针在植物细胞间游走的时间却略有增加，由对照的 169.68min 增长到 250.24min。口针不刺探（np）的时间和次数也显著增加，由对照组豌豆蚜 64.05min 到处理组豌豆蚜的 138.76min，几乎增加了一倍，由对照组豌豆蚜 8h 内每只 27.83 次增加到处理组豌豆蚜每只 42.62 次，大约增加了 50%。

干扰 *Armet* 基因的豌豆蚜在蚕豆上取食行为的 4 个主要 np 波、C 波、pd 波、E 波，分布情况发生了变化，分别由对照占总时间的 13%、35%、4%和 40%（图 17.4），转变为干扰后 30%、53%、4%和 10%。

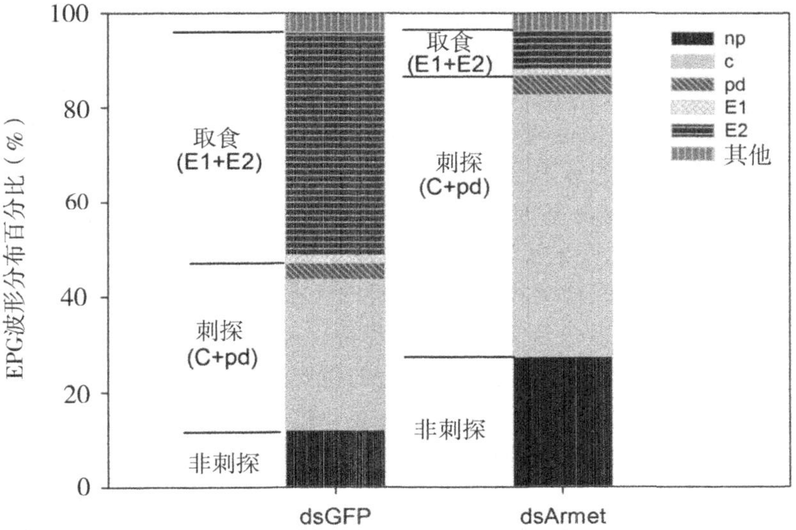

图 17.4 干扰 *Armet* 基因对豌豆蚜在蚕豆植株上取食 EPG 波形分布情况

总记录时间为 8h，G 波和 F 波并入一处标记为 other。对照组和处理组的有效数据分别为 23 个和 29 个

如图 17.5 所示，干扰 *Armet* 的豌豆蚜在某次取食行为里 E1 重复出现或者说 E2/E1 交替出现了 3 次。更多的 EPG 数据显示对照组呈现出这种 E1 重复出现或者说 E2/E1 的形式，在 11 只干扰 *Armet* 基因且记录到取食行为的豌豆蚜中有 5 只出现了该特性，E2/E1 的交替出现 2~6 次不等，而对照组中，记录到取食行为的 19 只豌豆蚜中，只有 5 只出现了该特性，且 E2/E1 的交替只出现一次。

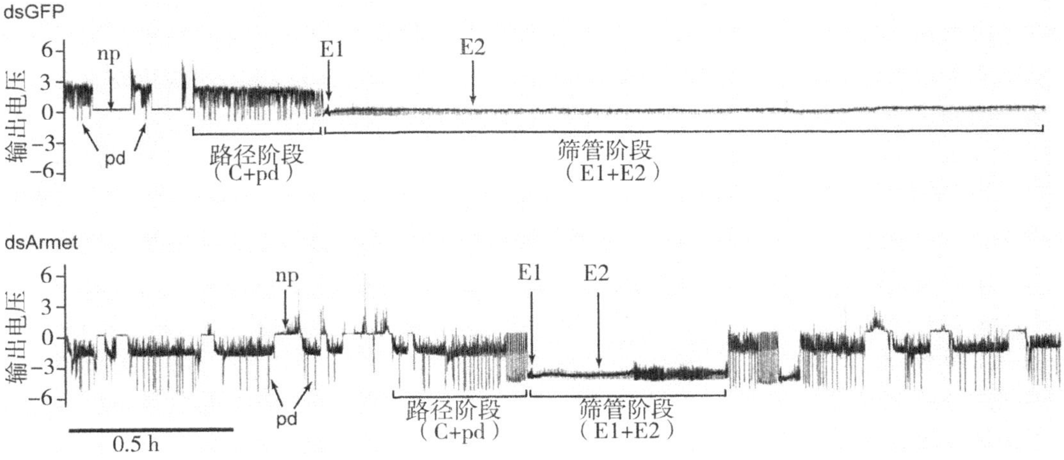

图 17.5 干扰 *Armet* 基因对豌豆蚜在蚕豆植株上取食行为的影响 EPG 波形

（Wang et al.，2005）

（四）讨论

干扰*Armet*基因后，豌豆蚜*Armet*基因整体转录水平下降了30%～40%，而唾液腺所在的头部转录水平下降了40%～50%，它影响豌豆蚜在寄主植物的存活率，却不会影响其在人工饲料的存活率，并且Western结果显示，*Armet*蛋白表达量的确在干扰后下降（Wang，2015），所以*Armet*蛋白是一种能够促进豌豆蚜取食寄主植物的重要蛋白。

在取食过程中，蚜虫的唾液会始终伴随着刺探行为而持续分泌，唾液是取食行为的重要影响因子。干扰*Armet*基因后，*Armet*表达量下降，导致豌豆蚜唾液成分可能发生变化，进而极大地干扰了蚜虫取食行为：口针第一次到达韧皮部的时间大大增加，取食波持续的时间和有效取食次数降低，同时口针在细胞内游走的路径波持续时间（C波）和蚜虫在植株上寻找适合刺探位置或者等待（np波）的时间增加。表明与对照相比，干扰*Armet*基因的豌豆蚜更难在植物适合的位置开始刺探，并且取食韧皮部汁液的时间大大缩短。干扰*Armet*基因后，豌豆蚜在蚕豆植株上的取食难度显著增加。

通常情况下，首先当唾液随着口针进入筛管中时，随之而来会有一段被动取食的阶段，在韧皮部进行的吸食行为会被记录为E1波形。而在E2波形时，反而因为口针插入韧皮内部的筛管，植物内压强大于外界，筛管内的植物汁液会冲入蚜虫口内，表现为蚜虫的被动取食，而此时唾液由于压力的原因不会到达植物中。因此，E1波形时会伴随唾液的进入，而E2时则没有。干扰*Armet*后，豌豆蚜短暂的取食行为中，被动取食极其容易被打断，且有效的被动取食数量下降。干扰*Armet*的豌豆蚜被动取食一段时间后，就需要不断地分泌唾液（E1）维持它在韧皮部的被动取食，当无法维持时，取食行为被迫停止，豌豆蚜重新开始刺探需要更适合取食的部位，这可能也是干扰*Armet*基因后豌豆蚜每次取食行为维持时间较短的原因之一。

目前最先鉴定的植物—昆虫效用因子是*C002*（Mutti，2008），与干扰*Armet*基因的结果类似，干扰*C002*基因也会导致豌豆蚜唾液成分的改变，唾液中起重要作用的蛋白含量降低从而导致豌豆蚜在蚕豆上取食更加困难，包括被动取食时间缩短，第一次到达筛管部取食的时间延长。但是两者又有很大差异，干扰*Armet*基因的豌豆蚜会花费更多的时间用于在植物细胞内刺探，而干扰*C002*的豌豆蚜则是花费更多时间等待。即使干扰*Armet*基因，豌豆蚜仍有短时间的取食，而干扰*C002*的豌豆蚜则基本不取食。干扰*C002*的豌豆蚜路径波中刺探的频率（rate of pd）减少，干扰*Armet*基因的豌豆蚜则不变。蚜虫的主唾液腺是由8个受到神经细胞调控不同形态和功能分泌细胞组成。结果显示*C002*定位于三型和四型细胞，*Armet*则位于八型细胞（Wang et al.，2015）。干扰*Armet*和*C002*的豌豆蚜取食行为改变既相似又有差异，可能是因为*Armet*和*C002*都是由主唾液分泌的蛋白，但是分泌它们的细胞却位于主唾液的不同细胞，蚜虫唾液腺不同的分泌细胞可能受神经信号的调控，分泌功能各异的蛋白质来响应寄主，抑制或干扰植物的先天免疫反应。

参考文献

[1] BELLAFIORE S, SHEN Z, ROSSO M N, et al. Direct identification of the *Meloidogyne incognitasecretome* reveals proteins with host cell reprogramming potential. *PLoS Pathogens*, 2008, 4: e1000192.

[2] CAROLAN JC, CARAGEA D, REARDON K T, et al. Predicted effector molecules in the salivary secretome of the pea aphid (*Acyrthosiphon pisum*) -A dual transcriptomic-proteomic approach. *Journal of Proteome Research*, 2011, 10: 1505-1518.

[3] DAVIS E L, HUSSEY R S, BAUM T J. Getting to the roots of parasitism by nematodes. *Trends in Parasitology*, 2004, 20: 134-141.

[4] DOUGLAS A E, PRICE D R, MINTO LB, et al. Sweet problems: insect traits defining the limits to dietary sugar utilisation by the pea aphid, *Acyrthosiphon pisum. The Journal of Experimental Biology*, 2006, 209: 1 395-1 403.

[5] DIXON, A. F. G. Aphid ecology: an optimization approach. 2nd edition, 1998.

[6] FIRE A, XU S, MONTGOMERY MK, et al. Potent and specific genetic interference by double-stranded RNA in Caenorhabdits elegans. *Nature*, 1998, 391: 806-811.

[7] FEBVAY G, DELOBEL B, RAHBE Y. Influence of the amino acid balance on the improvement of an artificial diet for a biotype of *Acythosiphon pisum* (Homoptera: aphididae) . *Canadian Journal of Zoology revue Canadienhe De Zoologie*, 1988, 66: 2 449-2 453.

[8] KARLEY A J, PARKER W E, PITCHFORD J W, et al. The mid-season crash in aphid populations: why and how does it occur? . *Ecological Entomology*, 2004, 29: 383-388.

[9] MUTTI N S, LOUIS J, PAPPAN L K, et al. A protein from the salivary glands of the pea aphid, *Acyrthosiphon pisum*, is essential in feeding on a host plant. *Proceedings of the National Academy of Sciences of the United States of America*, 2008, 105: 9 965-9 969.

[10] PONSEN M B. The site of potato leafroll virus multiplication in its vector, *Myzus persicae*: An anatomical study. Wageningen, the Netherlands: H. Veenman, 1972.

[11] TJALLINGII W F. Electrical recording of stylet penetration activities. Aphids, Their Biology, Natural Enemies and Control, Vol. 2B (ed. by AK Minks & P Harrewijn), Elsevier, Amsterdam, the Netherlands, 1998: 95-108.

[12] TJALLINGII W F. Continuous recording of stylet penetration activities by aphids. Aphid-Plant Genotype Interaction (ed. By RK Campbell & D Eikenbary) Elsevier, Amsterdam, the Netherlands, 1990: 89-99.

[13] WEI WANG, HUAIEN DAI, YI ZHANG. et al. Armet is an effector protein mediating aphid-plant interactions. *The FASEB Journal*, 2005, 29: 2 032-2 045.

[14] WILL T, TJALLINGII W F, THÖNNESSEN A, et al. Molecular sabotage of plant defense by aphid saliva. *Proceedings of the National Academy of Sciences of the United States of America*, 2007, 104: 10 536-10 541.

[15] ZAMORE P D, TUSCHL T, SHARP P A, et al. RNAi: double-stranded RNA directs the ATP-dependent cleavage of mRNA at 21 to 23 nucleotide intervals. *Cell*, 2000, 101: 25-33.

（罗　岚　王　炜）

第十八章　小麦抗蚜性研究

（一）研究背景

麦蚜是麦类作物的重要害虫，世界范围内为害小麦的蚜虫有 30 多种（叶炳元和郝纪华，1992），在我国主要发生的是麦二叉蚜 *Schizaphis graminum* Rondani，禾谷缢管蚜 *Rhopalosiphum padi*（L.）和麦长管蚜 *Sitobion avenae*（Fab.）。麦长管蚜喜光照，多分布在小麦植株上部叶片正面；麦二叉蚜畏光、耐干旱，多分布在植株的中下部叶片；禾谷缢管蚜畏光、喜湿，多在植株的茎秆和叶鞘部取食。麦蚜可吸食小麦叶片、茎秆和嫩穗的汁液，造成小麦灌浆不足、籽粒干瘪、千粒重降低、品质下降（王美芳等，2010），受害麦田轻者减产 10%～20%，重者减产 30%～40%。除直接为害外，麦蚜还是传播小麦黄矮病毒的重要媒介，从而造成更为严重的间接为害。我国目前麦蚜防治依然主要依靠化学农药，但化学防治容易造成环境污染，杀伤麦田天敌（刘爱芝和李世功，1999）。研究选育和种植抗麦蚜品种是控制麦蚜最经济有效的措施，符合有害生物可持续治理和农业可持续发展的要求，有利于环境保护（Dogimont et al.，2010）。

植物凝集素是一种抗蚜的重要物质，将植物凝集素加入人工饲料或通过转基因技术在植物体内表达，对包括蚜虫、灰飞虱和小绿叶蝉等刺吸式害虫均表现出了较高的抗性。在抗蚜转基因研究中应用较多的有雪花莲凝集素（GNA）（Hilder et al.，1995）、伴刀豆凝集素（ConA）（Gatehouse et al.，1999）、蒜叶凝集素（ASAL）（Dutta et al.，2005a；Dutta et al.，2005b）、半夏凝集素（PTA）（Yu et al.，2008）等。2002 年和 2005 年转雪花莲凝集素小麦已经进行了环境释放实验，不仅证明对蚜虫有控制效果而且还提高了小麦品质（徐琼芳等，2003，2005）。关于植物凝集素对蚜虫生理生化的影响已有较多的研究和报道，但转凝集素基因小麦对麦蚜取食行为的影响鲜有报道。本文利用 EPG 技术分析在同一寄主植物上麦长管蚜、麦二叉蚜和禾谷缢管蚜的取食行为，以及转雪花莲凝集素基因抗蚜小麦对三种麦蚜取食行为的影响。该研究有助于揭示三种麦蚜生态位分离的机制，转雪花莲凝集素基因抗蚜小麦对三种麦蚜抗性水平的差异以及可能抗性机制。

（三）材料与方法

1. 麦蚜来源于饲养　麦长管蚜、麦二叉蚜和禾谷缢管蚜均由中国农业科学院植物保护研究所提供，三个麦蚜种群均在人工气候箱即温度：(23±1)℃，相对湿度：(70±5)%，光照 L/D=16h/8h 中，用冬小麦（*Triticum aestivum* L.，品种名：郑麦 9023）苗饲养最少5代后选择日龄和体形大小一致的无翅成蚜用于实验。

2. 植物材料 转雪花莲凝集素基因小麦和转基因受体小麦品种郑麦9023均由中国农业科学院作物研究所提供，该转基因小麦品种已证实能过高剂量表达雪花莲凝集素基因并对麦长管蚜具有较高的抗性水平（薛琼芳等，2005）。转基因和非转基因小麦均用营养土种植在塑料培养钵（长、宽、高均为10cm）中，然后放置在人工气候箱即温度：(23±1)℃，相对湿度：(70±5)%，光照L/D=16h/8h中，出苗1周后用于实验。

3. EPG记录 八通道直流电型（DC-EPG Giga-8）昆虫刺探电位仪（荷兰瓦赫宁根大学制造）和随机附带的ACQ、ANA软件分别用于三种麦蚜在转基因抗蚜小麦和非转基因小麦上取食行为的记录和数据的分析统计。受试无翅成蚜测试前饥饿2h后，用导电银胶将一根2~3cm长的细金线（直径为10μm）的一端与中胸背板相连，另一端连接至EPG的昆虫电极。将植物电极插入种植有小麦苗的培养钵营养土中，接通电流后将悬挂的蚜虫轻轻地放置在小麦叶片的背面开始记录。实验在温度为（25±1)℃的室温下进行，整个记录系统置于法拉第笼中以防止外源噪声影响记录。每只蚜虫只用于一次测试，每次记录5h，每种麦蚜分别在两种小麦上最少要成功记录20头。

4. 数据分析 根据实际记录的EPG波形，依据各波形特征识别、判断各特征波的频次和时间，并做记录。根据Tjallingii（1990）的结论分析5类EPG波形，np波：非刺探波；C波：口针位于表皮和维管束之间；G波：木质部主动吸食；E1波：分泌水溶性唾液；E2波：韧皮部被动取食。SPSS 10.0（SPSS，Chicago，IL）软件用于数据间差异性比较，各处理间的显著差异性均设为$P<0.05$水平。

（三）结果

1. 三种麦蚜取食策略的差异 表18.1显示，5h的监测过程中，禾谷缢管蚜的非刺探时间最长，麦长管蚜次之，麦二叉蚜最短，并与禾谷缢管蚜之间差异显著（$P<0.05$)。从接触叶面到开始刺探所需时间禾谷缢管蚜也明显长于麦长管蚜和麦二叉蚜($P>0.05$)。第一次刺探的持续时间麦长管蚜明显比麦二叉蚜和禾谷缢管蚜的持续时间长（$P>0.05$)。三种麦蚜在取食时F波的持续时间为麦长管蚜>禾谷缢管蚜>麦二叉蚜，三者间差异显著（$P<0.05$)，说明麦长管蚜在刺探过程中口针遇到的机械阻力最大。禾谷缢管蚜在取食过程中G波出现的时间最早，第一次G波的持续时间、G波总的持续时间和每次刺探G波的平均持续时间都明显长于麦长管蚜和麦二叉蚜（$P<0.005$)。

麦长管蚜第一次E1波的时间、E1波总的持续时间和每次刺探E1波的平均持续时间都明显比麦二叉蚜和禾谷缢管蚜要长（$P<0.05$)，说明麦长管蚜在韧皮部取食之前需要向韧皮部筛管内分泌更多的水溶性唾液。韧皮部被动取食波E2出现的时间三种蚜虫间差异不显著，但E2波总的持续时间和每次刺探E2波的平均持续时间麦二叉蚜明显长于麦长管蚜和禾谷缢管蚜（$P<0.05$)。麦二叉蚜在韧皮部刺探的时间E波（E1+E2）达到了（163.11±13.69）min，占到总时间的（54.33±4.56)%，长于麦长管蚜和禾谷缢管蚜，与禾谷缢管蚜的差异达到了显著水平（$P<0.005$)，每次刺探E波的平均持续时间麦二叉蚜也明显长于禾谷缢管蚜（$P<0.05$）而与麦长管蚜差异不显著。

2. 转基因抗蚜小麦对三种麦蚜取食行为的影响 共统计20个主要的EPG参数，用于分析转基因抗蚜小麦对三种麦蚜取食行为的影响（表 18.2）。与对照相比，转雪花莲凝集素基因抗蚜小麦对禾谷缢管蚜取食行为影响较小，对麦长管蚜和麦二叉蚜的取食行为则具有显著影响。麦长管蚜和麦二叉蚜在转基因抗蚜小麦上取食时，非刺探时间（np 波）明显增加（$P<0.05$），第一个 E2 波的持续时间和 E2 波总的持续时间则明显缩短（$P<0.05$），韧皮部持续取食时间（E1+E2>10min）占总测试时间的比例明显低于对照（$P<0.05$）。麦长管蚜总的刺探次数和 C 波的持续时间也明显高于对照组（$P<0.05$）。

表 18.1　三种麦蚜的 EPG 参数（总记录时间为 5h）

	EPG 参数	麦长管蚜	麦二叉蚜	禾谷缢管蚜
1	总的刺探次数（*n*）	4.93±0.77a	3.26±0.38a	3.83±0.53a
2	到达韧皮部前的刺探次数（*n*）	2.29±0.45a	1.74±0.30a	1.75±0.57a
3	第一次刺探出现时间（min）	5.36±1.26b	5.92±1.18b	12.15±3.08a
4	第一次刺探时间（min）	34.18±17.47b	94.67±23.52a	71.29±31.23a
5	C 波时间（min）	73.32±10.69a	92.21±10.16a	70.32±13.02a
6	F 波时间（min）	27.57±7.42a	3.28±1.19c	11.69±3.16b
7	G 波时间（min）	39.98±10.59b	26.53±6.58b	94.37±22.42a
8	E1 波时间（min）	33.23±6.17a	3.41±1.05b	14.16±5.32b
9	E2 波时间（min）	100.63±18.69b	159.71±13.53a	82.36±21.07b
10	E 波（E1+E2）时间（min）	133.86±16.55ab	163.11±13.69a	96.53±22.03b
11	单次刺探 C 波平均持续时间（min）	25.57±4.82a	33.98±3.48a	27.20±8.44a
12	单次刺探 G 波平均持续时间（min）	10.23±3.79b	9.23±2.40b	45.62±18.82a
13	单次刺探 E1 波平均持续时间（min）	7.62±1.72a	1.51±0.54b	3.87±1.22b
14	单次刺探 E2 波平均持续时间（min）	28.97±8.44b	84.32±16.85a	28.97±8.37b
15	单次刺探 E 波(E1+E2）平均持续时间(min)	36.59±8.16b	85.83±17.03a	32.85±8.41b
16	np 波时间（min）	23.04±4.37ab	14.89±3.11b	27.35±3.65a
17	pd 波时间（min）	4.16±0.69a	5.50±0.60a	2.84±0.35b
	pd 波的数目	44.73±4.52ab	51.97±7.82a	32.99±4.22b
18	每个 pd 平均持续时间（s）	5.58±0.09b	6.35±0.25a	4.51±0.14c
19	第一次 G 波出现的时间（min）	141.19±33.97b	143.08±29.53b	98.11±27.26a
20	第一次 G 波持续时间（min）	28.47±9.51b	26.53±6.58b	66.81±19.48a
21	第一次 E1 波出现的时间（min）	102.82±17.63	109.28±12.61	138.72±29.93
22	第一次 E1 波持续时间（min）	15.29±2.76a	2.25±1.06b	1.47±0.49b
23	第一次 E2 波出现的时间（min）	150.77±21.67	113.14±12.22	145.66±29.63
24	第一次 E2 波持续时间（min）	89.00±19.94a	90.59±18.59a	35.69±16.78b
25	持续韧皮部取食（E>10min）时间（min）	132.02±16.82ab	161.63±13.69a	91.69±21.46b
26	韧皮部取食占总时间的%	44.54±5.57%ab	54.33±4.56%a	33.28±6.94%b

注：同一行内不同字母表示差异性显著（$P<0.05$）；下同。

表 18.2　转 GNA 抗蚜小麦对 3 种麦蚜取食行为的影响（记录时间为 5h）

EPG 参数	麦长管蚜 *S. avenae*		麦二叉蚜 *S. Graminum*		禾谷缢管蚜 *R. padi*	
	GNA	Non-GNA	GNA	Non-GNA	GNA	Non-GNA
总刺探次数	6.93±1.00a	4.93±0.77a	8.25±1.12a	3.26±0.38b	6.60±1.17a	3.83±0.95a
第一次刺探之前时间（min）	5.17±1.53a	5.36±1.26a	4.23±1.13a	5.92±1.17a	9.70±2.03a	12.10±3.08a
第一次刺探持续时间（min）	8.27±2.69b	34.18±8.04a	29.26±12.21b	94.67±23.52a	56.80±0.79a	71.30±0.68a
非取食时间（min）	54.37±6.25a	23.04±4.37b	51.02±10.89a	14.89±3.11b	46.10±0.83a	27.40±0.86a
C 波持续时间（min）	117.63±12.09a	101.9±10.68a	154.12±13.35a	95.49±10.16b	126.00±0.93a	81.30±0.94a
G 波持续时间（min）	40.72±8.71a	39.98±10.59a	33.02±13.41a	29.06±6.96a	61.50±6.93a	94.40±8.42a
第一个 E1 波持续时间（min）	10.88±1.96a	15.29±2.76a	1.58±0.53a	2.25±1.06a	2.14±0.67a	1.47±0.49a
E1 波总时间（min）	27.44±5.44a	25.36±3.68a	2.37±0.68a	3.41±1.05a	15.50±5.47a	14.20±5.32a
第一个 E2 波持续时间（min）	41.35±8.95b	89.00±12.49a	45.47±10.97b	90.59±17.31a	22.70±4.20a	35.70±8.78a
E2 波总时间（min）	55.38±10.12b	100.63±14.77a	65.83±14.24b	159.71±13.53a	51.60±9.71a	82.40±21.07a
持续韧皮部取食总时间 [E（pd）>10 min]	76.20±19.75b	132.02±16.82a	65.99±14.70b	161.63±13.69a	58.45±9.92a	91.69±21.46a
	26.34±7.18b	44.54±4.34a	22.70±4.84b	54.33±4.56a	23.36±3.11a	32.17±7.34a

(四) 讨论

昆虫的取食行为是昆虫在寄主植物上所进行的摄取食物以及与此相关的一系列活动，一般同种昆虫个体的取食行为既表现出种群的相似性又表现出种所特有的固定模式（Hsiao，1985）。虽然三种蚜虫都以小麦作为寄主植物，但在生活习性上存在一定的差异，自然状态下取食寄主部位不尽相同，表明三者之间存在一定的生态位分化。应用 EPG 技术研究三种麦蚜在小麦同一部位的取食行为，能够模拟在种间竞争强度较大的情况下，三种蚜虫取食行为的差异。结果表明，三种蚜虫中，麦长管蚜 E1 波的时间最长，而 E2 波的时间相对较短，说明麦长管蚜对食物的利用率可能高于二叉蚜和禾谷缢管蚜；麦二叉蚜 E1 波的时间很短，但韧皮部被动取食波（E2 波）持续时间最长；禾谷缢管蚜木质部取食时间则显著高于麦长管蚜和麦二叉蚜。以上的结果反映出，三种蚜虫各自拥有不同取食策略以满足自身对营养的需求。三种麦蚜通过不同的取食策略减小了种间竞争的压力，进而促进了生态位的分化，这是与寄主植物长期协同进化的结果，同时也是种间竞争的结果。

EPG 技术的应用之一就是通过比较昆虫在不同抗性品种上的刺探和取食行为，从而确定抗性因子存在的部位，继而运用其他化学分析手段，则可能确定抗性因子的化学成分。将蚜虫在各类抗蚜品种的 EPG 波形参数特征概括起来为：E2 波和 np 波是抗蚜的重要标志。如果出现较多短暂的不连续口针刺探、长时间的路径刺探以及长时间的非取食活动，可推断抗性因子存在于植物表面、表皮层和叶肉层；如果出现很长时间的唾液分泌，较短时间甚至没有韧皮部取食可推断抗性因子存在于韧皮部。将 EPG 技术作为快速筛选和评价小麦品种对蚜虫的抗性水平，以及抗性因子定位的一种生物测定手段，对于加快我国抗蚜育种工作具有重要的实践和理论意义。

转雪花莲凝集素基因小麦对麦长管蚜和麦二叉蚜均表现出了明显拒食效果，表明该转基因小麦对两种麦蚜具有较高的抗性水平。麦长管蚜和麦二叉蚜在转雪花莲凝集素基因小麦上取食时，刺探波首次出现的时间明显迟于转基因受体植物，表明转入雪花莲凝集素基因后可能改变了小麦叶片的表面性状，阻碍了它们的取食。同时，两种麦蚜在转基因小麦上取食时 E2 波的持续时间显著短于转基因受体小麦，表明这两个转基因株系的韧皮部可能存在抗性因子使麦长管蚜持续取食困难。邓青（2012）利用 EPG 技术分析了转半夏凝集素基因小麦和转天南星凝集素基因小麦对麦长管蚜取食行为的影响，也表现出了相同的结果。

禾谷缢管蚜在转雪花莲凝集素基因小麦上取食时，除 E2 波持续时间较转基因受体小麦略有减少外，对其取食行为无明显的影响。表明转雪花莲凝集素基因小麦对禾谷缢管蚜的抗性水平较低。同一个抗蚜小麦对三种麦蚜表现出不同的抗性水平，可能与三种蚜虫具有不同的取食策略有关。与麦长管蚜和麦二叉蚜相比，禾谷缢管蚜韧皮部取食时间短，摄取的植物凝集素的量较少，因此对其行为以及生理的影响也比较小。

参考文献

[1] DOGIMONT C, BENDAHMANE A, CHOVELON V, et al. Host plant resistance to aphids in cultivated crops: genetic and molecular bases, and interactions with aphid populations. *Comptes Rendus Biologies*, 2010, 333: 566-573.

[2] DUTTA I, MAJUMDER P, SAHA P, et al. Constitutive and phloem specific expression of Allium sativum leaf agglutinin (ASAL) to engineer aphid (*Lipaphis erysimi*) resistance in transgenic Indian mustard (*Brassica juncea*) . *Plant Science*, 2005a, 169: 996-1007.

[3] DUTTA I, SAHA P, MAJUMDER P, et al. The efficacy of a novel insecticidal protein, *Allium sativum* leaf lectin (ASAL), against homopteran insects monitored in transgenic tobacco. *Plant Biotechnology Joural*, 2005b, 3: 601-611.

[4] GATEHOUSE A M R, DAVISON G M, STEWART J N. Concanavalin A inhibits development of tomato moth (*Lacanobia oleracea*) and peach-potato aphid (*Myzus persicae*) when expressed in transgenic potato plants. *Molecular Breeding*, 1999, 5: 153-165.

[5] HILDER V A, POWELL K S, GATEHOUSE A M R. et al. Expression of snowdrop lectin in transgenic tobacco plants results in added protection against aphids. *Transgenic Research*, 1995, 4: 18-25.

[6] HSIAO TH. Feeding behavior, *in* Comprehensive Insect Physiology, Biochemistry and Pharmacology (eds. Kerkut, FA, Gilbert, LI), Vol. 9, New York: Pergamon Press, 1985, 471-512.

[7] TJALLINGII, W F. Continuous recording of stylet penetration activities by aphids, pp. 88D89. *In* R. K. Campbell and R. D. Eikenbary, Aphid - plant genotype interactions. Elsevier, Amsterdam, the Netherlands.

[8] YU Y, WEI Z. Increased oriental armyworm and aphid resistance in transgenic wheat stably expressing *Bacillus thuringiensis* (Bt) endotoxin and Pinellia ternate agglutinin (PTA) . *Plant Cell, Tissue and Organ Culture*, 2008, 94: 33-44.

[9] 邓青．转基因抗虫小麦抗麦长管蚜效果的评价技术研究．中国农业科学院研究生院博士论文，2012.

[10] 刘爱芝，李世功．四种农药对麦蚜的防治效果及其对天敌的影响．农药，1999，38（5）：29-30.

[11] 雷宏，徐汝梅．温室白粉虱取食行为的刺探电位（EPG）研究．昆虫学报，1998，41（2）：113-123.

[12] 王美芳，杨会民，刘进前，等．黄淮冬麦区小麦品种抗蚜性鉴定及蚜虫对小麦产量和品质的影响．河南农业科学，2010，7（4）：16-20.

[13] 徐琼芳，陈孝，田芳，等．转基因小麦京 411 的选育及其利用．全国作物细胞工程与分子技术育种学术研讨会．江苏苏州：陆维忠，郑企成，2003：343-346.

[14] 徐琼芳，田芳，陈孝，等．转 GNA 基因小麦新株系的分子检测和抗蚜虫性鉴定，麦类作物学报，2005，25（3）：7-10.

[15] 叶炳元，郝纪华．世界小麦蚜虫种类与鉴定．植物检疫，1992，6（1）：63-73.

（苗　进）

第十九章　褐飞虱在转 Bt 基因水稻上的取食行为

（一）研究背景

鳞翅目害虫如二化螟 *Chilo suppressalis*（Walker）、三化螟 *Tryporyza incertulas*（Walker）等是水稻生产中重要的限制因素之一。我国每年螟虫全国年发生面积达1 500万 hm^2，直接经济损失在100多亿元人民币（中国科学技术协会，2002）。由于水稻及其近缘种没有发现螟虫等害虫的抗性种质资源，不可能通过常规育种及分子标记辅助选择等技术方式培育抗虫水稻新品种，应用转基因培育抗虫水稻是现阶段及未来解决水稻虫害问题的主要策略（张启发，2009）。*Bacillus thuringiensis* Berliner（Bt）蛋白基因是目前世界上应用最广泛的抗虫基因，已经被转入多种作物中并创造了可观的经济效益（Wu et al.，2008；Clive，2009；Jaffe，2009；Kruger et al.，2009）。随着在1993年第一例转 Bt 基因水稻研制成功以来（Fujimoto et al.，1993），越来越多的转 Bt 基因水稻被成功转化，并表现出对二化螟、三化螟等鳞翅目害虫很高的抗性，高者可达100%的杀虫活性（Cheng et al.，1998；Tu et al.，2000；Ye et al.，2000；High et al.，2004；Chen et al.，2005；Tang et al.，2006；Wang et al.，2010）。在中国开展的大田试验还表明，和传统的水稻相比，种植 Bt 水稻每公顷可节省16.77kg 农药，相当于节省80%的农药使用量，同时还能提高6%~9%的产量（Huang et al.，2005）。

然而，转基因生物的安全性问题一直是公众争论的焦点。转 Bt 基因水稻最终能否商业化，很大程度上还取决于它们的环境安全性和食品安全性，其中非靶效应的研究是转基因作物安全性评价的一个必要组成部分（Andow and Hilbeck 2004；Bai et al.，2006；Chen et al.，2006；Li et al.，2007；Obonyo et al.，2010；Sun et al.，2015）。褐飞虱 *Nilaparvata lugens*（Stål）（Hemiptera：Delphacidae）是广泛分布于亚洲地区的一种水稻及野生稻上的专食性害虫（Gorman et al.，2008），它可刺伤茎叶组织，吸食水稻汁液，造成稻茎损伤，虫量大时极易引起稻茎倒伏，俗称“冒穿”。褐飞虱也是水稻病毒病的传播媒介（Chen et al.，1978；Velusamy and Heinrichs，1986；Gorman et al.，2008；邓飞等，2011）。为害严重时，水稻大面积枯死，呈现火烧状，颗粒无收。褐飞虱是 Bt 水稻的非靶标昆虫，Bt 水稻的种植可能会影响褐飞虱在稻田生态系统中竞争压力，因此，褐飞虱被认为是 Bt 水稻生态风险评价非靶标昆虫的一种代表性昆虫（Chen et al.，2011）。很多研究对不同转基因稻田中褐飞虱种群进行了调查，得到的结果有的比较一致，而有的则相互矛盾。如一个两年的田间观察结果表明，在水稻生长的前期和中期，转Bt基因汕优63水稻和非转基因对照水稻上的褐飞虱种群数量无差异，但在后期，Bt水稻上的褐飞虱高于对照田（Wang et al.，2010）；而其他多年的田间试验结果则发现，Bt水稻田和非Bt水稻田褐飞虱种群数量无差异（Chen et al.，2006b；Chen et al.，2007）。为此，建立在实验室可控条件下，更为科学的非靶标害虫影响效应评价技术非常有必要。

取食行为及效率是考察昆虫对宿主植物适应性的重要指标之一。刺探电位技术（electrical penetration graph，EPG）可以准确地记录刺吸式口器昆虫在寄主植物组织中的刺探行为和位置，并可以反映出影响昆虫取食的因素（闫凤鸣，2000；Seo et al.，2010；Ghaffar et al.，2011）。研究刺吸式昆虫在转Bt基因水稻上的取食行为，可以在一定程度上解释非靶标昆虫在转基因稻田种群的变化。为此，我们在建立褐飞虱EPG技术的基础上，分析了褐飞虱在转Bt基因水稻与其亲本上的取食行为（Sun et al，2017），以期明确转Bt基因水稻是否会对非靶标昆虫飞虱产生影响及转Bt水稻安全性评价提供的参考及技术支持。

（二）材料与方法

1. 褐飞虱与水稻

（1）供试水稻材料：本实验采用的两个转Bt基因水稻材料T1C-19和T2A-1分别表达*cry1C*和*cry2A*基因，转入的*cry1C*和*cry2A*基因对水稻鳞翅目害虫表现出明显的抗性（Chen et al.，2005；Tang，Chen et al.，2006）；对照为转Bt基因水稻亲本明恢63及感虫品种TN1水稻。试验用的转Bt基因水稻材料T1C-19、T2A-1、明恢63和TN1水稻均由华中农业大学农作物遗传和改良育种国家级重点实验室林拥军老师提供。

实验用的水稻栽培方法如下：首先在温室中把所用的水稻品种（T1C-19、T2A-1、明恢63和TN1）的种子撒到土中，待15d后将长成的水稻秧苗移入单独的盆中（直径10cm，高度12cm），整个水稻生长保持温度（26±4）℃以及湿度（60±10）%的环境中，待水稻生长的（70±15）d后用于EPG实验。

（2）供试昆虫：实验用的褐飞虱于2008年采自华中农业大学试验田中，后一直用TN1水稻饲养于温室的虫笼之中传代饲养。实验使用日龄1d之内的长翅褐飞虱雌虫。

2. EPG记录 实验采用购自荷兰瓦赫宁根大学的四通道刺吸电位仪进行记录。首先将实验用的褐飞虱的长翅雌虫饥饿30min，然后用3cm的直径18.5μm的金丝蘸取少量黏虫银胶黏结在褐飞虱的前胸背板上，待银胶凝固之后将固定好的褐飞虱放入待测的水稻上同时开始记录。整个实验过程保持温度（25±2）℃，湿度（60±10）%的环境，每天17：00开始记录，持续14h，其中前12h用于实验分析，整个记录过程处于黑暗中。记录过程使用随机附带的Style$^+$b软件进行记录，记录完毕后使用同样随机附带的Style$^+$a软件进行分析。

（三）结果

1. 褐飞虱取食过程中的EPG波形的描述 图19.1展示了褐飞虱取食水稻过程中

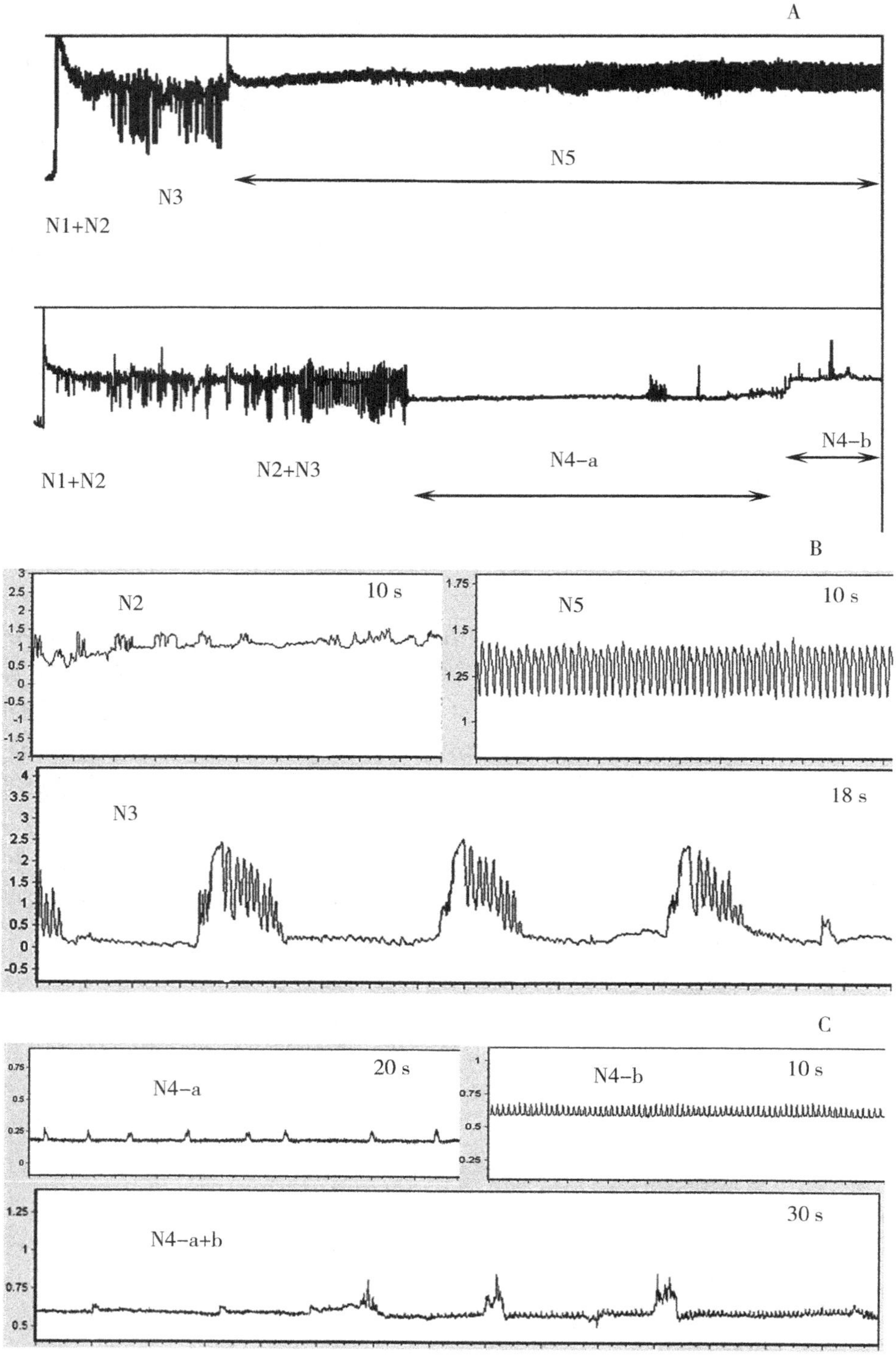

图 19.1　褐飞虱取食过程中典型的取食波形

A. 褐飞虱在水稻韧皮部和木质部取食过程中典型的整体波形；B. 褐飞虱取食过程中的典型波形包括路径波中的 N2 和 N3 波形及在木质部取食的 N5 波形；C. 褐飞虱韧皮部取食过程中的典型波形包括分泌水溶性唾液的 N4-a 波形、韧皮部取食的 N4-b 波形以及发生在 N4-a 和 N4-b 之间的过渡波形

典型的 DC-EPG 波形，其中，我们鉴定出 6 个典型的 EPG 波形：np 波（非刺探波形），刺探过程中的 3 个典型波形（N1、N2、N3），N4a 波（筛管中分泌唾液），N4b 波（韧皮部取食），N5 波（木质部取食）。鉴定出的各个波形与 Seo et al.（2010）和 Ghaffar et al.（2011）在各自文章中描述的褐飞虱取食的典型波形基本一致。并且我们发现在褐飞虱取食的 N4a 和 N4b 波形之间的过渡过程中存在一个未报道的褐飞虱取食波形。

2. EPG 记录结果分析　通过对 EPG 波形和其发生顺序的分析，我们筛选了一系列的指标用于对褐飞虱取食行为的分析（Xue et al.，2008），其中用路径波（ph）来统称褐飞虱刺探过程中的 E1、E2 和 E3 波形，便于对褐飞虱取食过程进行统计和数据计算。通过对采用的 12 个指标进行分析（表 19.1），除了褐飞虱筛管取食分泌唾液的时间这个指标出现差异之外，其他指标均未出现明显的差异，其中褐飞虱在 TN1 水稻上取食时在筛管取食时分泌唾液的时间显著长于褐飞虱在 T2A-1 水稻上取食时的时间，但是褐飞虱在转 Bt 基因水稻 T2A-1 和 T1C-1 及其亲本明恢 63 上取食时分泌唾液的时间没有差异。

（四）讨论

昆虫取食过程中植物内部的化学物质在刺吸式口器昆虫到达植物木质部、韧皮部和其他组织的过程中起到关键的作用。因此，利用 EPG 技术研究刺吸式口器昆虫的取食行为，插入到水稻基因组的 Bt 基因的表达对研究刺吸式口器害虫有十分重要的意义，并且利用 EPG 技术还能提供更多的在刺吸式口器昆虫取食过程中口针运动的细节（Tjallingii，2006）。目前，已有很多文献对褐飞虱取食过程中的 EPG 波形进行描述和功能分析（Velusamy and Heinrichs，1986；Kimmins，1989；Lösel and Goodman，1993；van Helden and Tjallingii，1993；Hattori，2001；Hao et al.，2008；Seo et al.，2009）。按照已有的文献报道，褐飞虱取食过程可以分为木质部取食和韧皮部取食两个过程，其中，木质部取食过程包括非刺探（np），路径波 ph（N1+N2+N3），木质部取食（N5）波形组成，而韧皮部取食则主要由 np 波、ph 波、在筛管中分泌水溶性唾液（N4-a）波和筛管中取食的（N4-b）波形组成。此外 Ghaffar 等人曾报道发现两个新波形：机械作用导致取食困难（N6）以及细胞内刺探（N7）波形。在本实验中我们研究发现了 7 个典型的 EPG 波形，并且与以往的利用 DC-EPG 的研究结果基本一致。此外，我们在研究中没有发现 N6 和 N7 波形，可能是因为这两个波形主要在褐飞虱取食高抗性水稻时出现。在一些 EPG 记录的波形中我们发现在 N4-a 波和 N4-b 波之间有一个新的亚波形，其波形的特征预示其可能是 N4-a 和 N4-b 的混合波形，因为其包括混合的 N4-b 和 N4-a 波形的特征，其生理学意义目前难以判定，但是其出现可能与褐飞虱的一个尝试取食过程相关。

表 19.1　褐飞虱取食不同水稻时 EPG 各指标的比较

指标	T1C-19（$n=11$）	T2A-1（$n=10$）	Minghui63（$n=8$）	TN1（$n=11$）
np 波持续时间	3 203.36±2 090.63 a	798.70±437.92 a	387.75±151.92 a	610.36±212.02 a
路径波（ph）持续时间	4 194.00±1 529.37 a	5 295.00±1 109.31 a	4 604.5.0±879.83 a	3 968.18±718.56 a
N4-a 波形持续时间	2 572.64±616.41 ab	1 650.778±325.29 b	3 235.00±532.35 ab	4 918.71±1 132.42 a
N4-b 波形持续时间	33 491.55±2 973.52 a	32 017.50±1 991.96 a	32 957.13±1 429.57 a	32 708.46±1 349.49 a
N5 波形持续时间	842.46±879.77 a	537.80±213.57 a	164.13±117.45 a	909.09±279.03 a
第一次刺探发生时间	163.182±69.97 a	282.20±101.93 a	206.88±76.91 a	136.18±65.34 a
路径波（ph）出现次数	4.00±1.52 a	5.90±0.79 a	5.38 ±0.65 a	6.73±1.13 a
第一次 N4a 波持续时间	1 925.73±253.85 a	1 979.30±151.14 a	2 374.125±542.31 a	2 023.00±373.16 a
第一次 N4 波出现前的刺探时间	3 426.73±1 564.42 a	1 476.70±164.53 a	1 251.25±478.86 a	1 591.91±264.17 a
第一次路径波（ph）持续时间	3 130.82±1 427.15 a	976.60±208.83 a	1 011.375±491.29 a	893.55±220.58 a
N4b 波持续时间	24 057.09±5 390.66 a	19 173.00±5 017.59 a	13 521.88±4 437.26 a	14 744.09±4 819.47 a
N4b 波的出现次数	1.55±0.25 a	2.00±0.39 a	2.38±0.63 a	1.82±0.24 a

注：数值采用平均值±标准误表示，相同字母标注表示差异不显著（$P>0.05$），不同字母标注表示差异显著（$P<0.05$）。

已有的文献报道，水稻对褐飞虱抗性的提高主要体现在褐飞虱取食过程中，其 np、ph、N5、N6 和 N7 波形持续时间在所测定的时间内比例升高，从而影响褐飞虱在韧皮部中的取食（N4-b）时间，而水稻对褐飞虱的抗性主要体现在褐飞虱能否在韧皮部中持续取食，而不是其口针刺探到韧皮部的过程中（Kimmins，1989；Ghaffar et al.，2011）。此外，我们的 EPG 实验结果也进一步证明褐飞虱主要选择在韧皮部中取食而不是在木质部中取食，这可能与植物韧皮部汁液中往往有更加丰富的营养物质有关（Ghaffar et al.，2011）。在我们使用的 12 个用于 EPG 波形分析的指标中，只有 N4-a 波形的持续时间这一个指标出现了明显的差异，其中褐飞虱在 TN1 水稻上取食过程中 N4-a 波形的持续时间比在 T2A-1 水稻上取食时显著延长。N4-a 波形的生理学意义在于将水溶性唾液注射到筛管之中，而水溶性唾液的功能主要体现在降低植物抗性，具体主要表现在与韧皮部中的钙离子结合阻止筛管阻塞；同时，减少植物抗性物质累积（Will et al.，2007）。但是 Ghaffar 等人的研究发现，当褐飞虱取食不同抗性的水稻时 N4-a 波形的出现在统计学上不会有明显的差异，所以 N4-a 波形持续时间上的差异，说明转 Bt 基因水稻 T2A-1 对褐飞虱的抗性可能低于 TN1 水稻。

从整体的 EPG 数据分析可以发现，Bt 基因的插入不会导致水稻对非靶标昆虫褐飞虱的取食行为产生影响，同时测定的两个转 Bt 基因水稻和其亲本明恢 63 在其生育期 70d 左右时是褐飞虱的易感品系，需要在减少鳞翅目靶标害虫的同时注意对刺吸式口器昆虫褐飞虱的控制。

参考文献

[1] ANDOW D A, HILBECK A. Science - based risk assessment for nontarget effects of transgenic crops. *Bioscience*, 2004, 54: 637-649.

[2] BAI Y Y, JIANG M X, CHENG J A, et al. Effects of CrylAb toxin on *Propylea japonica* (Thunberg) (Coleoptera: Coccinellidae) through its prey, *Nilaparvata lugens* (Homoptera: Delphacidae), feeding on Transgenic Bt rice. *Environmental Entomology*, 2006, 35: 1 130-1 136.

[3] CHEN H, TANG W, XU C, et al. Transgenic indica rice plants harboring a synthetic *cry*2A * gene of *Bacillus thuringiensis* exhibit enhanced resistance against lepidopteran rice pests. *Theoretical and Applied Genetics*, 2005, 111: 1 330-1 337.

[4] CHEN M, LIU Z, YE G Y, et al. Impacts of transgenic *cry*1Ab rice on non-target planthoppers and their main predator *Cyrtorhinus lividipennis* (Hemiptera: Miridae) —A case study of the compatibility of Bt rice with biological control. *Biological Control*, 2007, 42: 242-250.

[5] CHEN M, SHELTON A, YE G Y. Insect-resistant genetically modified rice in China: from research to commercialization. *Annual Review of Entomology*, 2011, 56: 81-101.

[6] CHEN M, YE G, LIU Z, et al. Field assessment of the effects of transgenic rice expressing a fused gene of *cry1Ab* and *cry1Ac* from *Bacillus thuringiensis* Berliner on nontarget planthopper and leafhopper populations. *Environmental Entomology*, 2006, 35: 127-134.

[7] CHEN M, ZHAO J Z, YE G Y, et al. Impact of insect-resistant transgenic rice on target insect pests and non-target arthropods in China. *Insect Science*, 2006, 13: 409-420.

[8] DESNEUX N, BERNAL J S. Genetically modified crops deserve greater ecotoxicological scrutiny. *Ecotoxicology*, 2010, 19: 1642-1644.

[9] GHAFFAR MBAB, PRITCHARD J, FORD-LLOYD B. Brown Planthopper (*N. lugens* Stal) Feeding Behaviour on Rice Germplasm as an Indicator of Resistance. *PloS ONE*, 2011, 6 (7): 22137.

[10] KIMMINS F. Electrical penetration graphs from *Nilaparvata lugens* on resistant and susceptible rice varieties. *Entomologia Experimentalis et Applicata*, 1989, 50: 69-79.

[11] KRUGER M, VAN RENSBURG J, VAN DEN BERG J. Perspective on the development of stem borer resistance to *Bt* maize and refuge compliance at the Vaalharts irrigation scheme in South Africa. *Crop Protection*, 2009, 28: 684-689.

[12] LI F F, YE G Y, WU Q, et al. Arthropod abundance and diversity in Bt and non - Bt rice fields. *Environment Entomology*, 2007, 36: 646-654.

[13] ROMEIS J, BARTSCH D, BIGLER F, et al. Assessment of risk of insect-resistant transgenic crops to nontarget arthropods. *Nature Biotechnology*, 2008, 26: 203-208.

[14] SEO B Y, KWON Y H, JUNG J K, et al. Electrical penetration graphic waveforms in relation to the actual positions of the stylet tips of *Nilaparvata lugens* in rice tissue. *Journal of Asia-Pacific Entomology*, 2009, 12 (2): 89-95.

[15] SUN X, YAN M J, ZHANG A, et al. Transgenic cry1C * gene rough rice line T1C-19 does not change the host preference of the non-target stored product pest, *Rhyzopertha dominica* (Fabricius) (Coleoptera: Bostrichidae), and its parasitoid wasp, *Anisopteromalus calandrae* (Howard) (Hymenoptera: Pteromalidae) . *Ecotoxicology and Environmental Safety*, 2015, 120: 449-456.

[16] SUN Z, YAN FM, WANG MQ. Transgenic expression of Bt in rice does not affect feeding behavior and population density of the brown planthopper, Nilaparvata lugens Stål (Hemiptera: Delphacidae). Entomologia Generalis, 2017, DOI: 10. 1127/entomologia/2017/0068.

[17] TANG W, CHEN H, XU C, et al. Development of insect-resistant transgenic indica rice with a synthetic *cry1C* gene. *Molecular Breeding*, 2006, 18: 1-10.

[18] TU J, ZHANG G, DATTA K, et al. Field performance of transgenic elite commercial hybrid rice expressing *Bacillus thuringiensis* δ-endotoxin. *Nature Biotechnology*, 2000, 18: 1 101-1 104.

[19] VELUSAMY R, HEINRICHS E. Electronic monitoring of feeding behavior of *Nilaparvata lagens* (Homoptera: Delphacidae) on resistant and susceptible rice cultivars. *Environmental Entomology*, 1986, 15 (3): 678-682.

[20] WANG Y, ZHANG G, DU J, et al. Influence of transgenic hybrid rice expressing a fused gene derived from *cry1Ab* and *cry1Ac* on primary insect pests and rice yield. *Crop Protection*, 2010, 29: 128-133.

[21] WILL T, TJALLINGII W F, THÖNNESSEN A, et al. Molecular sabotage of plant defense by aphid saliva. *Proceedings of the National Academy of Sciences*, 2007, 104: 10 536-10 541.

[22] WU K M, LU Y H, FENG H Q, et al. Suppression of cotton bollworm in multiple crops in China in areas with *Bt* toxin-containing cotton. *Science*, 2008, 321: 1 676-1 678.

（孙 泽 王满困）

第二十章　水稻对褐飞虱抗性机制研究

（一）研究背景

褐飞虱 *Nilaparvata lugens*（Stål）属于半翅目飞虱科（Hemiptera：Delphacidae），具有远距离迁飞习性，是我国和其他许多亚洲国家当前水稻上的首要害虫（Sogawa，1982；Velusamy & Heinrichs，1986）。该虫为单食性害虫，只能在水稻和普通野生稻上取食和繁殖后代。褐飞虱对水稻的为害主要表现在以下几方面：①直接吸食为害：以成虫、若虫群集于稻丛基部，刺吸茎叶组织汁液。虫量大、受害重时稻株瘫痪倒伏，俗称"冒穿"，导致严重减产或失收。②产卵为害：产卵时，刺伤稻株茎叶组织，形成大量伤口，促使水分由刺伤点向外散失，同时破坏疏导组织，加重水稻的受害程度。③传播或诱发水稻病害：褐飞虱不仅是传播水稻病毒病——草状丛矮病和齿叶矮缩病的虫媒，也有利于水稻纹枯病、小球菌核病的侵染为害。取食时排泄的蜜露，因富含各种糖类、氨基酸类，覆盖在稻株上，极易招致煤烟病菌的滋生（程遐年等，2003）。

由于主栽品种抗性差、褐飞虱抗药性增强等原因，从 2003～2006 年，全国稻飞虱发生面积不断扩大，每年造成稻谷损失都在数十亿千克。褐飞虱在我国南方稻区为害严重，尤其 2005～2007 年，为害面积近 5 亿亩。2005 年仅华东四省就因褐飞虱损失 25 亿元。化学防治是目前遏制褐飞虱猖獗发生的主要手段，但是也出现了褐飞虱的高抗药性、农药残留以及对天敌的杀伤（Matsumura et al.，2008）等问题。利用水稻品种自身的抗性、推广应用抗虫品种被认为是防治褐飞虱为害的最经济有效途径之一。然而，在育种家努力选育一个又一个抗虫品种以抵抗褐飞虱为害的同时，褐飞虱也同样通过生物型的改变一次又一次地克服了水稻品种的抗性，导致水稻抗虫性迅速衰退。据我们对广东省 71 个主栽品种进行苗期褐飞虱抗性测定结果表明，在 71 份材料中，只有 2 份表现为中抗，其余 69 份均对褐飞虱无抗性，抗性的衰退已经到了令人吃惊的程度（肖汉祥等，2011）。因此，加强对现有优良抗源抗性机制研究，利用中等抗性的抗虫品种或多基因控制的抗虫品种延缓抗性衰退，以及不断挖掘新的抗源材料，成为当前利用水稻品种自身抗性防治褐飞虱为害的首要任务。

EPG 是一种研究刺吸式口器昆虫取食行为的很好工具（McLean & Kinsey，1964；Tjallingii et al.，1978；Golawska et al.，2012；Trebicki et al.，2012），很多国内外学者利用 EPG 对褐飞虱的取食波形、褐飞虱取食抗虫、感虫水稻植株的表现、喷施内吸性新药剂吡蚜酮后的褐飞虱取食行为表现等进行研究（Velusamy & Heinrichs，1986；Khan & Saxena，1988；Lösel & Goodman，1993；Kimmins，1989；Hattori，2001；Hao et al.，2008；Seo et al.，2009，2010；He et al.，2011b；Ghaffar et al.，2011）。

麦黄酮（英文名：tricin；分子式：5，7，4′-trihydroxy-3′，5′-dimethoxyflavone）广泛存在于禾本科植物，已有的研究结果表明，植物体内的麦黄酮在植物遭受非生物胁迫（abiotic stresses）诸如干旱胁迫、盐胁迫、低温胁迫，以及生物胁迫（biotic stresses）诸如虫害、病害、杂草等时参与一些重要的生理响应和化学防御过程（Moheb et al.，2013），因此麦黄酮在禾本科作物如小麦、水稻的抗虫性方面受到广泛的关注。本研究在前人对褐飞虱波形研究的基础上，利用 EPG 技术，围绕水稻植株韧皮部重要次生化合物麦黄酮，研究褐飞虱对含有不同浓度的麦黄酮水稻品种以及添加有不同浓度麦黄酮的人工取食液的取食行为反应，试图建立水稻品种麦黄酮的含量与水稻抗虫性的联系，从另一个的角度阐述水稻抗虫性，为含麦黄酮的水稻抗虫品种选育提供理论支撑。

（二）材料与方法

1. 水稻材料　抗虫品种 Rathu heenati（RHT）和感虫对照水稻品种 Taichuang Native 1（TN1）均由国际水稻研究所（International Rice Research Institute，IRRI）提供。所有水稻用育苗盒育苗然后转移至小盆中。

2. 昆虫来源及饲养　生物型 2 褐飞虱采集自广东省农业科学院植物保护研究所白云试验基地稻田，在室内人工气候箱温度：25℃±2 ℃，相对湿度：75%±5%，光照 L/D=14h/10h，用 TN1 连续饲养 30 代以上。

3. 水稻植株上的 EPG 波形记录　采用四通道直流电型（DC-EPG 型）昆虫刺吸电位仪进行取食观察。实验过程中均采用短翅型雌成虫，记录前进行 1h 左右的饥饿处理。为了减少干冰、CO_2 气体或者乙醚气体处理对褐飞虱行为造成的影响，本实验采用了一个自制的昆虫负压固定装置（已获专利授权，专利号 ZL 2013 2 0620911.5）来固定昆虫。首先利用水流产生的负压通过固定装置把雌虫的腹部固定在 PVC 平板上，然后在褐飞虱前胸背板滴上一滴细微的银胶，将金线插入银胶滴待凝固后就连接到 EPG 上。植物电极插入水稻生长的土壤中，连接有金线的褐飞虱通常固定在土壤表面水层以上 3~5cm 处取食。实验在温度：(25±2)℃，相对湿度：(75±5)%的房间开展，通常连续记录褐飞虱取食 4h 以上，每个通常经常变化测定不同的水稻品种，每个处理有效记录 20 次以上。记录的数据利用 PROBE 3.4 软件（瓦赫宁根大学，荷兰）和 EXCEL 软件进行分析。

4. 水稻植株麦黄酮含量定量测定　为了研究水稻植株麦黄酮含量与褐飞虱取食行为之间存在的关联，褐飞虱记录实验完成后将水稻植株叶片进行收集、溶剂提取、SPE 萃取小柱纯化、浓缩等，最后利用液质联用分析系统（API4000 Q-Trap LC/MS/MS，Applied Biosystems，Carlsbad，CA，USA）进行麦黄酮的定量分析。

5. 褐飞虱液体饲料囊上的刺吸电生理记录　褐飞虱液体饲料囊（liquid-diet sac，LDS）由一个直径为 6cm、高 5cm 的中空、透明的有机玻璃筒和双层封口膜组成（图 20.1），褐飞虱取食液加入双层膜中间，EPG 植物电极的一端绑上细小的、长约 10cm 的铜丝插入双层膜的人工取食液当中。以 5%的蔗糖溶液作为对照，以添加了不同浓度

(10mg/L，100mg/L 和 1 000mg/L) 麦黄酮的蔗糖溶液作为处理。其他的实验方法同在水稻植株上的电生理记录。

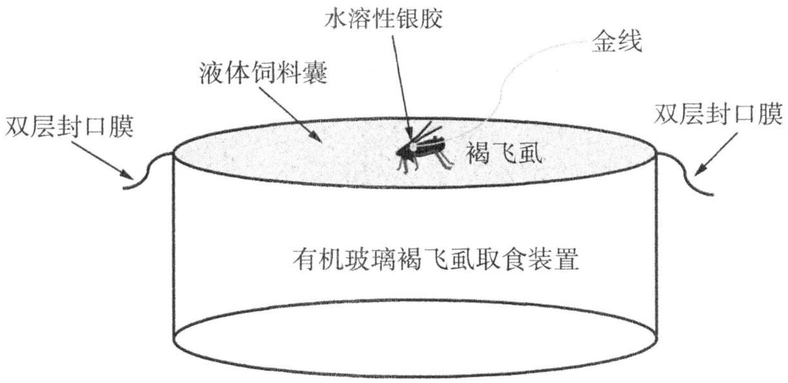

图 20.1 褐飞虱人工取食台（LDS）

（三）结果

1. 褐飞虱取食水稻的 EPG 波形 褐飞虱取食水稻的经典 EPG 波形如图 20.2 所示，与 Velusamy & Heinrichs (1986)，Kimmins (1989)，Losel &Goodman (1993)，Hattori (2001)，Hao et al. (2008)，Seo et al. (2009)，Ghaffar et al. (2011) 等对褐飞虱取食水稻植株的波形描述类似。根据相对电压、频率和波形性状大致可以将波形分为 6 类，分别是 np 波（口针非刺探状态）、N1 波（刺探开始）、N2 波（唾液分泌和口针活动）、N3 波（口针的细胞间隙刺探活动）、N4 波（韧皮部取食）、N5 波（木质部取食）。根据我们的分析，np 波往往和褐飞虱非取食状态相关联，表示口针尚未刺入水稻组织或从水稻组织拔出后的一种状态。根据 Seo 等人和 Ghaffar 等对褐飞虱波形的描述，在本项研究中，为了方便统计分析，我们也将 N1 波、N2 波和 N3 波归纳为褐飞虱取食过程波 PW 波，表示褐飞虱口针刺探后、韧皮部取食前口针在水稻茎秆组织中活动反映出的一种电生理状态（图 20.2A）。其中，N1 波往往比较难以鉴定，因为其持续时间通常非常短。N2 波在 NP 波出现后通常会立即出现，由一组不同性状和不同频率及振幅的波形组成，N2 波的频率大约 0.2Hz（图 20.2C）。N2 波之后紧接着出现的是 N3 波或者 N5 波，N3 波是一组性状连续、更高振幅（2.0～4.0V）、更高频率（0.3～1.0Hz）的波形组成（图 20.2E）。随后，N4 波出现，与路径波 PW 相比，N4 波具有更高的频率（通常 8～10Hz）和较低的振幅（0.8～2.4V）（图 20.2F、图 20.2G）。N5 波偶然出现在路径波发生过程中，也由一连串持续的高频率、高振幅波形组成，这类波经常出现在 N3 波之后（图 20.2H）。

2. 褐飞虱取食水稻植株的各类 EPG 波形持续时间、频率的比较 EPG 波形分析方法与 He 等（2011）、Sarria 等（2009）一致，褐飞虱在抗虫水稻品种 Rathu Heenati (RHT) 和感虫对照品种 TN1 上的取食波形如图 20.3 所示。本项目主要研究 3 个与抗虫品种抗性相关的非连续性变量参数（non-sequential variables）：总持续时长、出现频

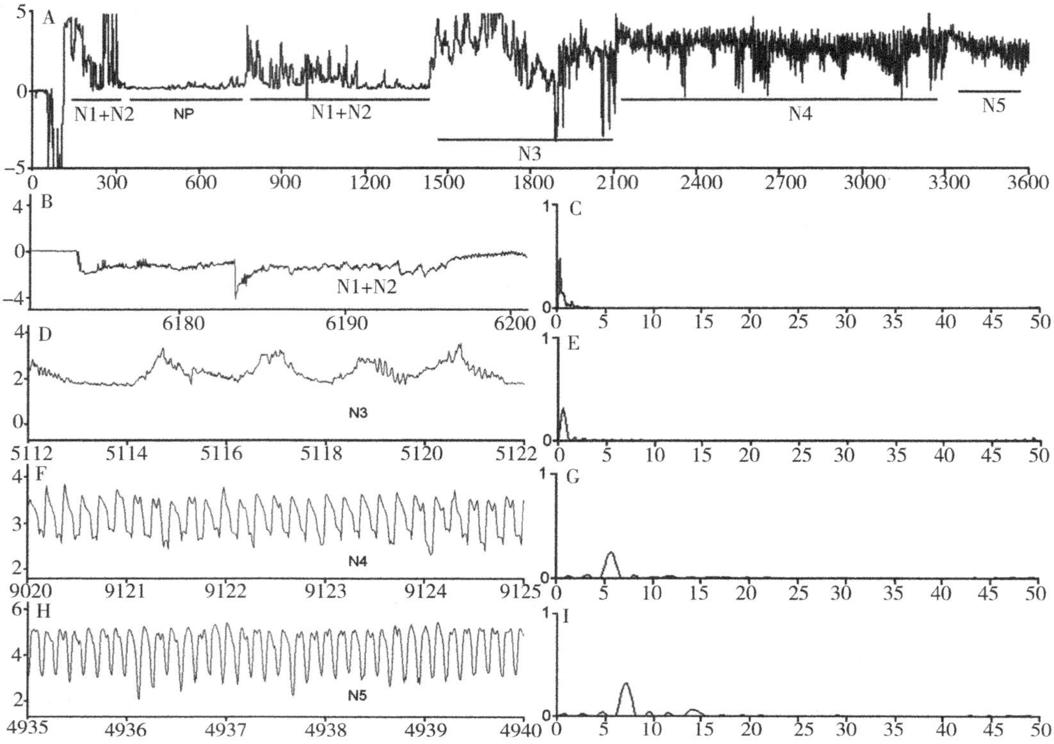

图 20.2　褐飞虱取食水稻 1h 内 EPG 波形图以及各种波形的振幅（B、D、F、H）和频率（C、E、G、I）

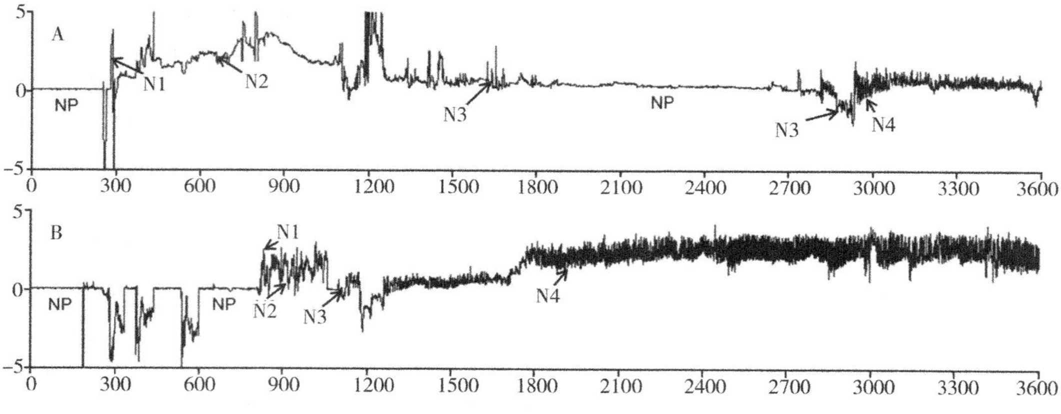

图 20.3　褐飞虱取食抗虫水稻品种 RHT（A）和感虫对照水稻品种 TN1（B）的 EPG 波形图比较

率、每次出现平均持续时间。我们分别记录 4h 内 np 波、PW 波、N4 波和 N5 波等上述参数值，经方差分析我们得知除 N5 波外，其他 3 类波形的总持续时间在抗虫和感虫品种上呈现显著性差异（图 20.4A）。RHT 水稻上 np 波和 PW 波显著长于 TN1 水稻，N4 波的总持续时间显著短于 TN1 水稻（图 20.4A）。RHT 水稻 np 波、PW 波和 N4 波的出现频率要高于 TN1 水稻（图 20. 4B）。4 种波形中，只有 RHT 水稻上的 N4 波的平均持续时间是显著低于 TN1 水稻上的（图 20. 4C）。

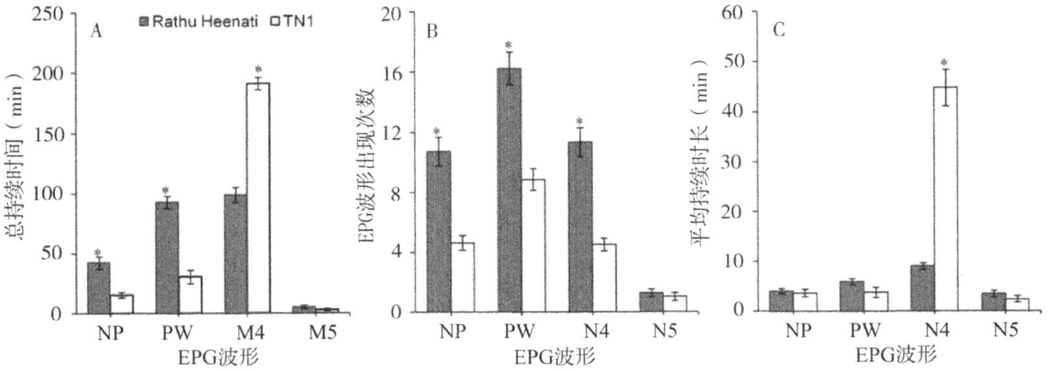

图 20. 4　褐飞虱取食抗、感水稻品种的各种波形比较

np 表示褐飞虱没有产生口针刺探行为；PW 表示路径波，是综合 N1 波、N2 波和 N3 波；N4 波表示褐飞虱在韧皮部的取食行为；N5 波表示褐飞虱在木质部的取食活动。（下同）

3. 不同抗性品种水稻麦黄酮含量的定量分析　新鲜 RHT 水稻叶片的麦黄酮质量浓度平均值为（23. 0±3. 0）$ng \cdot g^{-1}$，要显著高于 TN1 水稻叶片的麦黄酮质量浓度平均值（5. 4±0. 7）$ng \cdot g^{-1}$（ANOVA：$F_{1,15}=32.35$，$P<0.0001$）。

4. 褐飞虱取食行为与水稻植株体内麦黄酮含量的关联分析　水稻取食 RHT、TN1 各种 EPG 波形（np 波、PW 波、N4 波和 N5 波）的非持续性变量（总持续时间、出现频率、平均持续时间）与水稻植株体内麦黄酮浓度的关联利用佩尔森相关性分析法进行分析（图 20. 5～图 20. 7，表 20. 1），结果表明 PW 波的总持续时间和出现频率呈现随麦黄酮质量浓度的增加而增加的趋势，N4 波的出现频率也随麦黄酮质量浓度增加而增加（图 20. 6）。相反，N4 波总持续时间和平均持续时间随水稻筛管中麦黄酮的浓度增高而减少（图 20. 5～图 20. 7）。EPG 波形与麦黄酮浓度变量相关分析结果见表 20. 1。水稻植株体内麦黄酮浓度与 PW 波的总持续时间（$R^2=0.522$，$P=0.002$）、PW 波出现的频率（$R^2=0.494$，$P=0.002$）、N4 波出现的频率（$R^2=0.689$，$P<0.0001$）呈显著的线性正相关，而与 N4 波总波总持续时间和平均持续时间呈显著的线性负相关（$R^2=0.478$，$P=0.002$；$R^2=0.599$，$P<0.0001$）（图 20. 5～图 20. 7，表 20. 1）。

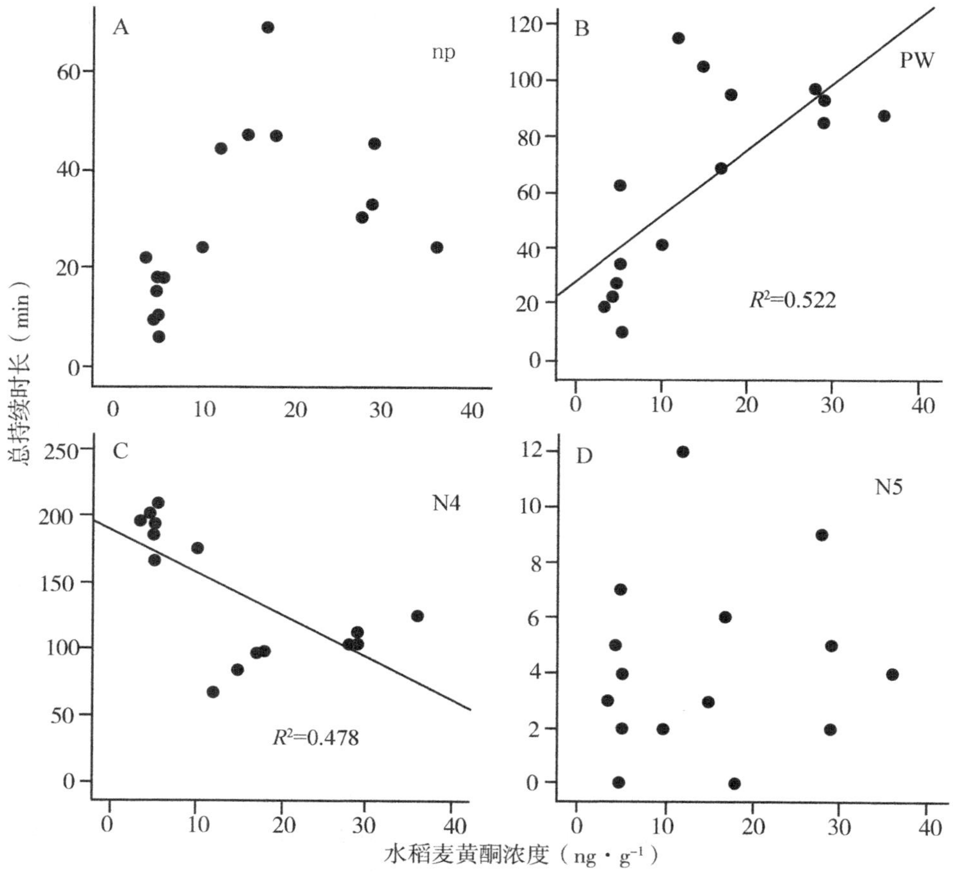

图 20.5 水稻麦黄酮含量与褐飞虱取食水稻各种波形的总持续时长的相关性分析

表 20.1 不同水稻品种（RHT 和 TN1）麦黄酮质量浓度与褐飞虱取食水稻植株的各种波形指标参数的相关性分析

	波形	R^2	P
总持续时间（min）	np	0.478	0.061
	PW	0.723	0.002
	N4	−0.691	0.003
	N5	0.184	0.50
波形出现频率	np	0.469	0.067
	PW	0.703	0.002
	N4	0.830	<0.001
	N5	0.134	0.62
平均持续时长（min）	np	0.115	0.67
	PW	0.358	0.17
	N4	−0.774	<0.001
	N5	0.255	0.34

np，未刺吸；PW，路径波。

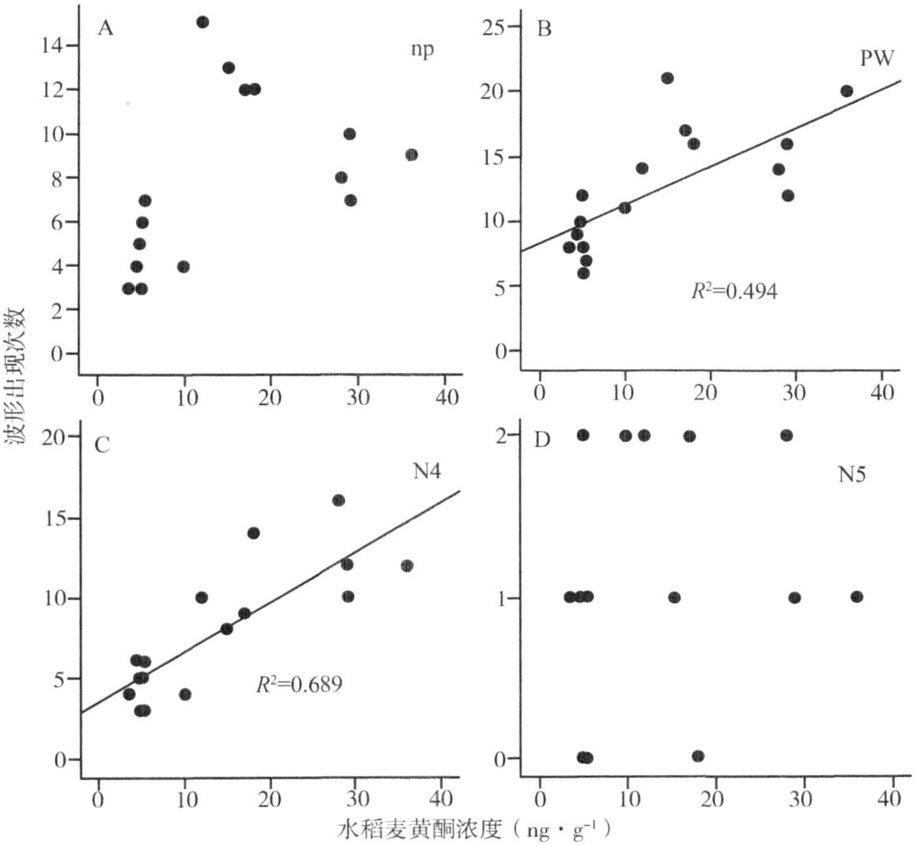

图 20.6 水稻麦黄酮含量与褐飞虱取食水稻各种波形的出现频率的相关性分析

5. 褐飞虱在液体饲料囊上取食的 EPG 波形特征 图 20.8 表示褐飞虱取食 LDS 内人工液体饲料的 EPG 波形图。我们根据波形的性状、振幅和频率，将这些 EPG 信号归为 4 类波形，分别是 np 波、N1 波、N2 波和 N4 波。本项目为了统计分析的方便将 N1 波和 N2 波统一归为 PW 波（图 20.8E）。这类波形常出现在 np 波之后，具有较低的频率（0.5~3Hz）和相对较低的振幅（0.3~1.5V）（图 20.8G）。取食人工饲料的 N4 波与取食水稻韧皮部产生的 N4 波非常相似，几乎拥有同样的频率范围（5~8Hz）和相对较高的振幅水平（3~5V）（图 20.8F、图 20.8H）。

6. 褐飞虱在经麦黄酮处理的 LDS 上取食行为的波形分析 记录褐飞虱在 CK（空白对照，5%蔗糖溶液）、不同麦黄酮浓度（10mg·L^{-1}、100mg·L^{-1}、1 000mg·L^{-1}）蔗糖溶液上取食的 EPG 波形（np 波、PW 波和 N4 波）的 3 种非连续性变量参数（总持续和平均持续时长、出现频率）（图 20.9A-I）。在 4h 记录时间内，褐飞虱在 CK 上取食的 np 波总时长和出现频率分别为（21.6±3.1）min 和（2.9±0.2）次，要显著短于和低于麦黄酮处理（ANOVA：所有的 P 均小于 0.01，图 20.9A、图 20.9D）。CK 与各个处理之间的 np 波平均持续时长并不显著（P 均大于 0.01，图 20.9G）。回归分析表明麦黄酮含量与 np 波的总持续时长和出现频率间存在显著的线性相关性（图 20.9A、

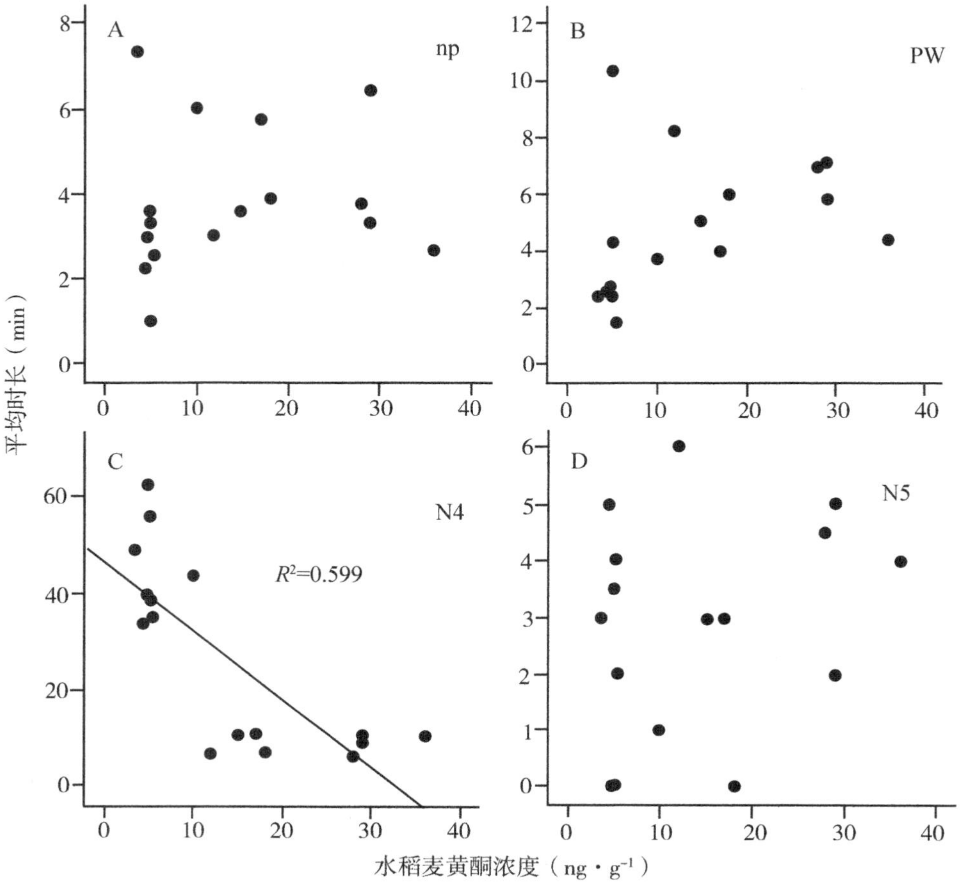

图 20.7　水稻麦黄酮含量与褐飞虱取食水稻各种波形的平均持续时长的相关性分析

图 20.9D）。CK 与各个处理间的 PW 波总时长间存在显著差异（图 20.9B），PW 波在 CK 上出现的频率也要显著低于各种麦黄酮浓度处理。回归分析表明麦黄酮含量与 PW 波的总持续时长和出现频率间存在显著的线性相关性（图 20.9A、图 20.9D）。PW 波在 CK 上的平均时长要显著低于各种麦黄酮浓度处理。回归分析表明麦黄酮含量与 PW 波的平均时长与麦黄酮浓度间存在显著的曲线相关（exponential decay）（图 20.9H）。CK 上的 N4 波总时长要显著长于个麦黄酮处理（图 20.9C），出现频率要显著低于麦黄酮处理（图 20.9F），平均持续时长要显著长于麦黄酮处理（图 20.9I）。回归分析表明麦黄酮含量与 N4 波的总持续时长和出现频率间存在显著的线性相关性（图 20.9C），与 N4 波出现的频率呈线性正相关（图 20.9F），与 N4 波形平均持续时长呈负的曲线相关（图 20.9I）。

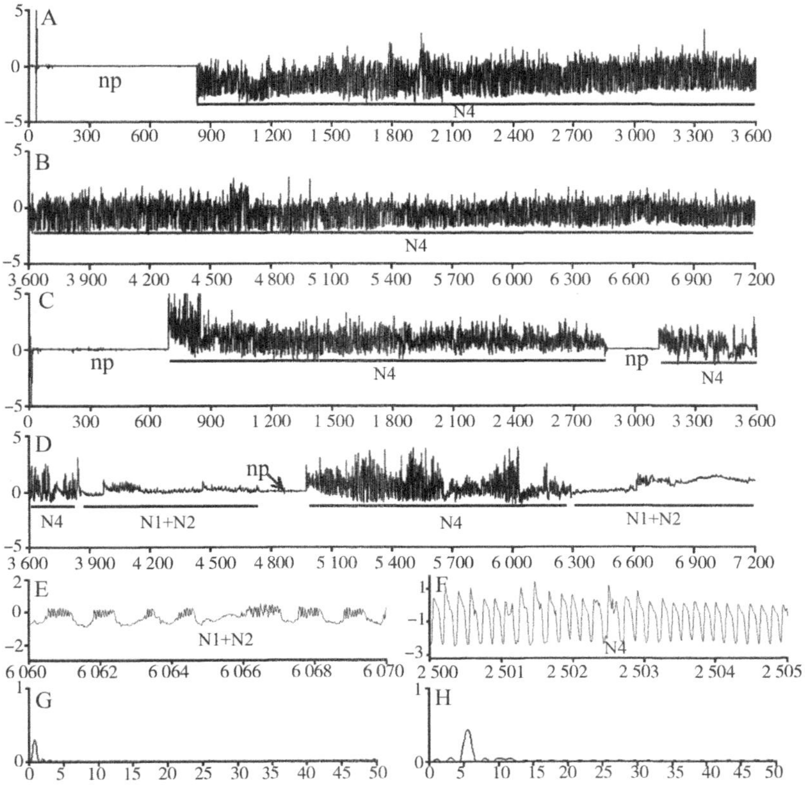

图 20.8 褐飞虱取食人工取食液 LDS 的波形及波形特征

（四）讨论

本研究利用 DC-EPG 描述了褐飞虱在抗虫水稻 RHT 和感虫对照水稻 TN1 上的取食行为，同时利用液质联用定量测定分析技术研究了麦黄酮在水稻植株中的含量。并在此基础上，利用人工取食液中添加麦黄酮的方法测定了褐飞虱的取食行为，以期验证麦黄酮在水稻抗虫过程中的生物学功能。

1. 褐飞虱在水稻植株上和 LDS 上取食波形的比较 我们在褐飞虱取食水稻植株过程记录到 6 种波形，这些波形的描述与分类基本采用 Seo 等（2009）和 Ghaffar 等（2011）对褐飞虱取食水稻波形的分类。一般而言，褐飞虱取食波形是从 np 波开始的，这类波形很好判断，一条直线表示褐飞虱尚未取食或者取食完毕口针离开水稻植株表皮。接着就是一些非常规的波形比如 N1 波、N2 波和 N3 波，表示褐飞虱口针在表皮细胞间或细胞内刺探、分泌唾液等生理行为。在本研究中，为了统计分析的需要，同时根据研究目的，我们将 N1 波、N2 波和 N3 波统一归为 PW 波，也就是路径波，这个与 Kimmins（1989）描述的 P2 波、P3 波，Velusamy & Heinrichs（1986）和 Hattori（2001）描述的 S 波应该是同一类波形。虽然命名不一样，但是根据波形的特点、性状等分析应该是同一类波形。N4 波是褐飞虱取食过程中非常重要的一类波形，代表褐飞

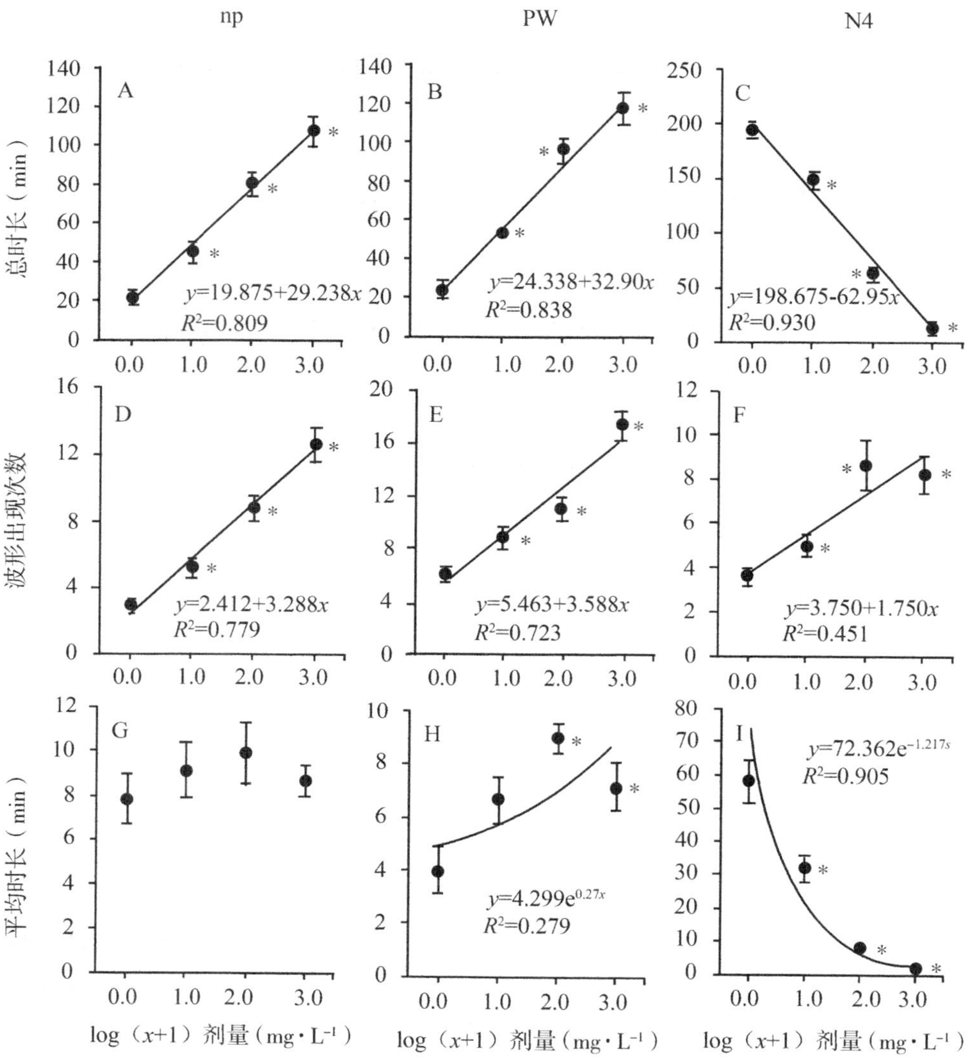

图 20. 9　褐飞虱取食添加麦黄酮的 LDS 波形与剂量之间的关系

虱在韧皮部取食汁液，这与蚜虫 EPG 研究中的 E 波很接近（Tjallingii，1978）。Kimmins（1989）把这类波形命名为 P4 波，Velusamy & Heinrichs（1986）和 Hattori（2001）把这类波命名为 I 波，但应该指的是同一类波形。N5 波出现在褐飞虱口针在木质部主动吸食水分时，Kimmins（1989）把这类波形命名为 P5 波。这类波形在褐飞虱取食水稻过程中不是每一次取食都会出现。

褐飞虱在 LDS 上的取食波波形相对简单，我们一共记录到 4 种波形，分别是 np 波、N1 波、N2 波、N4 波。N1 波和 N2 波与在水稻植株上记录到的这两种波形相比持续时间要短，波形图没有那么复杂。LDS 上的 N4 波与水稻上的 N4 波相比，振幅和频率几乎一致，LDS 上的 N4 波大小更加均匀，持续时间更长。

不管褐飞虱是在水稻上还是在 LDS 上取食，根据褐飞虱波形特征和生物学意义，我们基本可以将所有波形分成两大类：非取食行为波形和取食行为波形（Hattori，2001）。非取食行为波形从褐飞虱口针开始接触植物表皮、刺探、在水稻植物组织内除取食外的相关活动如唾液分泌等、木质部吸取水分等都属于这个范畴。取食行为波形仅仅发生在褐飞虱在韧皮部筛管吸食汁液的过程。而非取食行为过程与水稻植株对褐飞虱抗性、褐飞虱的传毒行为等密切相关（Khan & Saxena，1986，1988；Velusamy & Heinrichs，1986；Ghaffar et al.，2011；He et al.，2011a，b；Jin et al.，2011）。

2. 麦黄酮在 RHT 水稻对褐飞虱抗虫特性中的生态学功能　褐飞虱在不同抗性水平的水稻上有着不同的取食行为表现。褐飞虱取食抗虫品种如 RHT、Babawee、IR64、IR56 以及一些禾本科杂草，通常会有 70%左右的时间花在非取食行为波形如 np 波、PW 波上（Khan & Saxena，1986，1988；Velusamy & Heinrichs，1986；Kimmins，1989）。但是，在取食感虫对照水稻品种 TN1 时，褐飞虱可以停留在韧皮部筛管中停留吸食很长一段时间。Hattori（2001）和 Ghaffar 等（2011）研究结果表明，水稻对褐飞虱的抗性由韧皮部相关的特性决定，韧皮部抗虫性可能是由于韧皮部化学特性导致（Sogawa，1982；Chen et al.，2009）。比如韧皮部含有的游离氨基酸、硅酸、草酸、酚酸、黄酮类化合物等。麦黄酮已被报道为褐飞虱水稻韧皮部取食抑制剂和口针刺探刺激剂（Adjei-Afriyie et al.；2000a，b；Ling et al.，2007），但是并未有严格的实验证据尤其是电生理方面的证据。水稻植株体内麦黄酮含量和褐飞虱取食行为之间的关系也没有被详细研究。本文选择对褐飞虱具有广谱抗性的水稻材料 RHT 和感虫水稻对照材料 TN1 作为实验材料，我们发现水稻韧皮部麦黄酮含量与 PW 波的持续时间、PW 波出现频率和 N4 波出现频率间呈显著正相关。但是褐飞虱在韧皮部的吸食行为显示麦黄酮含量与 N4 波的总持续时间以及平均持续时间之间呈显著负相关。这些实验结果清楚地表明韧皮部筛管中存在的麦黄酮在阻止褐飞虱取食方面发挥了重要的作用。我们同样通过对 11 种不同抗性水平的水稻品种的抗虫鉴定和麦黄酮定量测定结果发现麦黄酮的含量水平和含有麦黄酮的水稻品种对褐飞虱的抗性级别之间也存在关联（张振飞，另文发表）。

我们对 EPG 分析结果还表明，褐飞虱取食经过麦黄酮处理的 LDS，取食行为会受到不同程度的抑制。np 波会随着麦黄酮浓度的增加延长。与对照相比，褐飞虱同时会有更长时间的休息以及更高的刺探频率。回归分析结果表明 np 波、PW 波和 N4 波的总持续时长和出现频率出现剂量反应，褐飞虱的取食行为（波形）将会随着 LDS 中的麦黄酮剂量的改变而改变。相类似的剂量效应也在 He 等（2011）、Jin 等（2011）的研究中存在。EPG 结合 LDS 技术来研究刺吸式口器昆虫对某种化合物的行为反应以及确定某些化合物的生态学功能正发挥着越来越重要的功能，也是 EPG 研究的一个热点。

在本项研究中，对麦黄酮生态作用的研究结果将有利于阐释抗虫水稻品种 RHT 以及其他一些含麦黄酮水稻品种的抗褐飞虱机制。基于 EPG 与 LDS 联合技术，我们研究发现褐飞虱口针不管是在水稻植株体内还是在人工液体饲料囊中，取食行为基本一致，都会出现一个相类似的取食波形。麦黄酮被发现可以抑制褐飞虱口针在筛管的取食，同时也在某种程度加快了刺探的频率。取食时间的缩短和刺探频率的加快在某种意义上来说正好可以作为抗虫水稻对刺吸式口器害虫抗性的重要生态学指标。下一步我们将从分子层面研究褐飞虱与水稻麦黄酮的互作机制。

参考文献

[1] ADJEI-AFRIYIE F, KIM C S, TAKEMURA M, et al. Probing stimulants from the rice plant towards the smaller brown planthopper, *Laodelphax striatellus* (Fallen) (Homoptera: Delphacidae). *Zeitschrift für Naturforschung*, 2000a, 55: 1 038-1 043.

[2] ADJEI-AFRIYIE F, KIM C S, TAKEMURA M, et al. Isolation and identification of the probing stimulants in the rice (*Oryza sativa*) plant for the white-back planthopper, *Sogatella furcifera* (Homoptera: Delphacidae). *Bioscience Biotechnology and Biochemistry*, 2000b, 64: 443-446.

[3] GHAFFAR M B A, PRITCHARD J, FORD-LLOYD B. Brown planthopper (*N. lugens* Stål) feeding behavior on rice germplasm as an indicator of resistance. *PLoS ONE*, 2011, 6 (7): e22137.

[4] GOLAWSKA S, LUKASIK I, KAPUSTA I, et al. Do the contents of luteolin, tricin, and chrysoeriol glycosides in alfalfa (*Medicago sativa* L.) affect the behavior of pea aphid (*Acyrthosiphon pisum*)? *Journal of Environmental Studies*, 2012, 21: 1 613-1 619.

[5] HAO P Y, LIU C X, WANG Y Y, et al. Herbivore-induced callose deposition on the sieve plates of rice: an important mechanism for host resistance. *Plant Physiology*, 2008, 146: 1 810-1 820.

[6] HATTORI M. Probing behavior of the brown planthopper, *Nilaparvata lugens* Stål (Homoptera: Delphacidae) on a non-host barnyard grass, and resistant and susceptible varieties of rice [J]. *Applied Entomology and Zoology*, 2001, 36: 83-89.

[7] HE Y P, CHEN L, CHEN J M, et al. Influence of pymetrozine on feeding behaviors of three rice planthoppers and a rice leafhopper using electrical penetration graphs. *Journal of Economic Entomology*, 2011, 104: 1 877-1 884.

[8] JIN L, YU X P, BIAN Y L, et al. Effects of mandelic acid on the feeding behavior of brown planthopper *Nilaparvata lugens* by electrical penetration graph (EPG) technique. *Journal of China University of Metrology*, 2011, 22: 95-101.

[9] KHAN Z R, SAXENA R C. Effect of steam distillate extracts of resistant and susceptible rice cultivars on behavior of *Sogatella furciferal* (Homoptera: Delphacidae). *Journal of Economic Entomology*, 1986, 79: 928-935.

[10] KHAN Z R, SAXENA R C. probing behavior of three biotypes of *Nilaparvata lugens* (Homoptera: Delphacidae) on different resistant and susceptible rice varieties. *Journal of Economic Entomology*, 1988, 81: 1 338-1 345.

[11] KIMMINS F M. Electrical penetration graphs from *Nilaparvata lugens* on resistant and susceptible rice varieties. *Entomologia Experimentalis et Applicata*, 1989, 50: 69-79.

[12] LING B, DONG H X, ZHANG M X, et al. Potential resistance of tricin in rice against brown planthopper *Nilaparvata lugens* (Stal). *Acta Ecologica Sinica*, 2007, 27: 1 300-1 307.

[13] LÖSEL PM, GOODMAN LJ. Effects on the feeding behaviour of *Nilaparvata lugens* (Stål) of sublethal concentrations of the foliarly applied nitromethylene heterocycle 2-nitromethylene-1,3-thiazinan-3-yl-carbarnaldehyde. *Physiological Entomology*, 1993, 18: 67-74.

[14] MATSUMURA M, TAKEUCHI H, SATOH M, et al. Species-specific insecticide resistance to imidacloprid and fipronil in the rice planthoppers *Nilaparvata lugens* and *Sogatella furcifera* in East and Southeast Asia. *Pest Management Science*, 2008, 64: 1 115-1 121.

[15] SEO B Y, KWON Y H, JUNG J K, et al. Electrical penetration graphic waveforms in relation to the actual positions of the stylet tips of *Nilaparvata lugens* in rice tissue. *Journal of Asia-Pacific Entomology*, 2009, 12: 89-95.

[16] SOGAWA K. Studies on the feeding habits of the planthopper, *Nilaparvata lugens* Stål (Homoptera: Delphacidae) Ⅳ. probing stimulant. *Applied Entomology and Zoology*, 1974, 9: 204-213.

[17] SOGAWA K. The rice brown planthopper: feeding physiology and host plant interactions. *Annual Review of Entomology*, 1982, 27: 49-73.

[18] TJALLINGII W F. Electronic recording of penetration behaviour by aphids. *Entomologia Experimentalis et Applicata*, 1978, 24: 721-730.

[19] TREBICKI P, TJALLINGII W F, HARDING R M, et al. EPG monitoring of the probing behaviour of the common brown leafhopper *Orosius orientalis* on artificial diet and selected host plants. *Arthropod-Plant Interactions*, 2012, 6: 405-415.

[20] VELUSAMY R, HEINRICHS E A. Electronic monitoring of feeding behavior of *Nilaparvata lugens* (Stål) (Hemiptera: Delphacidae) on resistant and susceptible rice cultivars. *Environmental Entomology*, 1986, 15: 678-682.

[21] WANG Y, GAO C, ZHU Y C, et al. Imidacloprid susceptibility survey and selection risk assessment in field populations of *Nilaparvata lugens* (Hemiptera: Delphacidae). *Journal of Economic Entomology*, 2008, 101: 515-522.

[22] YOSHIHARA T, SOGAWA K, PATHAK M D, et al. Soluble silicic acid as a sucking inhibitory substance on rice against the brown planthopper (Homoptera: Delphacidae). *Entomologia Experimentalis et Applicata*, 1979, 26: 314-322.

[23] YOSHIHARA T, SOGAWA K, PATHAK M D, et al. Oxalic acid as a sucking inhibitor of the brown planthopper (Homoptera: Delphacidae) in rice. *Entomologia Experimentalis et Applicata*, 1980, 27: 149-155.

[24] 程遐年，吴进才，马飞．褐飞虱研究与防治．北京：中国农业出版社，2003.

[25] 肖汉祥，李燕芳，张扬，等．不同抗性水稻品种对褐飞虱的抗性评价．广东农业科学，2011，22：64-65.

（张振飞）

第二十一章　利用 EPG 评价水稻品种对白背飞虱抗性的研究

（一）研究背景

白背飞虱 *Sogatella furcifera*（Hemiptera：Delphacidae）是水稻生产中的重要迁飞性害虫之一。南亚、东南亚、南太平洋及日本、朝鲜、中国、澳大利亚北部稻区均有该虫为害，在我国，其分布几乎遍布所有稻区。白背飞虱 1 年可发生 3~4 代，若虫分 5 龄，其成、若虫均刺吸取食水稻韧皮部汁液，造成稻株生长缓慢，分蘖延迟，瘪粒增加，甚至枯死，呈“虱烧”状（沈君辉等，2003a）。实践证明，培育和利用抗虫品种是控制白背飞虱为害最有效和环境友好的控制措施之一（Gamalath et al.，2012）。了解白背飞虱在不同水稻品种上的取食细节有助于明确水稻的抗虫机制。Khan 和 Saxena（1984）使用 EPG 记录了白背飞虱在抗虫、感虫水稻品种上的取食行为，结果表明，白背飞虱在感虫品种上更容易刺探，取食时间更长；相反，在抗虫品种上则产生更多短暂而重复的刺探，有效刺吸时间更短。限于当时的实验条件，白背飞虱的 EPG 波形与其具体的生物学意义的关联并不十分明确。Seo 等（2009）的研究明确了不同 EPG 波形的生物学意义，其结果为将 EPG 技术成功应用于白背飞虱取食行为的记录提供了基础。

不仅如此，白背飞虱还通过持久—循回—增殖型方式传播南方水稻黑条矮缩病毒（southern rice black-streaked dwarf virus，SRBSDV）（Zhou et al.，2008）。水稻秧苗期是感染该病毒的危险生育期（曹杨等，2011；Li et al.，2012）。人们利用 EPG 技术已探明，介体昆虫获得持久性病毒与吸食韧皮部汁液有关，而接毒与分泌水溶性唾液有关（Prado and Tjallingii，1994；Jiang et al.，2000）。因此，抗虫品种可能通过干扰介体昆虫这些与传毒相关的取食动作而减缓病毒病的流行（Jones，1987）。目前尚不清楚哪些抗虫品种可干扰与白背飞虱传播 SRBSDV 相关的吸食韧皮部汁液和分泌水溶性唾液的取食动作。

因此，本研究试图利用 EPG 技术探明白背飞虱在 6 个不同水稻品种上的取食行为差异，明确不同水稻品种对该虫的抗性水平及抗性部位，不仅为合理利用抗性品种减轻白背飞虱的直接为害提供参考依据，而且为进一步探讨利用抗虫品种延缓 SRBSDV 在稻苗间的扩散、蔓延提供理论依据。

（二）材料与方法

1. 稻苗和白背飞虱的准备　供试水稻品种名称为丰源优272、R9810-T、华恢1号、明恢63、麻糯谷和Rathu Heenati。水稻种子经保湿催芽后转移到带土的塑料盆（直径5cm）中单苗种植。盆栽稻苗在无虫的温室条件下培养，挑选发育基本一致的三叶期稻苗用于实验。白背飞虱在水稻品种丰源优272上连续饲养多代，饲养条件为人工气候箱，温度26℃±1℃，相对湿度85%±5%，光周期L/D=14h/10h。挑选生长发育基本一致的长翅型雌成虫用于实验。

2. EPG记录　在法拉第笼中用Giga 8 DC-EPG记录白背飞虱的取食行为，其输入电阻为10^9 Ω，输入偏压电流小于1 pA。昆虫在使用前饥饿、饲水处理2h，不采用任何麻醉措施，用水溶性银胶将金丝（直径20μm，长3cm）的一端直接连接到白背飞虱的前胸背板上，另一端与放大器相连。将连接好金丝的试虫轻轻放到水稻叶鞘上。将一根铜线（直径2mm，长10cm）插入土壤中作为植物电极。放大器的放大倍数为50×，调试植物电压使输出电压在-5~5V。每次同时记录6个水稻品种，6株稻苗在法拉第笼内随机排列。所有白背飞虱与稻苗只使用一次。每次实验在26℃±1℃和持续光照条件下记录8h。每个品种的有效重复数为15次。EPG数据记录和分析使用Stylet Version 1.2软件（Wageningen University，the Netherlands）。

3. 数据统计分析　EPG数据采用DPS软件进行统计分析，对6个水稻品种的EPG指标进行方差分析（one-way ANOVA）和多重比较（Duncan氏新复极差法，$P=0.05$）。

（三）结果

1. 白背飞虱的EPG波形　白背飞虱的EPG波形与He等（2011b）和Seo等（2009）报道的波形基本一致，分为以下几种波形（图21.1）：np代表非刺探，N1代表刺探的起始，N2代表分泌唾液和口针移动，N3代表口针在韧皮部附近的细胞外活动，N4-a代表分泌水溶性唾液，N4-b代表口针在韧皮部内吸食汁液，N4代表口针在韧皮部组织内活动（包括N4-a和N4-b），N5代表口针在木质部活动。本研究中将口针从叶鞘表面至韧皮部途中的N1、N2和N3波合并为Nc波（路径波）。EPG指标的选择参考相关文献（闫凤鸣，2000；Jiang et al.，2001；胡想顺等，2008）。

2. 白背飞虱口针到达韧皮部取食之前的刺探行为

本研究测定了白背飞虱在水稻上取食行为的31个EPG指标（表21.1），其中，前面9个指标反映的是口针到达水稻韧皮部前的刺探行为，后面22个指标反映了口针在水稻韧皮部内的刺探行为。从非刺探波np指标来看，该虫在品种Rathu Heenati上单次np的平均持续时间显著地长于其他品种。该虫在Rathu Heenati上非刺探波np的总时间为5 282.75s，显著长于在麻糯谷上的np总时间3 206.03s，而其他品种的np总时间都显著地比两者短（$P<0.05$）。可见，白背飞虱在抗虫品种Rathu Heenati上非刺探波np的总时间最长，其次为麻糯谷，然后为其余4个品种。相应地，该虫在各品种上的总刺探时间则表现为相反的趋势。

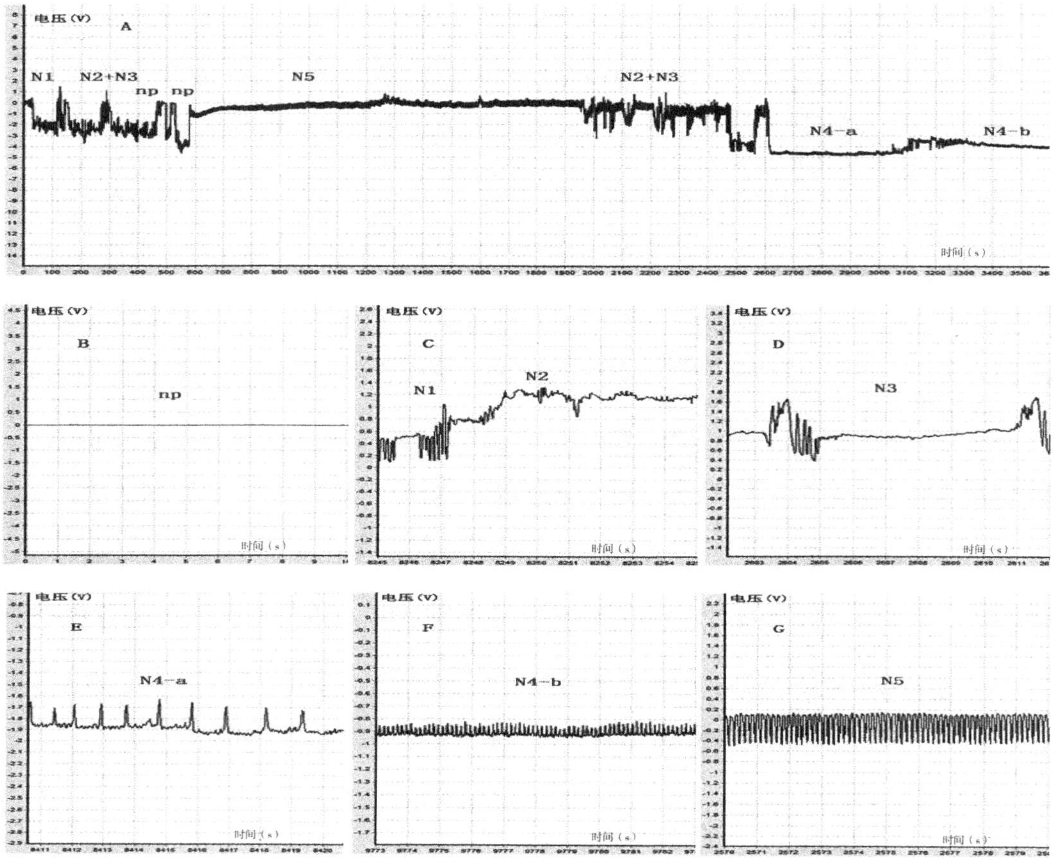

图 21.1　白背飞虱长翅型雌成虫在水稻苗上的典型 EPG 波形

A：记录 1h 的全图；B~G：白背飞虱在水稻上的 7 种典型 EPG 波形，显示 10s 内的细节；np：非刺探；N1：刺探起始；N2：口针移动；N3：韧皮部附近的细胞外活动；N4-a：在韧皮部分泌水溶性唾液；N4-b：吸食韧皮部汁液；N5：在木质部内活动。

表 21.1　白背飞虱在 6 个水稻品种上的 EPG 指标（总记录时间为 8 h）

指标	丰源优 272	R9810-T	华恢 1 号	明恢 63	麻糯谷	Rathu Heenati
非韧皮部指标						
单次 np 的平均持续时间（s）	62.87±9.97 b	72.83±13.96 b	83.74±20.01 b	101.13±18.79 b	183.23±35.49 b	378.90±93.21 a
np 的总时间（s）	1 101.71±255.46 c	1 501.66±564.30 c	1 291.65±497.01 c	974.86 ±297.44 c	3 206.03 ±651.00 b	5 282.75 ±895.99 a
第 1 次刺探的持续时间（s）	2 862.53±1 861.45 a	4 120.64±2 159.80 a	5 907.46±2 519.36 a	7 880.02±2 765.35 a	639.00±227.36 a	2 827.56±1 120.75 a
单次刺探的平均持续时间（s）	4 741.55±1 894.76 a	5 192.27±1 861.31 a	7 806.30±2 379.50 a	7 438.62±2 375.09 a	2 548.77±607.85 a	1 482.55±172.45 a
总刺探时间（s）	27 697.12 ±255.19 a	27 298.34±564.30 a	27 501.35±498.04 a	27 825.15 ±297.43 a	25 592.79 ±650.76 b	23 513.79 ±895.40 c
单次 Nc 的平均持续时间（s）	280.07±40.93 ab	221.13±25.37 ab	349.25±60.25 a	246.12 ±22.79 ab	176.26±19.33 b	279.17±34.42 ab
Nc 的总时间（s）	6 216.33 ±932.02 ab	4 361.43 ±519.65 ab	5 685.52 ±1，200.03 ab	3 807.99 ±543.95 b	4 101.49 ±523.10 ab	8 523.41 ±1 329.01 a
单次 N5 的平均持续时间（s）	429.44 ±137.04 a	612.29±113.87 a	639.37±131.31 a	1，010.67 ±406.85 a	584.54 ±131.64 a	439.71±69.37 a
N5 的总时间（s）	2 325.15 ±348.55 a	2 166.41 ±432.69 a	2 362.70±489.02 a	3 062.62 ±794.33 a	2 233.13 ±345.20 a	3 645.97 ±597.64 a
韧皮部指标						
单次单个 N4-a 的平均持续时间(s)	193.43±73.90 a	93.69±30.65 a	55.19±26.61 a	141.63±33.35 a	85.2±37.57 a	104.84±22.81 a
单个 N4-a 的总时间（s）	626.67±191.58 a	653.73±237.52 a	238.23±135.03 a	438.76±117.96 a	255.53±93.59 a	787.29±198.88 a
单次跟随 N4-b 的 N4-a 的平均持续时间（s）	1 887.69 ±358.73 a	1 153.43 ±166.51 b	674.19 ±111.20 b	755.23 ±125.47 b	678.91 ±91.97 b	879.47 ±174.40 b
跟随 N4-b 的 N4-a 的总时间（s）	2 379.09±474.41 a	2 311.33±409.66 a	1 254.29±247.65 a	2 002.34±329.51 a	1 745.61±197.14 a	1 629.73±311.21 a
单次 N4-a 的平均持续时间（s）	904.71±154.82 a	526.06±81.32 b	452.20±95.84 b	421.35±51.94 b	458.91±93.46 b	498.39±145.32 b
N4-a 的总时间（s）	3 005.76±562.58 a	2 965.05±503.93 a	1 492.49±267.89 a	2 441.10±378.07 a	2 001.13±254.22 a	2 416.99±309.76 a
到达第 1 次 N4-a 的时间（s）	4 751.85 ±794.84 a	3 423.54 ±400.96 a	3450.16 ±581.75 a	3 334.08 ±609.58 a	4 494.04 ±512.69 a	5 210.91 ±1 182.46 a
从第 1 次刺探到达第 1 次 N4-a 的时间（s）	4 677.18 ±805.38 a	3 386.75 ±397.02 a	3 424.82 ±581.54 a	3 303.35 ±607.83 a	4 452.23 ±513.31 a	5 076.35 ±1 165.18 a

续表

指标	丰源优 272	R9810-T	华恢 1 号	明恢 63	麻糯谷	Rathu Heenati
单次刺探到达第 1 次 N4-a 的平均时间（s）	450. 14 ±94. 85 a	440. 41±61. 63 a	400. 86±54. 92 a	316. 32±42. 37 a	363. 56±59. 22 a	394. 33±49. 77 a
单次刺探到达第 1 次 N4-a 的最短时间（s）	322. 74 ±96. 32 a	269. 69±50. 00 a	232. 29±52. 05 a	141. 30±24. 41 a	209. 55±37. 54 a	218. 61±53. 25 a
到第 1 次持久 N4-b 以前 N4-a 的总持续时间（s）	2 549. 35 ±456. 33 a	1 792. 03±385. 96 ab	948. 09±190. 90 b	1 112. 57 ±226. 92 b	1 079. 95 ±127. 67 b	1 724. 48 ±249. 28 ab
单次 N4-b 的平均持续时间（s）	15 100. 75±1 896. 15 a	11 564. 21±1 976. 28 ab	13 146. 71±2 480. 08ab	8 874. 54±1 693. 69 bc	8 356. 22±1 605. 50bc	5 295. 21±1 260. 51c
N4-b 的总时间（s）	16 149. 91±1 609. 76 a	17 805. 44±1 362. 14 a	17 960. 67 ±1 926. 69 a	18 513. 44 ±1 240. 94 a	17 257. 05 ±1 306. 83 a	8 927. 42 ±1 596. 28 b
单次持久 N4-b（>10 min）的平均持续时间（s）	15 477. 04±1 718. 44 a	12 937. 25±1 814. 98 a	13 942. 14±2 372. 35 a	11 343. 08±1 940. 00 a	10 977. 97±1 749. 92 a	5 851. 22 ±1 228. 45 b
持久 N4-b 的总时间（s）	16 059. 19 ±1 652. 57 a	17 758. 75 ±1 360. 52 a	17 899. 07 ±1 939. 69 a	18 359. 13 ±1 236. 38 a	17 089. 81 ±1 326. 95 a	8 817. 53 ±1 589. 16 b
到达第 1 次 N4-b 的时间（s）	8 288. 57±1 180. 90 a	5 306. 51±953. 77 bc	5 257. 33 ±999. 22 bc	3 702. 90 ±620. 92 c	4 637. 39 ±504. 32 c	7 742. 87 ±1 388. 85 ab
从第 1 次刺探到第 1 次持久 N4-b 的时间（s）	11 461. 99 ±1 505. 57 a	7 045. 96 ±980. 62 b	6 201. 03 ±1 049. 41 b	5 128. 04 ±673. 23 b	6 245. 24 ±597. 62 b	10 374. 82 ±1 468. 68 a
第 1 次 N4-a 到第 1 次 N4-b 的时间（s）	10 243. 27±1 329. 12 a	6 526. 72 ±1 015. 15 bc	5 981. 99±992. 81 bc	4 437. 29 ±743. 04 c	5 347. 61 ±509. 61 c	8 647. 61 ±1 398. 51 ab
从第 1 次 N4-a 到第 1 次持久 N4-b 的时间（s）	6 784. 82 ±1 533. 86 a	3 659. 20±820. 56 bc	2 776. 21±859. 88 bc	1 824. 69 ±432. 13 c	1 793. 01 ±403. 76 c	5 072. 09 ±942. 79 ab
单次 N4 的平均持续时间（s）	17 020. 22±1 870. 18 a	12 849. 81±1 951. 45 ab	13 832. 17±2 496. 22 ab	9 697. 46±1 716. 40 bc	9 565. 74±1 708. 98 bc	6 266. 95±1 323. 71 c
N4 的总时间（s）	18 528. 97±1 418. 63 a	20 116. 76±1 356. 23 a	19 214. 94±1 899. 06 a	20 515. 78±1 040. 70 a	19 002. 65±1 300. 36 a	10 557. 15±1 715. 66 b
从第 1 次刺探到第 1 次 N4 的时间（s）	10 317. 98 ±1 331. 47 a	6 563. 49 ±1 013. 93 bc	6 007. 33 ±995. 03 bc	4 468. 02 ±743. 88 c	5 389. 41 ±509. 72 c	8 782. 16 ±1 411. 23 ab

注：表中数据为均值±标准误，同行数据后不同字母表示不同品种间在 0. 05 水平上差异显著（Duncan 氏新复极差法）。每品种的有效重复数为 15 次。

从路径波 Nc 指标来看，该虫在 Rathu Heenati 上 Nc 总时间显著地长于在明恢 63 上的 Nc 总时间，而在其他品种的 Nc 总时间与两者相比都没有显著差异。

从口针在木质部活动指标 N5 来看，该虫在 6 个品种上无论是单次 N5 的平均持续时间还是 N5 的总时间均无显著性差异（$P \geq 0.05$）。

综上所述，白背飞虱在 6 个水稻品种上吸食木质部水分的时间（N5）没有显著差异，在品种 Rathu Heenati 上非刺探波 np 时间最长，其次为麻糯谷，然后是其他 4 个品种。白背飞虱在 Rathu Heenati 上路径波 Nc 的总时间为 8 523.41 s，是明恢 63 Nc 的总时间的 2.24 倍，这表明白背飞虱的口针在 Rathu Heenati 的韧皮部外组织停留的时间长，可能在不断地寻找取食位点。

3. 白背飞虱口针在韧皮部的刺吸行为　白背飞虱口针在韧皮部的刺吸行为（表 21.1）表明：在品种丰源优 272 上，单次跟随 N4-b 的 N4-a 的平均持续时间和单次 N4-a 的平均持续时间分别为 1 887.69s 和 904.71s，均显著地长于其他品种；到第 1 次持久 N4-b 以前 N4-a 的总持续时间在丰源优 272 上显著地长于华恢 1 号、明恢 63 和麻糯谷（$P<0.05$）；各品种间的 N4-a 的总时间无显著性差异（$P \geq 0.05$），其余与分泌水溶性唾液 N4-a 有关的 EPG 指标也呈现相同的情形。这说明在所有供试品种中该虫在丰源优 272 韧皮部内分泌水溶性唾液次数少但分泌时间长。

在品种 Rathu Heenati 上，N4-b 的总时间，N4 的总时间，单次持久 N4-b（>10min）的平均持续时间，持久 N4-b 的总时间都显著地短于其他品种（$P<0.05$），说明白背飞虱在此品种上更难持续地吸食韧皮部汁液。

单次 N4-b 的平均持续时间和单次 N4 的平均时间在丰源优 272 上最长，显著地长于明恢 63、麻糯谷和 Rathu Heenati（$P<0.05$）。可能受单次 N4-a 的平均持续时间长的影响，白背飞虱在丰源优 272 上到达第 1 次 N4-b 的时间、从第 1 次刺探到第 1 次持久 N4-b 的时间、第 1 次 N4-a 到第 1 次 N4-b 的时间、从第 1 次 N4-a 到第 1 次持久 N4-b 的时间、从第 1 次刺探到第 1 次 N4 的时间这些首次韧皮部取食的指标都为最长，显著地长于除 Rathu Heenati 外的其余 4 个品种，而且其他 4 个品种间没有显著差异（$P \geq 0.05$）。这说明白背飞虱更易持续地吸取丰源优 272 的韧皮部汁液。

（四）讨论

1. 白背飞虱的 EPG 波形特点　He 等（2011b）利用 EPG 技术记录了 3 种稻飞虱即褐飞虱、白背飞虱和灰飞虱的取食行为，结果表明，这 3 种稻飞虱共用相同的 EPG 波形，并对应相同的生物学意义。本研究中白背飞虱在水稻上的 EPG 波形与 Seo 等（2009）、Ghaffar 等（2011）和 He 等（2011a，2011b）报道的波形在形状、振幅和频率等方面都表现出相同的特征（图 21.1）。然而，Ghaffar 等（2011）选用短翅型雌成虫用于实验，观察到另外两种新波形，即 N6（代表口针遇到机械阻力，与蚜虫的 F 波类似）和 N7（代表细胞内刺探，与蚜虫的 pd 波类似）。本研究选用的是长翅型雌成虫，N6 和 N7 波极为罕见，这可能与供试昆虫的翅型有关，这与 Seo 等（2009）和 He 等（2011a）的报道类似。祝增荣等（2001）指出，白背飞虱长翅型成虫和短翅型成虫在若虫历期、产卵前期、繁殖动态、寿命及种群增长参数等方面都存在差异。

2. 基于 EPG 对水稻抗性的评估 如果寄主植物散发趋避剂，造成昆虫的不选择性，则刺探之前的非刺探（np）时间延长；若抗性因子位于表皮层，如口针不易刺入或有拒食剂的存在，则在刺探开始时有较高频率的中断；若抗性因子位于叶肉组织，则路径波波程发生变化；若韧皮部存在抗性物质，则与韧皮部取食有关的波形波程发生变化（胡想顺等，2008）。

本研究结果显示，白背飞虱在品种 Rathu Heenati 上非刺探波 np 和路径波 Nc 的总时间都为最长；相反，其在韧皮部取食波 N4－b 的总时间为最短。这表明，Rathu Heenati 可能存在忌避成分，并且在韧皮部外组织和韧皮部组织可能都存在抗性因子。

在品种麻糯谷上，非刺探波 np 的总时间也相对较长，显著地长于除 Rathu Heenati 外的其他品种。这表明，在麻糯谷上可能也存在不利于该虫刺探的忌避成分，这可能是麻糯谷在田间抗性鉴定中表现为高抗的重要因素之一（张克云，1991）。然而，飞虱的口针一旦刺入麻糯谷的叶鞘表皮以后，则与在感虫品种明恢 63 上的取食行为基本一致。因此，推测麻糯谷可能并不存在其他抗性因子。

明恢 63 既是感虫的杂交稻恢复系，也是转 Bt 水稻华恢 1 号的亲本（沈君辉等，2003b）。本研究中明恢 63 与华恢 1 号之间所有的 EPG 指标都没有显著性差异（$P \geq 0.05$），说明华恢 1 号中的 Bt 毒素对非靶标害虫白背飞虱的取食行为并没有产生显著的影响。R9810-T 是将茶陵野生稻的基因组 DNA 导入栽培稻 R9810 后得到的表型稳定的后代（蒋斌元等，2012）。R9810-T、华恢 1 号和明恢 63 三者间所有 EPG 指标都没有显著性的差异，由于明恢 63 是感虫的杂交稻恢复系（沈君辉等，2003b），所以推测导入野生稻基因组后的 R9810-T 也属感虫品种。

丰源优 272 是湖南地区广泛栽培的杂交稻品种，白背飞虱在此品种上单次分泌水溶性唾液的平均时间显著地长于其他品种，但分泌唾液的总时间与其他品种没有显著差异，单次韧皮部取食的时间在所有品种中最长。这表明，白背飞虱在丰源优 272 上分泌水溶性唾液的特点是少次而多量，并伴随着长时间的韧皮部取食，这可能有利于提高白背飞虱在该品种上的取食效率，该品种可能是比明恢 63 更为感虫的品种。

3. 抗性程度不同的品种对白背飞虱获毒和接毒能力的影响 SRBSDV 是一种水稻韧皮部限制性病毒（Zhou et al.，2008）。接毒时 SRBSDV 将随着染毒白背飞虱的水溶性唾液进入无毒水稻的韧皮部，无毒白背飞虱获毒时 SRBSDV 将随着染毒水稻韧皮部汁液进入该虫的消化道内。因此，白背飞虱在水稻韧皮部中能否顺利分泌水溶性唾液和吸食汁液将直接影响接毒成功率或获毒成功率。

本研究中白背飞虱在抗虫品种 Rathu Heenati 韧皮部汁液吸食的时间仅为感虫品种的一半左右，这意味着白背飞虱从该品种韧皮部吸取的汁液量较少，暗示白背飞虱取食带毒 Rathu Heenati 时成功获毒的概率变小。另外，白背飞虱在田间不必受实验中所用金丝的束缚，可能在刺探以前或口针到达韧皮部以前就已放弃取食抗虫品种，从而丧失成功获毒的机会。白背飞虱在感虫品种丰源优 272 上单次分泌水溶性唾液的时间长，可能会增加毒虫在此品种上成功接毒的概率。

从白背飞虱在水稻上的取食行为来看，田间种植抗虫品种肯定能降低白背飞虱从染毒稻苗上获毒的机会，而种植感虫品种将增加毒虫在健康稻苗上接毒的成功率。

参考文献

[1] GAMALATH N S, TUFAIL M, SHARMA P N, et al. Differential expression of vitellogenin mRNA and protein in response to rice resistance genes in two strains of *Nilaparvata lugens* (Hemiptera: Delphacidae) with different levels of virulence. *Applied Entomology Zoology*, 2012, 47: 9-16.

[2] GHAFFAR MBAB, PRITCHARD J, FORD-LLOYD B. Brown planthopper (N. lugens Stål) feeding behaviour on rice germplasm as an indicator of resistance. *PLoS ONE*, 2011, 6 (7): 1-13.

[3] HE Y P, CHEN L, CHEN J M, et al. Electrical penetration graph evidence that pymetrozine toxicity to the rice brown planthopper is by inhibition of phloem feeding. *Pest Management Science*, 2011a, 67: 483-491.

[4] HE Y P, ZHANG J F, CHEN J M, et al. Influence of pymetrozine on feeding behaviors of three rice planthoppers and a rice leafhopper using electrical penetration graphs. *Journal of Economic Entomology*, 2011b, 104: 1 877-1 884.

[5] JIANG Y X, DE BLAS C, BARRIOS L, et al. Correlation between whitefly (Homoptera: Aleyrodidae) feeding behavior and transmission of tomato yellow leaf curl virus. *Annals of the Entomological Society of America*, 2000, 93 (3): 573-579.

[6] JIANG Y X, NOMBELA G, MUŃIZ M. Analysis by DC-EPG of the resistance to *Bemisia tabaci* on a Mi-tomato line. *Entomologia Experimentalis et Applicata*, 2001, 99: 295-302.

[7] JONES A T. Control of virus infection in crop plants through vector resistance: a review of achievements, prospects and problems. *Annals of Applied Biology*, 1987, 111: 745-772.

[8] KHAN Z R, SAXENA R C. Electronically recorded waveforms associated with the feeding behavior of *Sogatella furcifera* (Homoptera: Delphacidae) on susceptible and resistant rice varieties. *Journal of Economic Entomology*, 1984, 77: 1 479-1 483.

[9] LI Y Z, CAO Y, ZHOU Q, et al. The efficiency of Southern rice black-streaked dwarf virus transmission by the vector *Sogatella furcifera* to different host plant species. *Journal of Integrated Agriculture*, 2012, 11: 621-627.

[10] PRADO E, TJALLINGII WF. Aphid activities during sieve element punctures. *Entomologia Experimentalis et Applicata*, 1994, 72: 157-165.

[11] SEO B Y, KWON Y H, JUNG J K, et al. Electrical penetration graphic waveforms in relation to the actual positions of the stylet tips of *Nilaparvata lugens* in rice tissue. *Journal of Asia-Pacific Entomology*, 2009, 12: 89-95.

[12] ZHOU G H, WEN J J, CAI D J, et al. Southern rice black-streaked dwarf virus: a new proposed Fijivirus species in the family Reoviridae. *Chinese Science Bulletin*, 2008, 53: 3 677-3 685.

[13] 曹杨，潘峰，周倩，等. 南方水稻黑条矮缩病毒介体昆虫白背飞虱的传毒特性. 应用昆虫学报，2011，48 (5): 1 314-1 320.

[14] 胡想顺，赵惠燕，胡祖庆，等. 麦长管蚜在3个小麦品种上取食行为的EPG比较. 中国农业科学，2008，41 (7): 1 989-1 994.

[15] 蒋斌元，张齐，康敏，等. 茶陵野生稻及其导入后代间的遗传差异RAPD分析. 湖南农业科学，2012，(13): 5-7.

[16] 沈君辉，尚金梅，刘光杰．中国的白背飞虱研究概况．中国水稻科学，2003a，17（增刊）：7-22.
[17] 沈君辉，刘光杰，陈爱辉，等．电子记录白背飞虱的取食和产卵行为．中国水稻科学，2003b，17（1）：73-76.
[18] 王强，周国辉，张曙光．南方水稻黑条矮缩病毒一步双重 RT-PCR 检测技术及其应用．植物病理学报，2012，42（1）：84-87.
[19] 闫凤鸣．芥子油苷在甘蓝蚜寄主部位选择行为中的作用．昆虫学报，2000，43（3）：297-304.
[20] 张克云．稻种资源抗病虫鉴定总结．湖南农业科学，1991（5）：33-35.
[21] 祝增荣，程家安，刘永军．白背飞虱长、短翅型成虫实验种群生物学比较．中国水稻科学，2001，15（3）：229-232.

（孙　凯）

第二十二章 柑橘木虱传播黄龙病的机制

（一）研究背景

柑橘黄龙病是目前世界柑橘生产上最严重的传染性病害，最早报道的病例发生在我国广东的潮汕地区（Reinking，1919），而疑似黄龙病症状早在18世纪的印度就有记载，当时认为是“梢枯病”（dieback）（Asana，1958）。华南农业大学林孔湘教授首次通过实验证明该病可以通过嫁接传播，属侵染性病害（林孔湘，1956），因此，1995年第13届国际柑橘病毒病学家会议决定以黄龙病（Huanglongbing，HLB）作为该病的正式名称（Bové，2006）。目前认为病原是限于韧皮部内寄生的革兰氏阴性细菌，属于变形菌门（Proteobacteria）α-变形菌纲（Alphaproteobacterial）根瘤菌目（Rhizobiales）根瘤菌科（Rhizobiaceae），候选的韧皮部杆菌属（*Candidatus* Liberibacter）（Jagoueix et al.，1996）。根据病原16S rDNA和β-操纵子基因的序列特征、传病媒介和病原性质，黄龙病菌分为亚洲种“*Ca.* L. *asiaticus*”、美洲种“*Ca.* L. *americus*”和非洲种“*Ca.* L. *africanus*”（Jagoueix et al.，1996；Teixeira et al.，2005）三种。其中亚洲种属于耐热型，分布最广，为害最大，在我国发生的黄龙病即为该种。

柑橘木虱 *Diaphorina citri* Kuwayama 属半翅目（Hemiptera），木虱科（Psyllidae），为黄龙病亚洲种和美洲种的媒介昆虫，为区别于传播黄龙病非洲种的柑橘木虱 *Trioza erytreae* Del Guercio，目前常称为亚洲柑橘木虱。尽管近年来已证实在我国云南发生的柚喀木虱 *Cacopsylla citrisuga* Yang & Li 也传播黄龙病亚洲种（Cen et al.，2012），柑橘木虱仍然为黄龙病最主要的媒介昆虫，在绝大多数黄龙病发生区域是唯一的媒介昆虫。许长藩等（1988）用电镜观察发现木虱体内唾液腺周围的脂肪体、中肠和后肠的细胞内有黄龙病病原，尤以唾液腺外皮层和中肠细胞病原密度最大，所以推测病原有可能在这些地方增殖。用分子生物学技术研究发现，木虱体内病菌浓度会随时间而增加，说明病菌在木虱体内可以不断增殖，而且木虱一旦获得病原体即能终身带菌，所以其传播黄龙病属于持久增殖型（Hung et al.，2004）。

刺吸电位（EPG）技术常用于媒介昆虫传播植物病毒病机制的研究（闫凤鸣，2003），但利用该技术对柑橘木虱传播黄龙病机制已开展的相关研究不多，现已初步确定了取食波形与其行为之间的关系（Bonani et al.，2010；杨成良等，2011）。在此基础上，我们应用EPG技术对柑橘木虱取食行为与其获菌和传菌的关系、柑橘木虱与黄龙病的互作关系开展了研究。

（二）材料与方法

1. 供试植物

（1）砂糖橘黄龙病苗：华南农业大学柑橘黄龙病研究室繁育，从感染黄龙病的砂糖橘园采集病芽嫁接到酸橘 *Citrus reticulata* Blanco‘Sunki’上，并通过 RT-qPCR 检测呈黄龙病阳性。

（2）酸橘黄龙病苗：华南农业大学柑橘黄龙病研究室繁育，从砂糖橘黄龙病树上采集病芽嫁接到健康酸橘上使其感染黄龙病，待酸橘显症后摘除病芽，并通过 RT-qPCR 检测呈黄龙病阳性。

（3）健康酸橘苗：为广东省博罗县柑橘研究所培育的脱毒苗，苗高 30cm 左右，单株移栽到塑料容器后放置在没有木虱和病苗的温室大棚中培育，在 EPG 实验前均通过 RT-qPCR 检测呈黄龙病阴性。移栽 1 个月后，长出嫩叶时用于获菌和进行 EPG 测定。

2. 供试虫源

（1）不带黄龙病菌的柑橘木虱：采自华南农业大学九里香 *Murraya exotica* L. 植株上，转移到网室内的盆栽健康酸橘苗上繁育多代，选择成虫进行实验。

（2）带菌的柑橘木虱：采自广东省惠州市龙门县一个全部显示黄龙病病症的失管砂糖橘果园，转移到华南农业大学网室内培育的黄龙病酸橘苗上繁育多代，选择成虫进行实验。

3. 柑橘木虱刺吸电位的测定方法

（1）前期准备：将待测定的木虱装在 2mL 离心管中饥饿处理 2h；将铜丝焊接在铜钉上，另外剪取 1~2cm 的金丝，用银胶（使用前用涡旋混合器将管内银胶混匀）缠粘在铜丝的另一端，并拉直金丝以免影响木虱活动。

（2）粘连木虱：使用冰冻法（汤清波等，2011）将木虱冻晕，同时将金丝另一端粘连银胶成小珠状，在木虱即将恢复活动时，快速将金丝粘到其背部中央。粘连好后，静置 1~2min 使金丝与木虱完全粘连。粘连过程中避免金丝粘在木虱其他部位，以免木虱不能正常活动和取食。

（3）调整植物：由于木虱偏好在柑橘叶片背中脉取食，故将叶片翻转，并用 Parafilm 膜固定叶片，防止实验过程中木虱从叶片上掉落；将植物电极插在土壤中，整个装置置于法拉第笼中以屏蔽外界电磁波的干扰。

（4）放置木虱：将粘连着木虱的铜钉插入设备孔中，并通过微量调节器调整木虱位置使其尽量自然地与叶背中脉接触；启动电脑记录软件。

实验使用的 EPG 信号放大器是荷兰瓦赫宁根大学昆虫实验室研制的 Giga-4 型直流放大器。EPG 波形的电信号通过转换器（DI-720，Dataq Instruments）转换为数字信号，分析软件为 Stylet v01. 20-111011。

4. 柑橘木虱取食行为与其获菌关系的研究 将健康成虫与 EPG 装置连接在感染黄龙病的砂糖橘不同成熟度部位（嫩芽、嫩叶、成熟叶）上进行获菌和取食行为记录，每头测定时间 8h，排除由于接线不好导致波形不清晰的木虱个体，每个部位各获得 20 个有效重复，雌雄虫各占一半。测定时室内温度约为 25℃，每头成虫只用于 1 个记录，每株植物分别用于三个部位的记录各 1 次。测定后将每头木虱编号，单独放置在装有 75%酒精的 0.5mL 离心管中保存，用 TIANGEN 血液组织 DNA 提取试剂盒提取 DNA 用于 RT-qPCR 检测，分析 EPG 参数与木虱成功获菌的关系。使用的 PCR 仪为 Bioscience qPCR MasterMix（SYBR Green），引物由英竣生物工程公司合成。

5. 柑橘木虱取食行为与其传病关系的研究　将可能携带黄龙病菌的柑橘木虱成虫与健康酸橘苗通过 EPG 装置连接，记录时间为 24h，每头木虱都在同一实验条件下进行，光照周期为 L/D=14h/10h，荧光照射，室内温度约为 25℃，每头成虫和 1 株植物只用于 1 次记录。传病和 EPG 测定结束后，将每头木虱单独放置在装有 75%酒精的 0.5 mL 离心管中保存，提取 DNA 并进行 RT-qPCR 检测。实验共使用雌雄成虫各 40 头，排除由于接线不好导致波形不清晰以及实验后传菌木虱检测为不带菌的植株，将饲菌后的苗木放到没有病树和木虱存在的温室中继续培育。6 个月后用 OMEGA 植物 DNA 提取试剂盒提取单株苗木的 DNA 并进行 RT-qPCR 检测。

6. 柑橘木虱感染黄龙病菌对其取食行为的影响　将携带和不携带黄龙病菌的柑橘木虱成虫（分雌雄虫）与健康酸橘苗通过 EPG 装置连接，实验条件同上，每头成虫和 1 株植物只用于 1 个记录。测试结束后将可能带菌的木虱编号保存在 75%酒精中，提取 DNA 并进行 RT-qPCR 检测，排除不带菌的木虱个体，与不带菌木虱个体的 EPG 参数进行比较。带菌和不带菌木虱各获得 20 个有效重复，雌雄各 10 头。

7. 寄主感染黄龙病对柑橘木虱取食行为的影响　不同发病程度酸橘植株的获取：从感染黄龙病的砂糖橘上采集病芽，嫁接到一年生大小一致的健康酸橘实生苗上，每株嫁接 3~4 个病芽。每个月嫁接一批苗，每批 30 株，共嫁接 3 个批次，另保留 20 株健苗作为对照。自嫁接第一批苗起 6 个月后对所有苗木进行黄龙病菌 RT-qPCR 检测，选择严重、中度、轻微发病的酸橘植株各 20 株作为本实验的材料，这三种发病程度的病原浓度分别为 397.54fg/μL ± 23.62fg/μL、46.73fg/μL ± 12.97fg/μL、78.73fg/μL ± 28.83fg/μL，差异显著，测定的木虱为健康的成虫。

（三）结果

1. 柑橘木虱取食柑橘的 EPG 波形

（1）柑橘木虱共产生 7 种不同的刺探取食波形以及非刺探波（图 22.1），与蚜虫取食波形比对分析可知，柑橘木虱的波形种类、形状特征与采用 DC-EPG 记录的蚜虫取食波形基本一致，本文采用相同的命名法对木虱取食波进行命名，即 A 波、B 波、C 波、E1 波、E2 波、G 波以及非刺探波 np 波，此外，还发现一种尚未报道过的波形 D 波，可能与韧皮部刺探活动有关（表 22.1）。

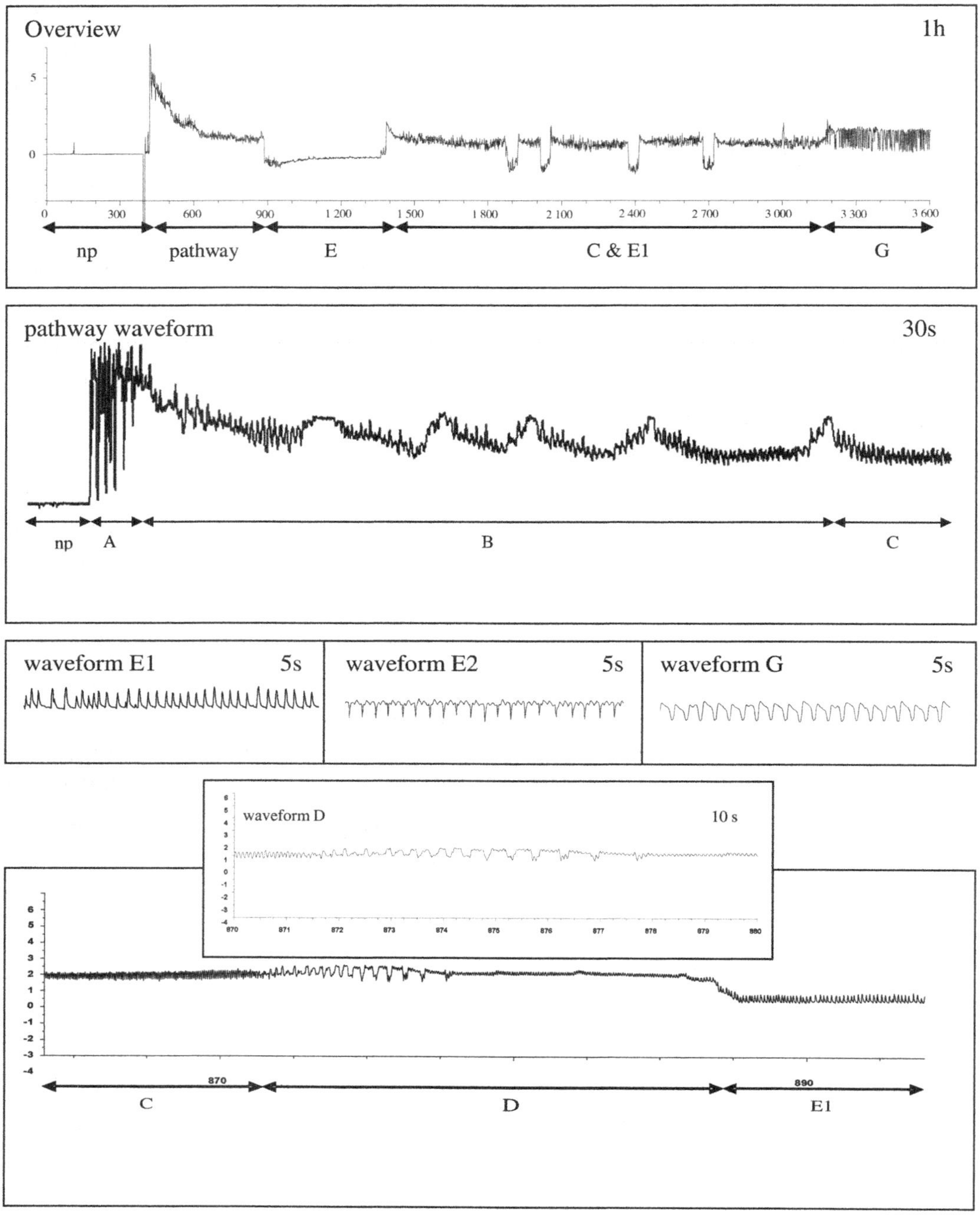

图 22.1　柑橘木虱取食 EGP 基本波形

注：1h EPG 记录的全貌图（上）；30s 路径波（中）；某些特征波谱细节图（下）。

表 22.1　柑橘木虱 EPG 波形特征及其与昆虫行为的关系

EPG 波形名称	波形特征			波形与昆虫行为的关系	
	频率（Hz）	电压水平	来源	口针所在部位	柑橘木虱行为
A	0.4	e	R	表皮	口针接触植物
B	0.5	e	R	表皮/叶肉	分泌凝胶性唾液
C	15～19	e	R	所有组织	路径行为
D	1～6	e	R	筛管	未知
E1	5～10	i	emf	筛管	分泌水溶性唾液
E2	4～9	i	emf/R	筛管	被动吸食
G	4～9	e	emf/R	木质部	主动吸食

注：e（extraceller level）：胞外水平；i（interceller level）：胞内水平；R：由电阻变化引起的电势波动；emf：由昆虫与寄主植物本身电生理变化引起的电势波动。

2. 柑橘木虱取食行为与其获菌的关系　EPG 记录的 60 头健康木虱在黄龙病苗上取食 8h 过程中共有 24 头出现韧皮部吸食波（E2 波），这 24 头中有 20 头还出现 G 波，在成熟叶、嫩叶、嫩芽上出现 E2 波的木虱数量分别为 11 头、7 头、6 头（表 22.2）。对所有出现 E2 波的木虱进行 RT-qPCR 检测结果显示，仅在成熟叶出现 1 头雄虫检测为黄龙病菌阴性，其余均为阳性，显示阳性（即成功获菌）个体的最短 E2 时间仅为 2min；同样使用 RT-qPCR 对未出现 E2 波的木虱进行检测，36 头木虱中有 34 头为黄龙病菌阴性，2 头为阳性，检测为阳性的 2 头均为雌虫，分别于嫩芽、嫩叶上取食，且波形中均出现韧皮部分泌唾液波（E1 波），但历时均较短。

木虱在病树成熟叶、嫩叶、嫩芽上取食 8h 后分别有 10 头、8 头、7 头显示黄龙病阳性（表 22.2），获菌率分别为 50%、40%、35%，说明在不同成熟度叶片上的获菌率有一定差异，可能随寄主的成熟度的增加而增加。

表 22.2　健康柑橘木虱取食行为与其获取黄龙病菌的关系

EPG 波形	仅出现此类波形的个体比例			木虱黄龙病菌检测结果	
	成熟叶	嫩叶	嫩芽	阳性（+）	阴性（-）
C		1/20 （5%）			1/1 （100%）
C、G	3/20 （15%）	7/20 （35%）	3/20 （15%）		13/13 （100%）
C、E1	1/20 （5%）		2/20 （10%）		3/3 （100%）
C、E1、G	5/20 （25%）	5/20 （25%）	9/20 （45%）	2/19 （10.5%）	17/19 （89.5%）

续表

EPG 波形	仅出现此类波形的个体比例			木虱黄龙病菌检测结果	
	成熟叶	嫩叶	嫩芽	阳性（+）	阴性（-）
C、E1、E2	3/20 (15%)		1/20 (5%)	4/4 (100%)	
C、E1、E2、G	(8/20) (40%)	7/20 (35%)	5/20 (25%)	19/20 (95%)	1/20 (5%)

注：EPG 测定和获菌时间共 8h；获菌试验后即对柑橘木虱进行 RT-qPCR 检测。

3. 柑橘木虱取食行为与其传病关系 在测试的 80 头柑橘木虱成虫中，排除 EPG 波形不清晰和传病试验后检测为不带菌的木虱重复数据后，一共得到 62 头成虫有效重复数据。这些成虫在 EPG 实验后通过 RT-qPCR 检测全部为黄龙病阳性，其中雄虫 28 头，雌虫 34 头。6 个月后，由这些木虱传病的部分健康柑橘苗叶片出现黄龙病特有的斑驳症状，并能通过 RT-qPCR 检测出带菌。17.7%（11 头）的成虫口针在韧皮部分泌了唾液但没有进行吸食（即仅有 E1 波而无 E2 波），61.3%（38 头）的木虱在韧皮部既分泌了唾液也进行被动吸食（即出现 E1 和 E2 波），88.7%（55 头）的木虱在木质部取食（出现 G 波），3.2%（2 头）的个体仅有刺探波（C 波）。6 个月后，检测到 14 株酸橘苗为黄龙病阳性，说明 62 头带菌木虱吸食 24h 后成功传病的共有 14 头，成功传病的个体达到 22.6%，其中有 1 头木虱没有韧皮部波形，4 头木虱仅在韧皮部分泌了唾液，另外 9 头木虱既分泌了唾液也有韧皮部被动取食（表 22.3）。

在这 14 头成功传病的木虱中，具有韧皮部波形的雌虫最短韧皮部分泌唾液可以传病的时间（即 E1 时间）为 5.1min，见图 22.2 中的 8 号；雄虫的最短韧皮部刺吸为 10.3min，见图 22.2 中的 2 号；雌虫和雄虫的最短传病时间都只有 E1 波出现，没有出现韧皮部被动吸食波 E2。另外，在测定的 24h 中，在植物上取食（C+G+E1+E2）总时间最短而可以传病的木虱个体实际取食时间为 88.8min，为图 22.2 中雄虫 2 号。

表 22.3 带菌柑橘木虱取食行为与其传播黄龙病菌的关系

EPG 波形	仅出现此类波形的个体比例	植物黄龙病菌检测结果	
		阳性（+）	阴性（-）
C	2/62 (3.2%)		2/2 (100%)
C、G	11/62 (17.7%)	1/11 (9.1%)	10/11 (90.9%)
C、E1	1/62 (1.6%)	1/1 (100%)	

续表

EPG 波形	仅出现此类波形的个体比例	植物黄龙病菌检测结果	
		阳性（+）	阴性（-）
C、E1、G	10/62 (16.1%)	3/10 (30%)	7/10 (70%)
C、E1、E2	4/62 (6.5%)	1/4 (25%)	3/4 (75%)
C、E1、E2、G	34/62 (54.8%)	8/34 (23.5%)	26/34 (76.5%)

注：EPG 测定和传毒时间共 24h；传毒 6 个月后对健康柑橘苗木进行黄龙病 RT-qPCR 检测。

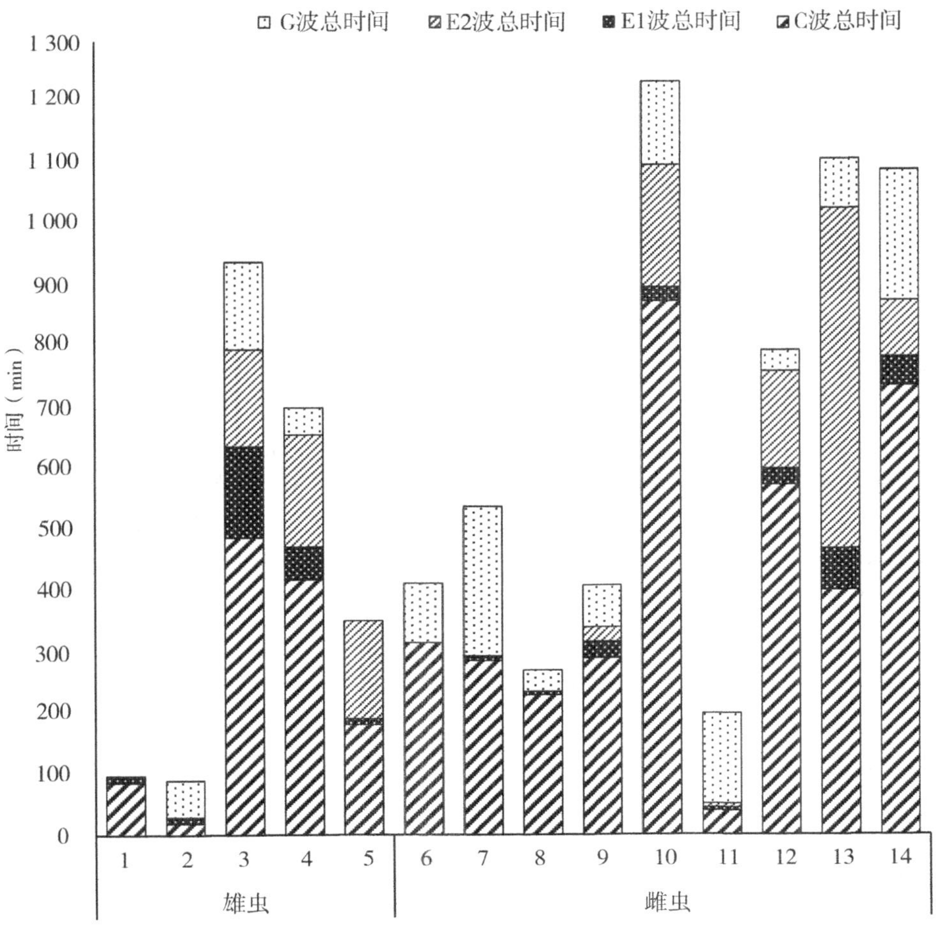

图 22.2　成功传病的 14 头木虱在 24h EPG 测定中各种取食波形的持续时间

4. 感染黄龙病菌的柑橘木虱对其取食行为的影响 在测定的20头带菌、20头不带菌木虱成虫中，带菌与不带菌成虫分别有2头和4头未进行韧皮部刺探与取食；有4头带菌成虫在植物韧皮部分泌了唾液，但未进行被动吸食（仅出现E1），14头带菌木虱成虫既有唾液分泌又有韧皮部被动吸食（出现E1+E2），而不带菌木虱15头具有韧皮部波形的都分泌了唾液并被动取食植物韧皮部汁液（出现E1+E2）（表22.4）。说明不带菌和带菌木虱成功取食的个体比例很接近。

表22.4 带菌与不带菌柑橘木虱成虫取食的EPG波形比较

EPG特征波形	仅有此波形的木虱比例			
	带菌		不带菌	
	雌虫	雄虫	雌虫	雄虫
C			1/10	
C，G	1/10	1/10	3/10	1/10
C，E1，G	2/10	2/10		
C，E1，E2		1/10	1/10	1/10
C，E1，E2，G	7/10	6/10	5/10	8/10

注：带菌和不带菌木虱分别测定雌虫、雄虫各10头；带菌木虱在EPG测定后均单头进行RT-qPCR检测确定带有黄龙病菌。

与不带菌木虱雄虫相比，带菌雄虫的第一次刺探、第一次E1、第一次E2出现的时间，以及刺探次数、E1和E2次数都无显著差异，E1波、E2波、C波、G波总时间、E1占韧皮部过程时间比例、口针在韧皮部的时间占总刺探过程比例等EPG参数也无显著差异，表明其取食并未受到影响（图22.3）。但雌虫分泌唾液时间显著推迟（图22.4），说明携带黄龙病菌对木虱雌成虫取食行为具有不利影响，可能导致其味觉敏感能力降低，在寻找合适取食位点过程中变得缓慢而迟钝。

5. 寄主感染黄龙病对柑橘木虱取食行为的影响 在8h记录过程中，与健康寄主植物相比，木虱口针在感染黄龙病叶片取食的np波时间无显著差异，但在下述几个方面有显著差异：①在重度发病的叶片上C波时间显著延长，表明口针在叶肉组织中穿刺寻找取食位点所需要的时间显著延长（图22.5）；②E波总时间显著减少，而G波时间显著增加，并且差异随发病程度的严重性而增大（图22.5）；③木虱口针首次到达韧皮部分泌唾液时间点和首次在韧皮部持续吸食时间点随发病程度的增加显著向后延迟（图22.6）；④在感病叶片韧皮部取食时，出现较多没有E2波跟随的E1波，唾液分泌次数和时间显著增加，而韧皮部持续吸食次数和时间显著减少，因此唾液分泌时间占总韧皮部取食时间的比例显著增加，而且差异随发病程度的严重性而增大（图22.7、图22.8）；⑤木虱在健树叶片上很少在木质部中取食，但在感病叶片上木质部中取食的个体比例和时间显著增加，发病严重时木质部成为木虱主要的取食部位（图22.9）。

（四）讨论

柑橘木虱在柑橘上取食共产生8种不同的波形，有非刺探波（np波）、路径波（A

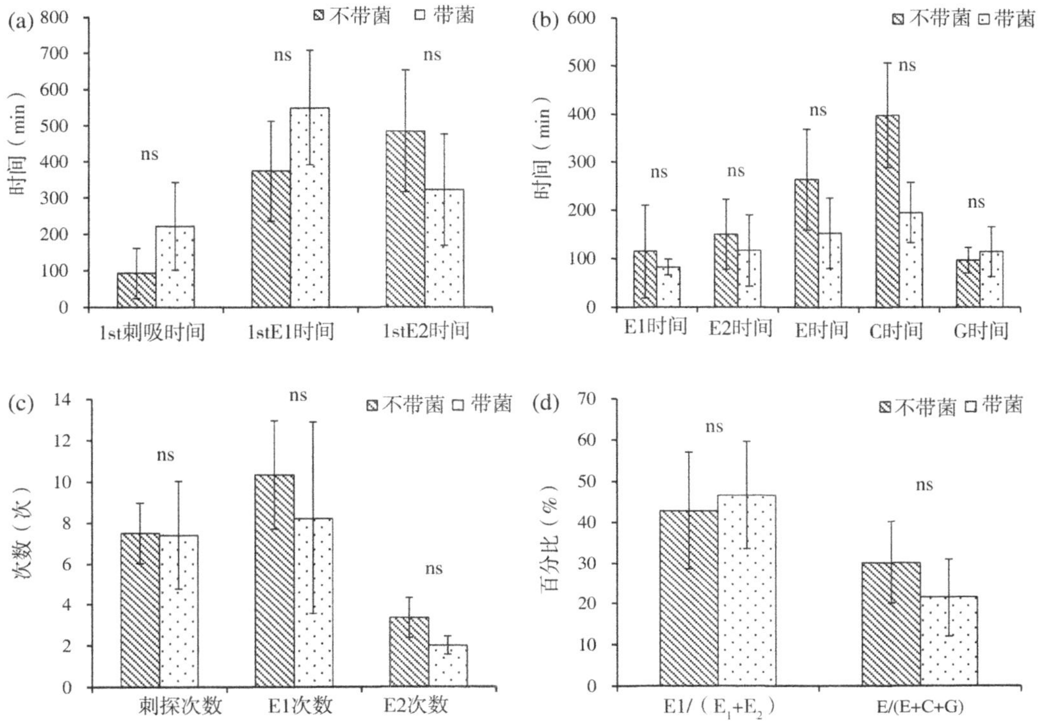

图 22.3　带菌与不带菌木虱雄虫 EPG 特征波形参数比较

注：图中相同类型数据均运用独立样本 t 检验进行比较，显著性水平分别用 $P>0.05$（ns），$P<0.05$（＊），$P<0.01$（＊＊），$P<0.001$（＊＊＊）表示。(a) 第一次刺吸、E1、E2 出现的时间；(b) E1 波、E2 波、E（E1+E2）波、C 波、G 波出现的总时间；(c) 刺探和 E1、E2 出现的次数；(d) E1 波占韧皮部总时间的比例，E 波占总刺探取食时间的比例。图 22.4 同。

波、B 波、C 波)、韧皮部分泌唾液波（E1 波)、韧皮部被动吸食波（E2 波）以及木质部主动吸食波（G 波)，此外，还发现在其他胸喙亚目刺吸式口器昆虫刺探取食波中没有的 D 波。D 波总是出现在 C 波之后、E1 波之前，推测为柑橘木虱口针刺探筛管相关的波形。木虱与蚜虫取食波形（Tjallingii，1985）最主要的区别是在 C 波中没有电势降落（pd 波）出现，表明其口针是从细胞间隙直接进入韧皮部取食。另外一个区别是蚜虫没有出现 D 波。

本研究结果表明，不带菌木虱获得黄龙病菌主要与韧皮部吸食（E2 波）相关，而带菌木虱传病主要与韧皮部分泌唾液（E1 波）相关，与黄龙病菌为韧皮部限制性细菌相吻合。不带菌木虱在病树成熟叶、嫩叶、嫩芽上取食 8h 后获菌率分别为 50%、40%、35%，成功获菌个体的最短 E2 时间仅为 2min；单头带菌木虱吸食 24h 后成功传病率达到 22.6%，成功传病的带菌木虱最短取食时间约为 1.5h，最短 E1 时间仅为 5.1min，说明柑橘木虱传播黄龙病菌的效率很高。Pelz-Stelinski 等（2010）的研究结果也表明，单头带菌木虱的传病率为 20%，与本研究结果相似。许长藩等（1988）也报道，单头木虱成虫传病 5h 和 7h 的效率分别 30% 和 20%。马铃薯木虱

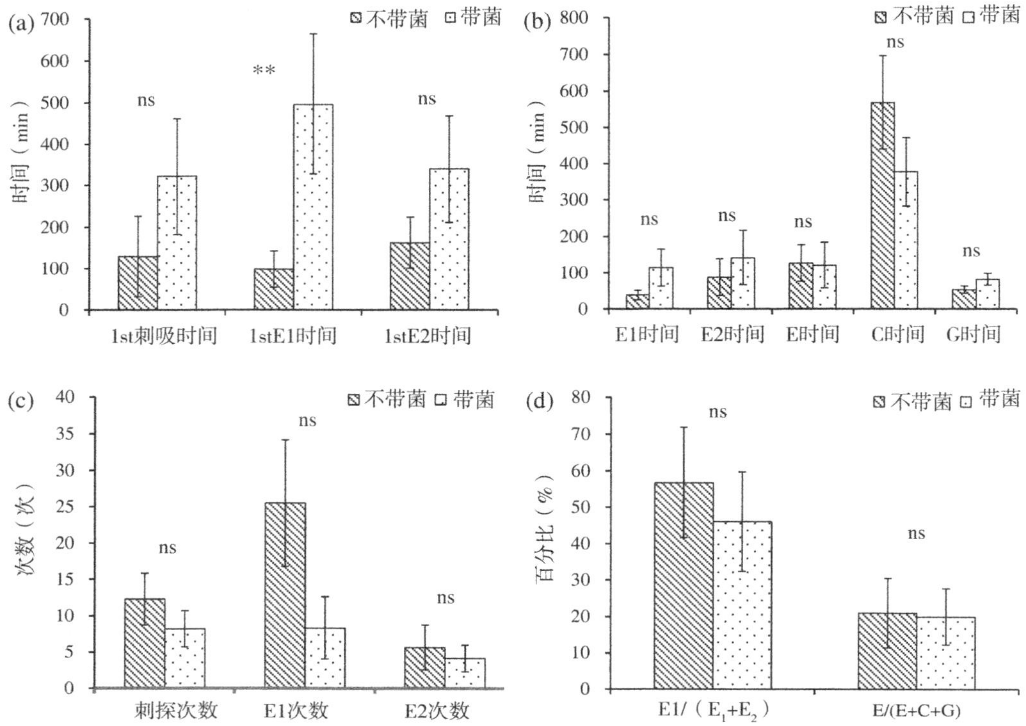

图 22.4　带菌与不带菌木虱雌虫 EPG 特征波形参数比较

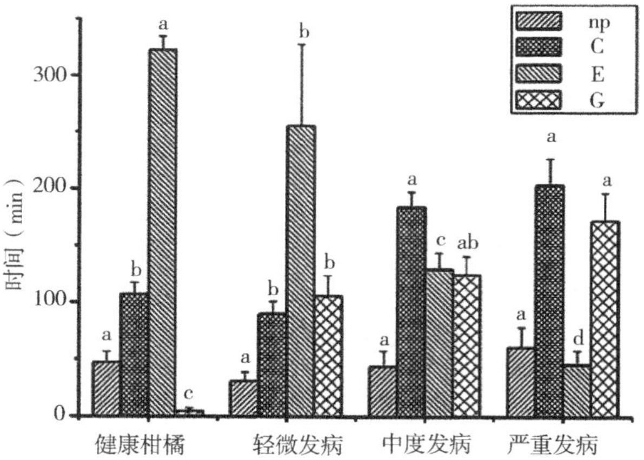

图 22.5　柑橘木虱在健康和黄龙病不同发病程度柑橘上取食各波形出现的时间

Bactericera cockerelli（Sulc）传播与黄龙病同属的马铃薯斑马病“*Ca.* L. *solanacearum*”，Sandanayaka 等（2014）用 EPG 监测了其取食与获菌、传菌的关系，发现单头木虱成功传病的最短取食时间为 2h 左右，也与本实验成功传病的最短时间相似。但是在本研究中，成功获菌的木虱有 2 头没有出现 E2 波，成功传病的有 1 头没有出现 E1 波，这种现象在马铃薯木虱传播马铃薯斑马病（Sandanayaka et al.，2014）和烟粉虱

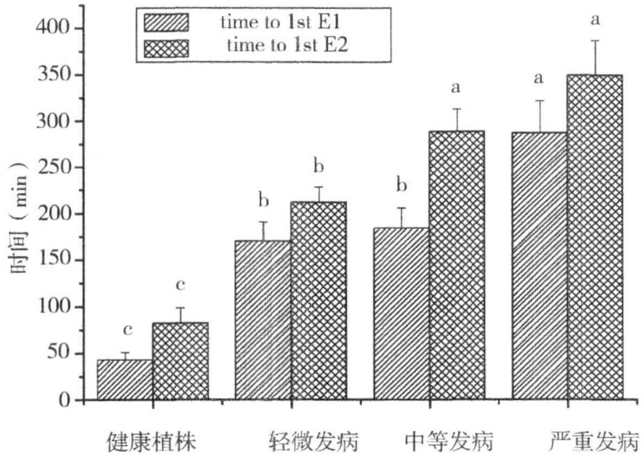

图 22.6　口针首次到达筛管分泌唾液和首次在筛管中持续吸食所需要的时间

注：测定时间：8h。np（non-probing）：非刺探波总时间；C（pathway phase）：路径波总时间（A+B+C）；E（phoem phase）：韧皮部阶段总时间（E1+E2）；G（xylem phase）：木质部吸食总时间。图中相同图案数据经 DMRT 法检验，有相同字母表示在 0.05 水平差异不显著。图 22.6~22.9 同。

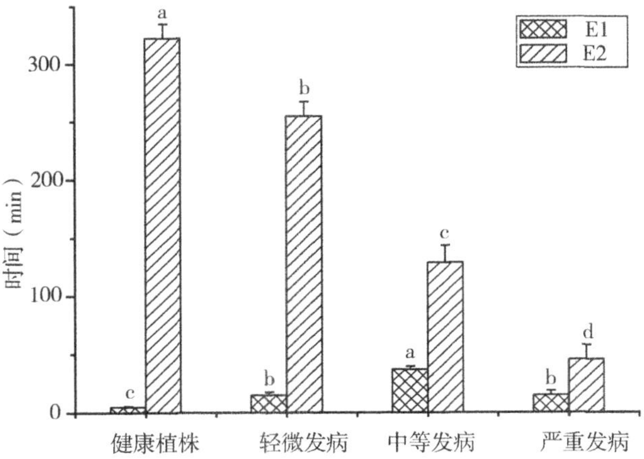

图 22.7　韧皮部分泌唾液和持续吸食平均持续总时间比较

*Bemisia tabaci*传播番茄黄化曲叶病毒（Jiang et al.，2000）的 EPG 测定中也有报道，一种可能是这些个体在传播韧皮部病菌的过程中时间很短，所以 EPG 未能检测到，其真正原因值得进一步研究。

本研究结果还表明，柑橘感染黄龙病后对木虱取食行为具有显著影响，主要表现在增加了取食障碍，而木虱雌虫携带黄龙病菌后取食行为也发生改变，分泌唾液时间显著推迟，表明病菌与寄主植物、媒介昆虫之间具有互作关系。关于黄龙病菌对柑橘木虱生理学特性的影响已有报道。黄龙病菌侵染会降低木虱体内相关酶的活性，从而增加其对不同作用机制杀虫剂的敏感性（Tiwari et al.，2010）。带菌木虱成

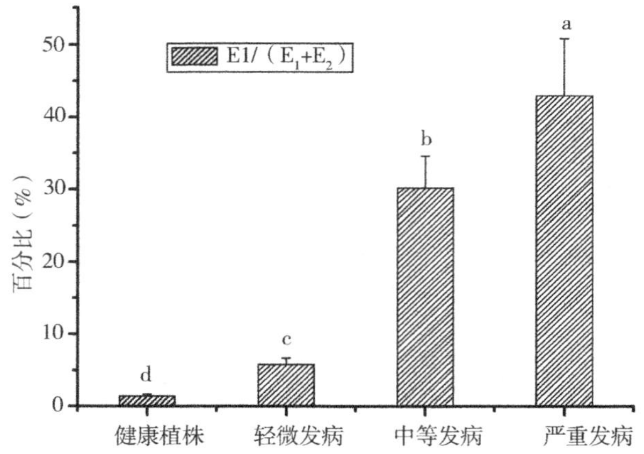

图 22.8　韧皮部分泌唾液平均总时间占韧皮部阶段平均总时间的比例

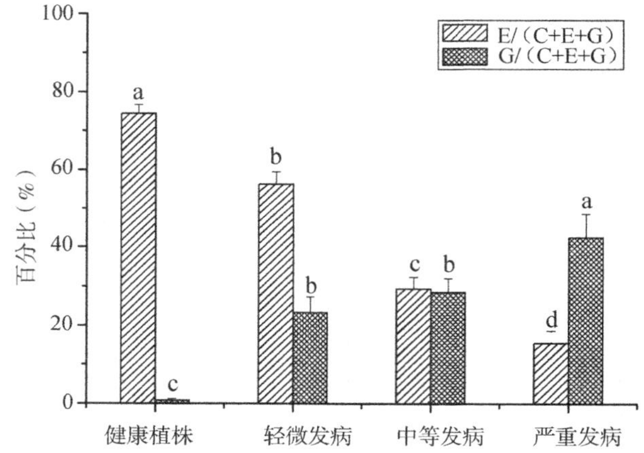

图 22.9　木虱在各部位取食平均总时间占取食总时间的比例

虫谷胱甘肽转移酶和细胞色素酶 P450 活性显著低于不带菌木虱，进一步对细胞色素酶 P450 的基因及其表达开展研究，鉴定出 5 个 *CYP*4 基因（*CYP*4*C*67，*CYP*4*DA*1，*CYP*4*C*68，*CYP*4*DB*1 和 *CYP*4*G*70），不带菌雄虫有 4 个基因水平显著高于带菌木虱，雌虫有 1 个基因水平显著高于带菌木虱（Tiwari et al.，2011）。

尽管 EPG 常被应用于媒介昆虫的传病机制研究，但应用该技术仅能解释媒介昆虫取食行为与传病的关系，更深层的传病机制，如病菌在虫体内的侵入途径、增殖部位、增殖方式等还需要用分子生物学方法开展进一步的研究。

参考文献

[1] ASANA R D. The citrus dieback problem in relation to cultivation of citrus fruits in India. *Indian Journal*

of Horticulture, 1958, 15: 283-286.

[2] Cen Y J, Zhang L N, Xia Y L, et al. Detection of '*Candidatus* Liberibacter asiaticus' in *Cacopsylla* (*Psylla*) *citrisuga* (Hemiptera: Psyllidae). *Florida Entomologist*, 2012, 95: 304-311.

[3] BONANI J P, FERERES A, GARZO E, et al. Characterization of electrical penetration graphs of the Asian citrus psyllid, *Diaphorina citri*, in sweet orange seedlings. *Entomologia Experimentalis et Applicata*, 2010, 134: 35-49.

[4] BOVÉ J M. Huanglongbing: A destructive, newly-emerging, century-old disease of citrus. *Journal of Plant Pathology*, 2006, 88: 7-37.

[5] HUNG T H, HUNG S C, CHEN C N, et al. Detection by PCR of *Candidatus* Liberibacter asiaticus, the bacterium causing citrus huanglongbing in vector psyllids: application to the study of vector-pathogen relationships. *Plant Pathology*, 2004, 53: 96-102.

[6] JAGOUEIX S, BOVÉ J M, GARNIER M. PCR detection of the two '*Candidatus*' liberobacter species associated with greening disease of citrus. *Molecular and Cellular Probes*, 1996, 10: 43-50.

[7] JIANG Y X, D E BLAS C, BARRIOS L, et al. Correlation between whitefly (Homoptera: Aleyrodidae) feeding behavior and transmission of tomato yellow leaf curl virus. *Annals of the Entomological Society of America*, 2000, 93: 573-579.

[8] PELZ-STELINSKI K S, BRLANSKY R H, EBERT T A, et al. Transmission parameters for *Candidatus* Liberibacter asiaticus by Asian citrus psyllid (Hemiptera: Psyllidae). *Journal of Economic Entomology*, 2010, 103: 1 531-1 541.

[9] REINKING O A. Diseases of economic plants in South China. *Philippine Agriculture*, 1919, 10: 109-135.

[10] SANDANAYAKA W R M, MORENO A, TOOMAN L K, et al. Stylet penetration activities linked to the acquisition and inoculation of *Candidatus* Liberibacter solanacearum by its vector tomato potato psyllid. *Entomologia Experimentalis et Applicata*, 2014, 151: 170-181.

[11] TEIXEIRA D C, DANET J L, EVEILLARD S, et al. Citrus huanglongbing in Sao Paulo, Brazil: PCR detection of the "*Candidatus* Liberibacter" species associated with the disease. *Molecular and Cellular Probes*, 2005, 19: 173-179.

[12] TIWARI S, PELZ-STELINSKI K, STELINSKI L L. Effect of *Candidatus* Liberibacter asiaticus infection on susceptibility of Asian citrus psyllid, *Diaphorina citri*, to selected insecticides. *Pest Management Science*, 2010, 67: 94-99.

[13] TIWARI S, GONDHALEKAR A D, MANN R S, et al. Characterization of five *CYP*4 genes from Asian citrus psyllid and their expression levels in *Candidatus* Liberibacter asiaticus-infected and uninfected psyllids. *Insect Molecular Biology*, 2011, 20: 733-744.

[14] TJALLINGII W F. Electrical nature of recorded signals during stylet penetration by aphids. *Entomologia Experimentalis et Applicata*, 1985, 38: 177-186.

[15] 林孔湘. 柑橘黄梢（黄龙）病研究：关于病原的探讨. 植物病理学报, 1956, 2 (1): 13-42.

[16] 汤清波, 张大山, 姬琨, 等. 刺吸电位技术应用中的几个问题. 应用昆虫学报, 2011, 48 (5): 1 519-1 527.

[17] 许长藩, 夏雨华, 李开本, 等. 柑橘木虱传播黄龙病的规律及病原在虫体内分布的研究. 福建省农科院学报, 1988, 3 (2): 57-61.

[18] 杨成良, 岑伊静, 梁广文, 等. 亚洲柑橘木虱的刺吸电位图谱研究. 华南农业大学学报, 2011, 32 (1): 49-52.

[19] 闫凤鸣. 化学生态学. 2 版. 北京：科学出版社，2011.

（岑伊静　杨成良　乌天宇　罗孝竹）

第二十三章 应用 EPG 技术研究 TYLCCNV 与烟粉虱互作关系的行为机制

（一）研究背景

刺吸式口器昆虫是作物上一大类重要害虫，其主要为害是植物病毒病的传播媒介，因此，携带病毒的昆虫的取食行为及其扩散对植物病毒的传播至关重要。媒介昆虫与植物病毒相互作用包括直接作用和通过寄主植物的间接作用两个方面。直接作用指病毒侵染虫体后调控昆虫的寄主选择和取食行为等从而影响媒介昆虫对病毒的传播（Ingwell et al.，2012；Rajabaskar et al.，2014）；间接作用是指病毒对植物的侵染改变植物的生理生化特性，进而影响媒介昆虫对寄主植物的选择、取食以及其在寄主植物上的寿命、产卵量等特性，而这些特性的改变会对病毒的传播产生影响（Belliure et al.，2005；Jiu et al.，2007；Mauck et al.，2010；Bosque-Pérez & Eigenbrode，2011）。由寄主植物介导的媒介昆虫与病毒之间的相互作用是影响田间病毒病流行和昆虫种群增长的重要因素（Stout et al.，2006）。

烟粉虱是刺吸式口器害虫，通过口针在植物细胞间刺探到达韧皮部取食韧皮部汁液。它作为重要的媒介昆虫能传播超过 200 种植物病毒，严重威胁农业经济作物的生产（Navas-Castillo et al.，2011）。烟粉虱是一个包含至少 36 个在形态上无法区分的隐种组成的物种复合体（De Barro et al.，2011；Liu et al.，2012；Firdaus et al.，2013；Boykin & De Barro，2014；刘银泉和刘树生，2012）。其中 Middle East-Asia Minor 1（MEAM1）和 Mediterranean（MED）两个隐种（之前分别被称为“B”和“Q”生物型烟粉虱）由于繁殖能力强、寄主适应范围广、抗药性强等特点，使得它们随着国际贸易的频繁往来而迅速扩散到世界各地（De Barro & Ahmed，2011；De Barro et al.，2011；Hu et al.，2011）。烟粉虱暴发之后通常伴随着病毒病特别是双生病毒病的流行（Navas－Castillo et al.，2011）。其中，双生病毒中的菜豆金黄花叶病毒属（*Begomovirus*）的病毒全部由烟粉虱特异性传播（Czosnek et al.，2002）。近年来，菜豆金黄花叶病毒—烟粉虱—寄主植物三者之间的相互作用对烟粉的入侵和病毒病流行的影响成为研究的热点（Colvin et al.，2006；Luan et al.，2014）。

双生病毒以持久性可循环的方式由烟粉虱特异性传播（Czosnek et al.，2002）。病毒在烟粉虱取食植物韧皮部汁液时进入烟粉虱体内，经过肠道、血淋巴、唾液腺的循环，随烟粉虱取食分泌的唾液而再次被传到新的植株上（Hogenhout et al.，2008）。病毒与烟粉虱虫体长时间的接触，也会影响其体内生理生化性质而对其行为产生直接影响（Moreno-Delafuente et al.，2013；Hemayet Jahan et al.，2014；王昱超等，2013）。此外，一些研究显示，部分双生病毒高度适应烟粉虱的细胞和组织，并能经卵传播给下一代，表明双生病毒还保留了许多昆虫病原菌的特点（Rubinstein & Czosnek，1997；Bosco et al.，2004；Luan et al.，2011；Czosnek & Ghanim，2012）。

像许多其他持久性传播的病毒一样，双生病毒也可以通过改变寄主植物特性而间接影响烟粉虱的生理和行为（Fang et al.，2013；Liu et al.，2013）。病毒感染植物后，引起植物生理生化水平的变化，这些变化又进一步影响烟粉虱在感毒植物上的行为和表现。首先，双生病毒感染引起的植物叶片卷曲、褶皱以及表皮毛的变化等会影响烟粉虱在植株上的取食和移动（Moreno-Delafuente et al.，2013）。其次，病毒感染还会引起植物防御途径的变化，从而间接影响烟粉虱在寄主植物上的表现（Luan et al.，2013b）。双生病毒通过寄主植物对昆虫种群增长的影响比较复杂，与植物、病毒及烟粉虱种类关系密切，其作用对昆虫可能是有利的，也可能是不利的，也可能影响不大（Mauck et al.，2012；Luan et al.，2014）。

对中国番茄黄化曲叶病毒（Tomato yellow leaf curl China virus，TYLCCNV）、烟粉虱以及寄主植物烟草的研究发现，尽管病毒感染对烟粉虱本身不利而使得烟粉虱的寿命和产卵量都降低，但烟粉虱在带毒烟草上成虫寿命和产卵量都显著地增强，使得烟粉虱种群增长显著加快（Jiu et al.，2007）。进一步研究发现，TYLCCNV 与烟粉虱之间的间接互惠是由于病毒感染抑制了植物的茉莉酸信号通路介导的植物防御反应，尤其是萜类代谢产物的合成（Zhang et al.，2012；Luan et al.，2013b；Li et al.，2014）。而且，烟草感染 TYLCCNV 病毒后，其对烟粉虱吸引作用显著强于未携带病毒的烟草（潘登等，2014）。对植物与昆虫营养的研究发现，病毒的感染对烟草氨基酸组分、必需氨基酸含量以及糖类与氨基酸的比例都没有显著的影响；然而烟粉虱在感毒烟草上能达到更为均衡的营养吸收而获得更高比例的必需氨基酸，尽管病毒对烟粉虱感染本身导致烟粉虱营养吸收更为不均衡而吸收更多的糖类（Wang et al.，2012）。烟粉虱与双生病毒直接和间接相互作用的行为机制值得我们更为深入地研究。

本研究中，我们首先通过 EPG 对带毒和不带毒烟粉虱在非病毒寄主植物棉花上取食行为进行测定，探讨病毒感染对烟粉虱本身的直接影响。然后，我们对带毒和不带毒烟粉虱在健康烟草上的取食行为以及带毒烟粉虱在带毒烟草上的取食行为进行测定，探讨病毒对烟粉虱取食行为直接和间接的影响。

（二）材料与方法

1. 植物、昆虫及病毒 实验用棉花（*Gossypium hirsutum* L.），品种为浙棉 1793；普通烟（*Nicotiana tabacum*），品种为 NC89。

中国番茄黄化曲叶病毒（TYLCCNV）DNA-A（GenBank Acc. No.：AJ319675.1）和 DNA-β 农杆菌侵染性克隆，由浙江大学生物技术研究所构建和提供。

烟粉虱 MEAM1 隐种（mtCOI GenBank Acc. No.：GQ332577.1），2010 年 8 月采集于浙江省温州市农科院蔬菜所，在人工气候室内以棉花作为寄主建立实验种群，并续代维持。

2. 带毒昆虫、植物的获取及病毒检测 选取 3~4 片真叶期、长势较为一致的烟草植株为材料，使用一次性无菌注射器取 TYLCCNV 农杆菌侵染性克隆菌液，将菌液注射到每株植株的茎部和叶柄韧皮部中。注射后植物置于自然光的控温温室内培养。15d 后获得发病明显的烟草，随机选取检测植物带毒情况。

取棉花上饲养的烟粉虱，分别在感染 TYLCCNV 烟草和健康烟草上饲养，繁殖两代后抽样检测烟粉虱体内的带毒情况，获得带毒和不带毒烟粉虱。

提取植物与烟粉虱的 DNA 后，参照 Cui 等（2004）的方法对病毒进行检测。

3. EPG 记录 首先用玻璃小管扣取供试烟粉虱，并置于冰上 10s 左右后，倒在装有冰的培养皿表面，使烟粉虱静止不动。然后取长度为 1.5~2.0cm 的金线重复蘸取银胶，将金丝粘到烟粉虱的中胸背板，粘连后静置 1min 左右，保证金线垂直稳定粘连在烟粉虱中胸背板上。再将连接好烟粉虱的金线用银胶粘到铜线上，并将粘连金线的烟粉虱放置在叶片的背面一个合适位置，以便烟粉虱能够在上面爬动。每头烟粉虱记录 8h，每次记录结束后，更换新的植物和烟粉虱。实验在 25~28℃，相对湿度 50%~70%的安静房间内进行，用 Giga-8 DC-EPG 系统及 STYLET 软件（荷兰瓦赫宁根大学）记录、分析波形。参考 Jiang 等（1999，2000）将烟粉虱 EPG 波形分为 7 种，分别为非路径波（np）、路径波（C）、唾液分泌波（E1）、韧皮部取食波（E2）、电势降落波（pd）、障碍波（F）、木质部波（G）。

4. 利用非寄主植物棉花研究 TYLCCNV 对烟粉虱取食行为的直接影响 为了研究 TYLCCNV 对 MEAM1 取食行为的直接影响，本研究比较了带毒和不带毒烟粉虱在 TYLCCNV 非寄主植物棉花上的取食行为。将数据分为两类参数进行分析，即非韧皮部取食参数及韧皮部取食参数。

5. TYLCCNV 通过寄主植物烟草对烟粉虱取食的综合影响 用 EPG 系统记录了以下三组处理：不带毒烟粉虱取食健康烟草、带毒烟粉虱取食健康烟草，以及带毒烟粉虱取食 TYLCCNV 感染烟草。各处理重复 24 次以上。统计分析取食行为的非韧皮部参数及韧皮部参数。

（三）结果

1. 利用非寄主植物棉花研究 TYLCCNV 对烟粉虱取食行为的直接影响 表 23.1 中带毒和不带毒烟粉虱在 TYLCCNV 非寄主植物棉花上的非韧皮部参数数据显示，带毒烟粉虱总的刺探次数显著增多（参数 2），路径波数量也显著增加，为不带毒烟粉虱的 2 倍多（参数 4）。而平均路径波时间与不带毒烟粉虱相比显著减少（参数 6）。此外，带毒烟粉虱从开始刺探到第一次接触韧皮部所用时间超过不带毒烟粉虱所用时间的 1 倍以上（参数 10）。而从记录开始到达第一次刺探的时间和总的刺探时间（参数 1 和参数 3）、总的路径波时间（参数 5）、烟粉虱到达韧皮部之前的刺探数（参数 7）、第一次开始韧皮部取食前的刺探数（参数 8）、第一次达到韧皮部所用的时间（参数 9）以及木质部波 G 波和障碍波 F 波总时间（参数 11，参数 12）在两处理间没有显著差别。

表 23.1　带毒与不带毒烟粉虱在棉花上取食行为的非韧皮部参数 (He et al.，2015)

参数	不带毒（n=25）	带毒（n=29）	P 值[a]
1. 开始到第一次刺探的时间（min）	34.64±13.32[b]	25.21±6.32	0.476
2. 总刺探数	30.95±5.69	68.03±12.29	**0.005**
3. 总刺探时间（min）	107.40±10.95	99.53±8.81	0.507
4. 路径波（C 波）数量	32.94±5.83	69.06±6.17	**0.012**
5. 路径波总时间（min）	72.49±9.48	52.46±4.84	0.132
6. 路径波平均时间（min）	2.64±0.41	0.98±0.08	**<0.001**
7. 到达韧皮部之前的刺探数	22.36±8.94	25.38±8.33	0.507
8. 第一次韧皮部取食前的刺探数	23.45±8.92	26.38±8.34	0.590
9. 第一次到达韧皮部所用时间（min）	249.50±41.66	310.8±28.84	0.386
10. 开始刺探到第一次接触韧皮部所用时间（min）	98.84±27.98	231.5±37.29	**0.018**
11. 木质部波（G 波）总时间（min）	23.36±8.86	38.91±8.42	0.257
12. 障碍波（F 波）总时间（min）	27.90±10.75	45.06±7.90	0.139

[a] P 值以 Mann-Whitney U-检验计算，粗体显示差异显著（α=0.05）。

[b] 表中数据为平均值±标准误。

韧皮部取食参数显示，带毒与不带毒烟粉虱两个处理之间在唾液分泌波 E1 波数量、E1 波平均时间、总的 E1 波时间、第一次唾液分泌到第一次持续的取食时间、E2 波数量、E2 波总时间、E 波总数量以及 E 波总时长相关的参数都没有显著差别（表 23.2）。然而，带毒烟粉虱从第一次 E1 波到第一次 E2 波的时间（参数 4）是不带毒烟粉虱需要时间的 2 倍以上，而其平均取食时间（参数 7）则只有不带毒烟粉虱的 1/3 左右。

表 23.2　带毒与不带毒烟粉虱在棉花上取食行为的韧皮部参数 (He et al.，2015)

参数	不带毒	带毒	P 值[a]
1. 唾液分泌波（E1 波）数	1.64±0.31[b]	3.25±0.96	0.178
2. 唾液分泌波平均时间（min）	2.49±1.08	1.95±0.51	0.528
3. 唾液分泌波总时间（min）	4.09±1.92	7.74±2.55	0.145
4. 第一次唾液分泌波到第一次韧皮部取食波（E2 波）时间（min）	1.71±0.46	3.92±0.64	**0.026**
5. 第一次唾液分泌到第一次持续的取食波时间（min）	4.40±1.47	6.39±2.02	0.230
6. 韧皮部取食波数	1.18±0.12	2.75±0.80	0.098

续表

参数	不带毒	带毒	P 值[a]
7. 韧皮部取食波平均时间（min）	88.31±15.70	27.60±8.54	**0.005**
8. 韧皮部取食波总时间（min）	100.30±14.59	75.88±17.63	0.529
9. 韧皮部 E 波总数	2.82±0.35	6.00±1.74	0.193
10. 韧皮部 E 波总时间（min）	88.24±16.15	74.08±15.75	0.650

[a]P 值以 Mann-Whitney U-检验计算，粗体显示差异显著（$\alpha=0.05$）。
[b]表中数据为平均值±标准误。

带毒和不带毒烟粉虱在 TYLCCNV 非寄主植物棉花上的取食行为结果显示，烟粉虱带毒后总的刺探次数增多，路径波数量也显著增加，但每次刺探用时减少，从开始刺探到第一次接触韧皮部所用时间延长。到达韧皮部后，从唾液分泌到开始取食所用时间增加，且平均取食时间显著降低，这表明携带病毒对烟粉虱取食行为有不利影响，但这种取食行为的变化有利于病毒的传播。

2. TYLCCNV 通过寄主植物烟草对烟粉虱取食的综合影响 实验设置的三组处理：不带毒烟粉虱取食健康烟草、带毒烟粉虱取食健康烟草，以及带毒烟粉虱取食 TYLCCNV 感染烟草分别进行了 30、28 和 24 个重复。

带毒及不带毒烟粉虱非韧皮部参数双因素方差分析结果显示，烟粉虱带毒状况只对 C 波数有着显著的影响；而植物带毒状况对 8 个参数中的 6 个都有显著影响；烟粉虱和植物带毒状况互作显著影响总刺探数和 C 波数（表 23.3）。数据分析结果表明植物带毒对烟粉虱取食行为的影响显著强于病毒感染烟粉虱本身对烟粉虱取食行为的影响。

表 23.3 带毒与不带毒烟粉虱在感病及健康烟草上取食行为的非韧皮部参数分析（He et al.，2015）

参数	P 值		
	烟粉虱	烟草	烟粉虱 * 烟草
a. 到达第一次刺探的时间	0.746	**<0.001**	0.573
b. 总刺探数	0.100	**<0.001**	**0.044**
c. 路径波（C 波）数	**0.002**	**<0.001**	**0.007**
d. 总刺探时间	0.419	0.305	0.484
e. 路径波平均时间	0.604	**0.031**	0.750
f. 第一次到达韧皮部所用时间	0.096	0.761	0.084
g. 到达韧皮部之前的刺探数	0.601	**0.010**	0.732
h. 第一次韧皮部取食前的刺探数	0.433	**0.004**	0.437

注：以双因素方差分析方法检验差异显著性。主要影响因子是烟粉虱（是否带毒）和番茄植株（是否带毒）及其互作。P 值粗体表示差异显著（$\alpha=0.05$）。

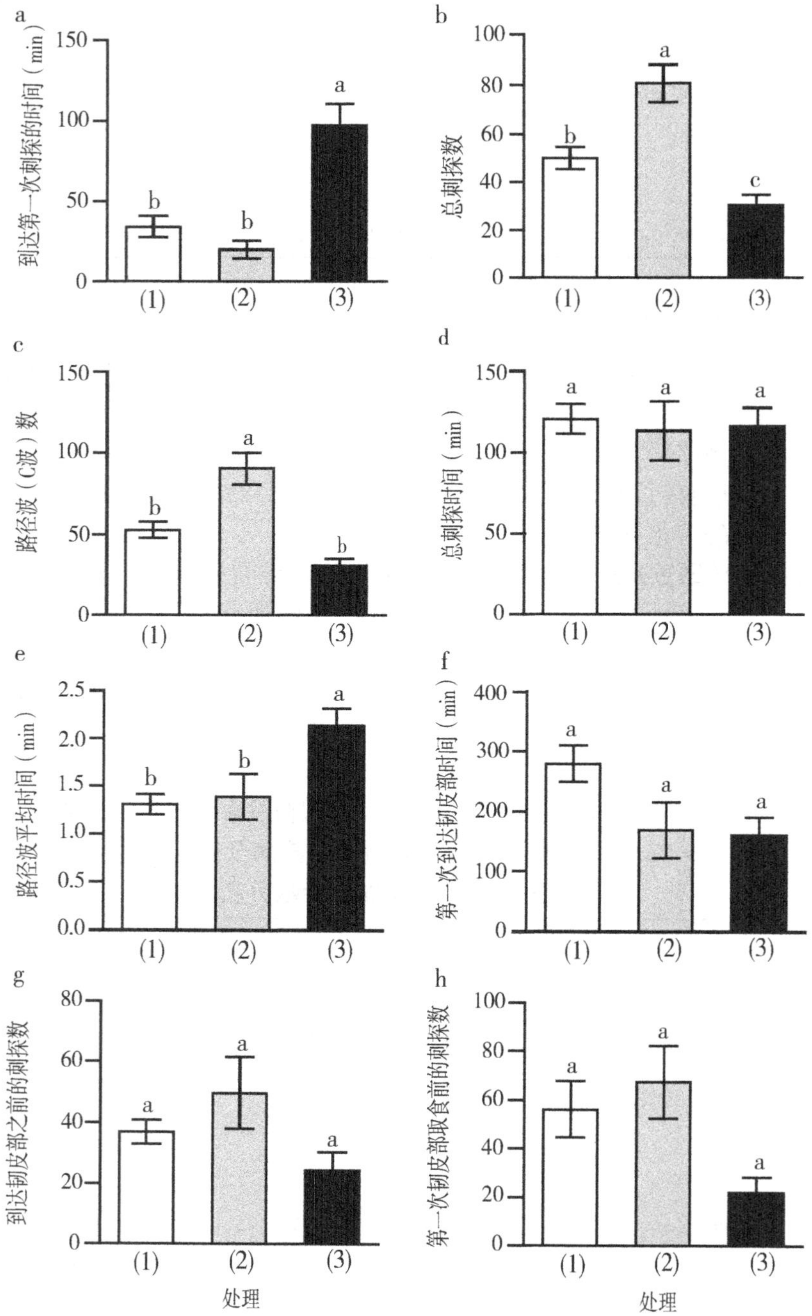

图 23.1　TYLCCNV 病毒对 MEAM1 烟粉虱取食烟草的非韧皮部参数影响

（1）不带毒烟粉虱在健康烟草上取食；（2）带毒烟粉虱在健康烟草上取食；（3）带毒烟粉虱在感病烟草上取食。

带毒烟粉虱在感毒烟草上取食时，烟粉虱第一次刺探所用的时间是带毒或不带毒烟粉虱在健康烟草上所用时间的 3~4 倍（图 23.1a）。在健康烟草上，带毒烟粉虱比不带毒烟粉虱刺探数显著增多；而带毒烟粉虱在带毒烟草上取食时比不带毒烟粉虱在健康烟草上取食时刺探数显著减少，并且其刺探数还不到带毒烟粉虱在健康烟草上刺探数的一半（图 23.1b）。在健康烟草上，带毒烟粉虱 C 波数比不带毒烟粉虱显著增多；而带毒烟粉虱在带毒烟草上的 C 波数比在健康烟草上显著减少（图 23.1c）。带毒烟粉虱在带毒烟草上的 C 波平均时间也比其他两组处理显著延长（图 23.1e）。此外，虽然差异未达到显著水平，带毒烟粉虱在带毒烟草上到达韧皮部和开始韧皮部取食前的刺探数都明显少于烟粉虱在健康烟草上的刺探数（图 23.1g，图 23.1h）。

表 23.4　带毒与不带毒烟粉虱在感病及健康烟草上取食行为的韧皮部参数分析

(He et al., 2015)

参数	P 值		
	烟粉虱	烟草	烟粉虱 * 烟草
i. 唾液分泌波（E1 波）数	0.702	0.358	0.593
j. 韧皮部取食波（E2 波）数	0.468	0.339	0.354
k. 唾液分泌波（E1）平均时间	0.427	0.993	0.355
l. 韧皮部取食波（E2）平均时间	0.446	**0.004**	0.465
m. 唾液分泌波（E1）总时间	0.753	0.776	0.693
n. 韧皮部取食波（E2）总时间	0.074	**0.010**	0.064
o. 第一次唾液分泌波（E1）到第一次韧皮部取食波（E2）时间	0.067	**0.009**	**0.040**
p. 唾液分泌波（E1）占韧皮部总取食时间的百分比	0.157	0.394	0.077

注：以双因素方差分析方法检验差异显著性。主要影响因子是烟粉虱（是否带毒）和番茄植株（是否带毒）及其互作。*P* 值粗体表示差异显著（$\alpha = 0.05$）。

烟粉虱在烟草上取食行为韧皮部参数的双因素方差分析结果显示，烟粉虱带毒状况对各参数没有显著影响，尽管其对韧皮部取食 E2 波总时间和第一次 E1 到第一次 E2 所用时间的影响接近显著水平（表 23.4，参数 n，o）。植物带毒状况对平均 E2 波时间、总的 E2 波时间和第一次 E1 波到第一次 E2 波时间有着显著的影响。这两个参数的相互作用对第一次 E1 波到第一次 E2 波所用时间有着显著影响，并且对 E2 波总时间的影响接近显著水平。与非韧皮部参数结果相似的是，病毒感染植物对烟粉虱取食行为有着更为显著的影响。

图 23.2 结果也显示，与不带毒烟粉虱在健康烟草上取食相比较，TYLCCNV 病毒对烟粉虱虫体本身的感染虽然增加了其在健康烟草上的韧皮部取食的平均时间和总时间，但差异未达显著水平（图 23.2l，图 23.2n）。当烟粉虱和植物同时感毒时，烟粉虱在带毒烟草上的韧皮部取食平均时间和总时间均显著延长（图 23.2l，图 23.2n），尽管

与带毒烟粉虱在健康烟草上的取食参数也没有达到显著水平。同样的，病毒对烟粉虱本身的感染使得其第一次到达韧皮部时间距第一次取食时间间隔也位于不带毒烟粉虱在健康烟草上取食与带毒烟粉虱在带毒烟草上取食之间。而后两者差异显著，带毒烟粉虱在带毒烟草上所需时间显著要短（图 23. 2o）。其他参数在三者之间无显著差异。

（四）讨论

本研究共设置了两组实验。第一组实验中，我们观察了带毒和不带毒烟粉虱在 TYLCCNV 病毒非寄主棉花上的取食行为。这组实验能排除病毒感染植株的影响，因而能够直观展示病毒对烟粉虱感染本身的影响。第二组实验中，我们观察了带毒和不带毒烟粉虱在健康烟草上的取食行为以及带毒烟粉虱在感毒烟草上的取食行为。因为烟粉虱在带毒烟草上获毒时间较短（Jiu et al.，2006），本研究中因此未设置不带毒烟粉虱在带毒烟草上取食的处理，以避免设置重复。

对两组实验的综合分析结果表明，病毒对烟粉虱本身的感染主要对烟粉虱取食有着直接的不利影响；相反，病毒对寄主植物的感染主要对烟粉虱的取食行为有着间接的有利影响，并且带毒寄主植物对烟粉虱取食行为间接的有利影响超过病毒对烟粉虱本身不利的影响而在带毒植物上表现出综合的有利影响（表 23. 1 和表 23. 2；图 23. 1 和图 23. 2）。例如，在病毒非寄主植物棉花的实验中，当烟粉虱带毒时，其刺探次数和到达韧皮部的时间都显著增加，路径波 C 波数目是不带毒烟粉虱的 2 倍。这表明烟粉虱带毒后在植物细胞间刺探的过程中不顺利，遇到更多的阻碍（表 23. 1）。此外，在棉花上的带毒烟粉虱比不带毒烟粉虱的韧皮部取食波平均时间显著缩短（表 23. 2，参数 7），这表明 TYLCCNV 不利于烟粉虱在棉花韧皮部吸取汁液。带毒对烟粉虱细胞间刺探的不利影响在健康烟草上的取食过程中也表现出来，其刺探次数和路径波数都比不带毒烟粉虱在健康烟草上要显著增多（图 23. 1b，图 23. 1c）；而这种阻碍在感毒的烟草上取食时就不存在了，反而进一步表现为刺探数和路径波数都显著减少（图 23. 1b，图 23. 1c）。在韧皮部取食的表现也说明病毒通过植物介导的影响作用有利于烟粉虱，取食波 E2 波的平均时间和总时间在带毒和不带毒烟粉虱取食健康烟草时没有显著差别，但当带毒烟粉虱取食感毒烟草时，该取食时间明显延长，且与不带毒烟粉虱取食健康烟草间达到显著水平（图 23. 2l，图 23. 2n）。不过，由于烟草植物带毒导致烟粉虱开始第一次刺探的时间显著延迟（图 23. 1a，表 23. 3），这说明烟粉虱在带毒烟草上寻找刺探位点时间较长，这一点意味着病毒对烟粉虱间接作用也有不利的一面。这可能是由于植物感毒后叶片表面结构如叶毛密度变化等引起，需要进一步验证（潘登等，2013）。

TYLCCNV 对 MEAM1 取食的直接的不利影响，说明 TYLCCNV 表现出了昆虫病原菌的某些特性而不利于烟粉虱的取食及生长繁殖等（Rubinstein & Czosnek，1997；Jiu et al.，2007；Czosnek & Ghanim，2012）。TYLCCNV 表现出的这些昆虫病原菌的特点与前人研究的结果也相吻合。TYLCCNV 对烟粉虱本身的感染显著缩短了烟粉虱的寿命并且减少了其平均产卵量（Jiu et al.，2007），并且使得烟粉虱营养吸收更为不均衡（Wang et al.，2012）。另外，对 TYLCCNV 感染烟粉虱和不带毒烟粉虱转录组分析结果表明，病毒感染

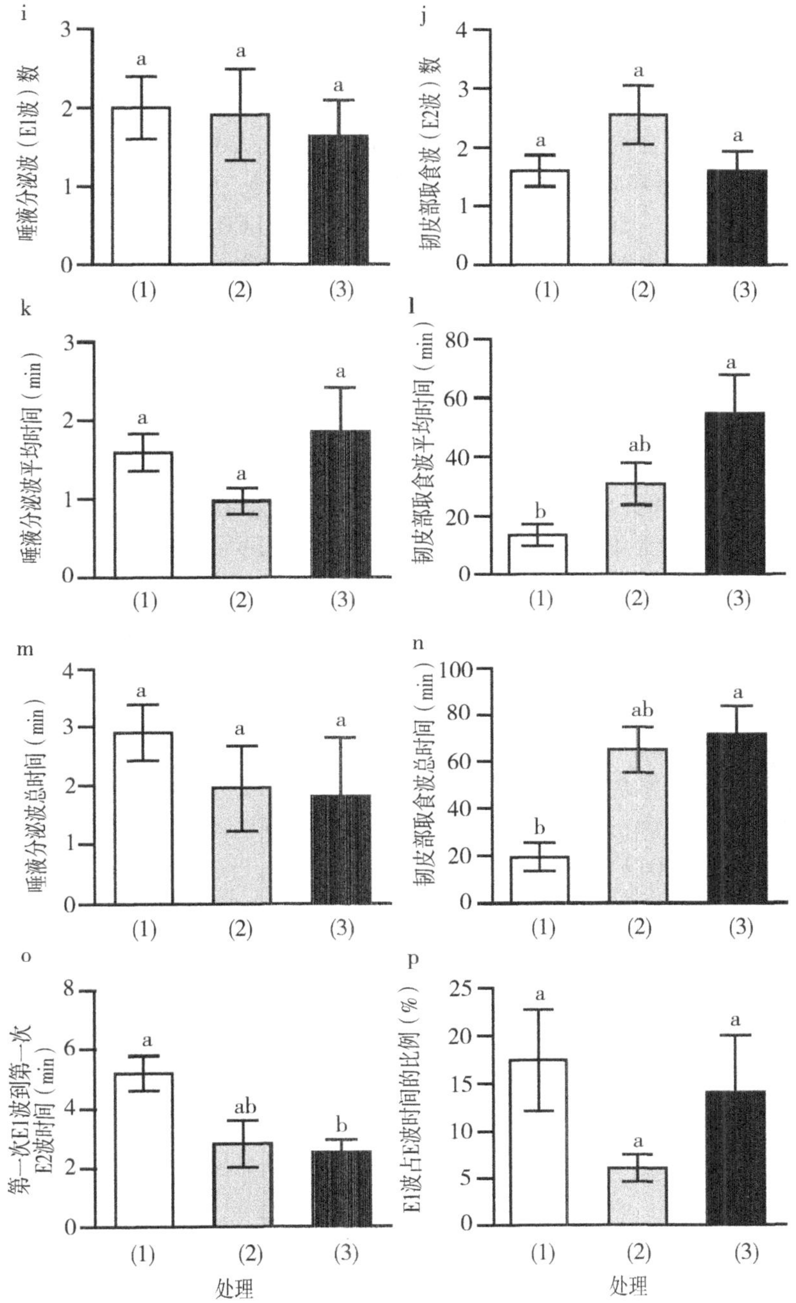

图 23.2　TYLCCNV 病毒对烟粉虱取食烟草的韧皮部参数影响

（1）不带毒烟粉虱在健康烟草上取食；（2）带毒烟粉虱在健康烟草上取食；

（3）带毒烟粉虱在感病烟草上取食。

也激活了许多与烟粉虱抗病毒过程有关的免疫相关反应（Luan et al.，2011）。

TYLCCNV 通过感染烟草对烟粉虱取食行为表现出更强的间接有利影响。这一结果也与前人的许多关于烟粉虱种群、生理及行为研究相吻合。这些结果表明，当烟粉虱在感毒烟草上比其在健康烟草上有着更高的存活率和产卵量以及更高的种群增长速率（Jiu et al.，2007；Guo et al.，2010；Wang et al.，2010；Guo et al.，2012）；带毒植株比不带毒植株对烟粉虱有着更强的吸引力（潘登等，2014）；烟粉虱在带毒烟草上比其在健康烟草上有着更为均衡的营养吸收（Wang et al.，2012）。带毒植株对烟粉虱间接的有利影响被证实与病毒对寄主植物抗性的抑制有关（Zhang et al.，2012；Luan et al.，2013b；Li et al.，2014）。另外，对烟粉虱在 TYLCCNV 感毒植株及健康植株上取食后转录分析结果也表明，烟粉虱在感染的烟草上取食时其解毒活动显著降低（Luan et al.，2013a）。

病毒对媒介昆虫取食行为直接和间接的影响在植物病毒及动物病毒中都有报道，这些对其行为的修饰被认为是病毒进化出的帮助自身传播的机制（Mauck et al.，2010；Stafford et al.，2011；Ingwell et al.，2012；Moreno-Delafuente et al.，2013；Kenney & Brault，2014）。通过 EPG 技术探究双生病毒和烟粉虱相互作用关系的研究目前还为数不多。Liu 等 2013 年的研究结果表明，MEAM1 烟粉虱在 TYLCV 感染的植株上取食活动更为活跃，并且带毒烟粉虱唾液分泌时间更长（Liu et al.，2013）。另有研究报道 TYLCV 对 MED 烟粉虱取食行为的影响，结果表明病毒的感染总体上有利于烟粉虱持续的唾液分泌和韧皮部汁液取食从而有利于病毒本身的传播（Liu et al.，2013；Moreno-Delafuente et al.，2013；Hemayet Jahan et al.，2014；王昱超等，2013）。然而各个研究之间也有不一致的情形发生。如 Moreno-Delafuente 等（2013）发现携带 TYLCV 病毒的 MED 烟粉虱呈现被束缚行为，比不带毒烟粉虱行动缓慢。而 Hemayet Jahan 等（2014）发现带毒烟粉虱更显得焦躁不安。这种不一致也预示着植物—病毒—媒介昆虫相互关系的复杂性。病毒和烟粉虱的互作对烟粉虱可能有利、不利或中性的（Mauck et al.，2012；Luan et al.，2014）。

综上所述，本研究利用 EPG 技术观察烟粉虱的取食行为，揭示了病毒 TYLCCNV 对 MEAM1 烟粉虱取食行为直接和间接影响的不同及其行为机制。研究结果表明，病毒对烟粉虱本身的感染不利于烟粉虱的取食，而病毒对寄主植物的感染间接有利于烟粉虱的取食，并且间接的有利影响超过其直接的不利影响从而提高媒介昆虫烟粉虱的适应性。研究结果有助于更好地理解双生病毒所保留的昆虫病原菌的特性以及其通过寄主植物与媒介昆虫之间的间接互惠。

参考文献

[1] BELLIURE B，JANSSEN A，MARIS P C，et al. Herbiovore arthropods benefit from vectoring plant viruses. *Ecology Letters*，2005，8：70-79.

[2] BOSCO D，MASON G，ACCOTTO G. TYLCSV DNA，but not infectivity，can be transovarially inherited by the progeny of the whitefly vector *Bemisia tabaci*（Gennadius）. *Virology*，2004，323：276-283.

[3] BOSQUE-PÉREZ N A，EIGENBRODE S D. The influence of virus-induced changes in plants on aphid vectors：insights from luteovirus pathosystems. *Virus Research*，2011，159：201-205.

[4] BOYKIN L M, DE BARRO P J. A practical guide to identifying members of the *Bemisia tabaci* species complex: and other morphologically identical species. *Frontiers in Ecology and Evolution*, 2014, 2: doi: 10. 3389/fevo, 2014, 00045.

[5] COLVIN J, OMONGO C A, GOVINDAPPA M R, et al. Host-plant viral infection effects on arthropod-vector population growth, development and behaviour: management and epidemiological implications. *Advances in Virus Research*, 2006, 67: 419-452.

[6] CUI X F, TAO X R, XIE Y, et al. A DNAβ associated with *Tomato yellow leaf curl China virus* is required for symptom induction. *Journal of Virology*, 2004, 78: 13 966-13 974.

[7] CZOSNEK H, GHANIM M. Back to basics: Are begomoviruses whitefly pathogens? *Journal of Integrative Agriculture*, 2012, 11: 225-234.

[8] CZOSNEK H, GHANIM M. The circulative pathway of begomoviruses in the whitefly vector *Bemisia tabaci*—insights from studies with *Tomato yellow leaf curl virus*. *Annals of Applied Biology*, 2002, 140: 215-231.

[9] DE BARRO P, AHMED M Z. Genetic networking of the *Bemisia tabaci* cryptic species complex reveals pattern of biological invasions. *PLoS ONE*, 2011, 6: e25579.

[10] DE BARRO P J, LIU S S, BOYKIN L M, et al. *Bemisia tabaci*: a statement of species status. *Annual Review of Entomology*, 2011, 56: 1-19.

[11] FANG Y, JIAO X G, XIE W, et al. *Tomato yellow leaf curl virus* alters the host preferences of its vector *Bemisia tabaci*. *Scientific Reports*, 2013, 3: 2 876.

[12] FIRDAUS S, VOSMAN B, HIDAYATI N, et al. The *Bemisia tabaci* species complex: additions from different parts of the world. *Insect Science*, 2013, 20: 723-733.

[13] GUO J Y, DONG S Z, YANG X L, et al. Enhanced vitellogenesis in a whitefly via feeding on a begomovirus-infected plant. *PloS ONE*, 2012, 7: e43567.

[14] GUO J Y, YE G Y, DONG S Z, et al. An invasive whitefly feeding on a virus-infected plant increased its egg production and realized fecundity. *PLoS ONE*, 2010, 5: e11713.

[15] HE W B, LI J, LIU S S. Differential profiles of direct and indirect modification of vector feeding behaviour by a plant virus. *Scientific Reports*, 2015, 5: 7 682.

[16] HEMAYET JAHAN, S M, LEE G S, et al. Upregulation of probing- and feeding-related behavioural frequencies in *Bemisia tabaci* upon acquisition of *Tomato yellow leaf curl virus*. *Pest Management Science*, 2014, 70: 1 497-1 502.

[17] HOGENHOUT S A, AMMAR E D, WHITFIELD A E, et al. Insect vector interactions with persistently transmitted viruses. *Annual Review of Phytopathology*, 2008, 46: 327-359.

[18] HU J, DE BARRO P, ZHAO H, et al. An extensive field survey combined with a phylogenetic analysis reveals rapid and widespread invasion of two alien whiteflies in China. *PLoS ONE*, 2011, 6: e16061.

[19] INGWELL L L, EIGENBRODE S D, BOSQUE-PÉREZ N A. Plant viruses alter insect behavior to enhance their spread. *Scientific Reports*, 2012, 2: 578.

[20] JIANG Y, LEI H, COLLAR J, et al. Probing and feeding behavior of two distinct biotypes of *Bemisia tabaci* (Homoptera: Aleyrodidae) on tomato plants. *Journal of Economic Entomology*, 1999, 92: 357-366.

[21] JIANG Y X, DE BLAS C, BARRIOS L, et al. Correlation between whitefly (Homoptera: Aleyrodidae) feeding behavior and transmission of *Tomato yellow leaf curl virus*. *Annals of the Entomolog-

ical Society of America, 2000, 93: 573-579.

[22] JIU M, ZHOU X P, LIU S S. Acquisition and transmission of two begomoviruses by the B and a non-B biotype of *Bemisia tabaci* from Zhejiang, China. *Journal of Phytopathology*, 2006, 154: 587-591.

[23] JIU M, ZHOU X P, TONG L, et al. Vector-virus mutualism accelerates population increase of an invasive whitefly. *PLoS ONE*, 2007, 2: e182.

[24] KENNEY J, BRAULT A. The role of environmental, virological and vector interactions in dictating biological transmission of arthropod-borne viruses by mosquitoes. *Advances in virus research*, 2014, 89: 39-83.

[25] LI R, WELDEGERGIS B T, LI J, et al. Virulence Factors of Geminivirus Interact with MYC2 to Subvert Plant Resistance and Promote Vector Performance. *The Plant Cell*, 2014, 26: 4 991-5 008.

[26] LIU B M, PREISSER E L, CHU D, et al. Multiple forms of vector manipulation by a plant-infecting virus: *Bemisia tabaci* and *Tomato yellow leaf curl virus*. *Journal of Virology*, 2013, 87: 4 929-4 937.

[27] LIU S S, COLVIN J, DE BARRO P J. Species concepts as applied to the whitefly *Bemisia tabaci* systematics: how many species are there? *Journal of Integrative Agriculture*, 2012, 11: 176-186.

[28] LUAN J B, LI J M, VARELA N, et al. Global analysis of the transcriptional response of whitefly to *Tomato yellow leaf curl China virus* reveals the relationship of coevolved adaptations. *Journal of Virology*, 2011, 85: 3 330-3 340.

[29] LUAN J B, WANG X W, COLVIN J, et al. Plant-mediated whitefly-begomovirus interactions: research progress and future prospects. *Bulletin of Entomological Research*, 2014, 104: 267-276.

[30] LUAN J B, WANG Y L, WANG J, et al. Detoxification activity and energy cost is attenuated in whiteflies feeding on *Tomato yellow leaf curl China virus*-infected tobacco plants. *Insect Molecular Biology*, 2013a, 22: 597-607.

[31] LUAN J B, YAO D M, ZHANG T, et al. Suppression of terpenoid synthesis in plants by a virus promotes its mutualism with vectors. *Ecology Letters*, 2013b, 16: 390-398.

[32] MAUCK K, BOSQUE-PÉREZ N A, EIGENBRODE S D, et al. Transmission mechanisms shape pathogen effects on host-vector interactions: evidence from plant viruses. *Functional Ecology*, 2012, 26: 1 162-1 175.

[33] MAUCK K E, DE MORAES C M, MESCHER M C. Deceptive chemical signals induced by a plant virus attract insect vectors to inferior hosts. *Proceedings of the National Academy of Sciences*, 2010, 107: 3 600-3 605.

[34] MORENO-DELAFUENTE A, GARZO E, MORENO A, et al. A plant virus manipulates the behavior of its whitefly vector to enhance its transmission efficiency and spread. *PloS ONE*, 2013, 8: e61543.

[35] NAVAS-CASTILLO J, FIALLO-OLIVÉ E, SÁNCHEZ-CAMPOS S. Emerging virus diseases transmitted by whiteflies. *Annual Review of Phytopathology*, 2011, 49: 219-248.

[36] RAJABASKAR D, BOSQUE-PÉREZ N A, EIGENBRODE S D. Preference by a virus vector for infected plants is reversed after virus acquisition. *Virus Research*, 2014, 186: 32-37.

[37] RUBINSTEIN G, CZOSNEK H. Long-term association of *Tomato yellow leaf curl virus* with its whitefly vector *Bemisia tabaci*: effect on the insect transmission capacity, longevity and fecundity. *Journal of General Virology*, 1997, 78: 2 683-2 689.

[38] STAFFORD C A, WALKER G P, ULLMAN D E. Infection with a plant virus modifies vector feeding behavior. *Proceedings of the National Academy of Sciences*, 2011, 108: 9 350-9 355.

[39] STOUT M J, THALER J S, THOMMA B P H J. Plant-mediated interactions between pathogenic mi-

croorganisms and herbivorous arthropods. *Annual Review of Entomology*, 2006, 51: 663-689.

[40] WANG J, BING X L, LI M, et al. Infection of tobacco plants by a begomovirus improves nutritional assimilation by a whitefly. *Entomologia Experimentalis et Applicata*, 2012, 144: 191-201.

[41] WANG J, ZHAO H, LIU J, et al. Low frequency of horizontal and vertical transmission of two begomoviruses through whiteflies exhibits little relevance to the vector infectivity. *Annals of Applied Biology*, 2010, 157: 125-133.

[42] ZHANG T, LUAN J B, QI J F, et al. Begomovirus-whitefly mutualism is achieved through repression of plant defences by a virus pathogenicity factor. *Molecular Ecology*, 2012, 21: 1 294-1 304.

[43] 刘银泉，刘树生．烟粉虱的分类地位及在中国的分布．生物安全学报，2012，21：247-255.

[44] 潘登，王岚岚，刘树生，等．番茄感染双生病毒对叶毛密和海氏桨角蚜小蜂搜寻行为及适合性的影响．昆虫学报，2013，56：644-651.

[45] 潘登，李元喜，栾军波，等．烟粉虱及海氏桨角蚜小蜂对中国番茄黄化曲叶病毒感染烟草的嗅觉反应．应用昆虫学报，2014，51：60-66.

[46] 王昱超，穆常青，郭晓军，等．番茄黄化曲叶病毒对烟粉虱MED隐种取食行为的影响．环境昆虫学报，2013，35：415-421.

(刘银泉　李　杰　何文波　刘树生)

第二十四章　麦蚜在人工饲料上的取食波形研究

（一）研究背景

小麦蚜虫是我国乃至世界上小麦的重要害虫，主要种类有麦长管蚜 *Sitobion avenae*（Fabricius）、禾谷缢管蚜 *Rhopalosiohum padi*（L.）、麦二叉蚜 *Schizaphis graminum*（Rondani）、麦无网长管蚜 *Metopolophium dirhodum*（Walker）4 种。它具有分布广、为害严重等特点；不仅使植株生长受阻，分蘖减少而产生直接损失；而且分泌蜜露干扰小麦的光合作用，引起霉菌寄生，刺吸小麦汁液后严重影响小麦的生长发育，造成小麦品质下降（杨益众等，1995，2005；李巧丝等，2003；袁浩，2013）；并且麦蚜作为大麦黄矮病毒（BYDV）的传播介体（张向才等，1985），造成小麦严重损失。

植物抗虫品种选育与应用是害虫综合防治的重要途径。植物基因工程为培育农作物抗虫新品种开辟了新的方向，这不仅减少了农药使用、降低农药残留，而且有利于保护害虫的天敌，维持生态平衡。植物抗性机制分为三个方面：不选择性（或拒趋性）、抗生性和耐害性。在与蚜虫长期协同进化选择的过程中，小麦主要形成了抗生性和不选择性两种主要抗性类型（陈巨莲等，1997）。植物次生物质在寄主植物抗虫作用中具有重要贡献。

植物次生物质主要来源于三类代谢途径的产物，包括酚类化合物、萜类化合物和含氮化合物，分为挥发性和不挥发性两类，前者在蚜虫降落时对寄主植物定向嗅觉反应起到重要作用；在蚜虫探测寄主表面和内部组织时是两类次生物质共同作用完成，而最后接受寄主则主要是植物内部因素通过蚜虫味觉来实现，即不挥发性次生物质起主导作用。与植物抗蚜性有关的次生物质主要有单宁酸、酚类化合物、黄酮类化合物、氧肟酸、生物碱、芥子油苷、非蛋白氨基酸。植物凝集素是一类重要的抗蚜物质，具有特异糖结合特性的蛋白质。研究报道，抗蚜效果明显的植物凝集素主要有雪花莲凝集素（GNA）、伴刀豆凝集素（Con A）、蒜叶凝集素（ASAL）、半夏凝集素（PTA）等。在抗蚜转基因研究中应用较多的有雪花莲凝集素（GNA）（Hilder et al.，1995）、伴刀豆凝集素（ConA）（Gatehouse et al.，1999）、蒜叶凝集素（ASAL）（Dutta et al.，2005a；Dutta et al.，2005b）、半夏凝集素（PTA）（Yu et al.，2008）等。

为明确植物次生物质对蚜虫的影响，本研究利用 EPG 结合液体人工饲料囊技术，研究了次生物质单种或几种混合、植物凝集素对麦蚜取食行为的影响，以期通过麦蚜取食波形参数与抗蚜作用的关联分析，获得抗虫组分剂量，从而建立合适的抗蚜评价指标。

（二）材料与方法

1. 麦蚜的来源及饲养　采用陈巨莲等（2000）的麦蚜人工饲料的配方及饲养方法。在饲料中加入不同量的次生物质，配制成不同的浓度，用微孔滤膜（孔径 0.45μm，规格 φ25mm）过滤灭菌，配制的饲养液分装于-20℃冰柜保存备用。

2. 实验仪器　采用的昆虫刺吸行为仪为 DC-EPG 系统，包括放大器（GIGA-4，为四通道；GIGA-8，为八通道）（荷兰瓦赫宁根大学）、探头、法拉第笼（自行设计）、记录仪（XWTD-204 型，上海大华仪表厂）和计算机等。放大器与插入计算机中的 A/D 卡（Metraeyte，Dash/8）连接，电信号经放大后转换为数值信号，GIGA-4 系统采用 STYLET2.2 软件系统将数值信号转化为波谱图，进行数据采集储存于计算机硬盘；GIGA-8 系统采用 Probe data acquisition 3.4 软件采集与保存数据。

植物电极插在位于取食小囊石蜡膜之间的饲料中。当蚜虫口针刺入植物组织时，回路接通。实验在室温连续光照下进行。

3. 测定方法　昆虫电极是一段长 3～5cm 银丝（直径 0.025mm）或 2～3cm，直径 18μm 的金丝，一端用水溶性导电银胶（美国产）粘在昆虫前胸背板上，另一端与放大器的探头相连。将粘好的蚜虫置于供试人工饲料取食小囊（双层石蜡膜拉制而成，液体人工密封其中，模拟小麦叶片）上任其取食。植物电极为细铜丝，插于石蜡膜所夹的饲料中。当蚜虫口针刺入植物组织时，回路接通，回路电流经 AD 转换器转换为数值信号被电脑记录下来。

蚜虫在添加次生物质人工饲料上取食行为监测：采用 DC-EPG Giga-4 刺探电位图谱记录仪。将全纯人工饲料添加单宁酸和没食子酸，配制成含有 0.015%、0.03%、0.04%、0.05%、0.06%等浓度梯度；对单宁酸、总酚、丁布、门布 4 种次生物质采用正交旋转组合实验设计，选取对麦长管蚜存活率抑制作用最强的两个组合（17 号、4 号）进行蚜虫刺探电位研究，17 号组合中单宁酸为 0.06%、总酚为 0.03%、丁布为 0.01%、门布为 0.02%；4 号组合单宁酸为 0.045%、总酚为 0.045%、丁布为 0.005%、门布为 0.01%。以未加次生物质的全纯人工饲料为对照。实验在室温 22～24℃条件下进行，每一处理重复 8 次（即供试蚜虫数为 8 头），每一重复连续记录 8h。

麦长管蚜在添加凝集素人工饲料上取食行为监测：采用 DC-EPG Giga-8 刺探电位图谱记录仪，取大小和日龄基本一致的无翅成蚜饥饿 2h 后连入刺探电位放大器的昆虫电极。每一株系重复 8 次，每一重复连续记录 8h，每头蚜虫只用于测试一次。

本实验主要统计 np 波、C 波（包括 A 波、B 波和 C 波）、E1 波、E2 波、G 波和 pd 波。

4. EPG 波形的记录　将储存于计算机硬盘中的波谱图，依据波形特征进行分析。7 种特征波形：np（非刺探波形）、C（A+B+C）（路径刺探波形）、pd（口针尖端在细胞内穿刺波形）、E1（韧皮部分泌唾液）、E2（韧皮部吸食）、F（路径中的机械阻碍）、G（木质部吸食）。对下列 3 类波形参数进行统计，并用 Duncan's 法测验和多重比较：①各种波形历时占总历时的比例；②出现各种波形的次数；③E 波细节，包括 E2 波跟随 E1 波而出现事件，E2 波持续时间超过 10min 的次数［E2（10）］。

（三）结果

1. 特征波形与分析 麦长管蚜和禾谷缢管蚜在人工饲料中的取食波形相对较简单（图 24. 1），可记录到非刺探波形 np，刺探波 C 波，分泌唾液的波形 E1 波和被动取食的波形 E2 波，蚜虫的吸水过程 G 波，但是没有细胞间穿刺的 pd 波。

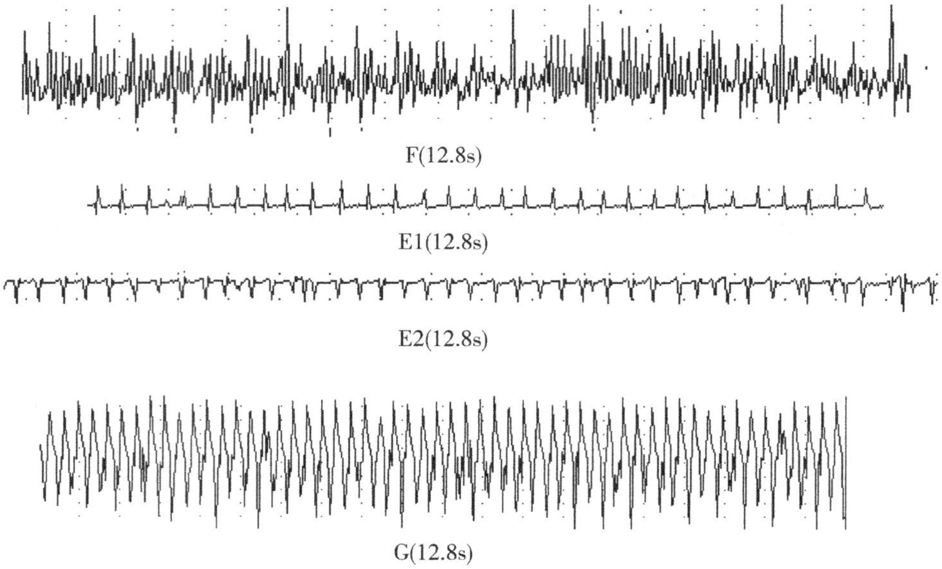

图 24. 1 麦长管蚜与禾谷缢管蚜取食的特征波形

2. 单宁酸对麦蚜取食行为的影响

（1）单宁酸对蚜虫取食波形历期比例的影响：由表 24. 1 可知，麦长管蚜非刺探波历期比例（np%）、刺探波历期比例（C%）和 E2 波历期比例（E2%）存在极显著差异（$F>F_{0.01}$）。在单宁酸浓度>0. 03%，np% 显著高于对照，C%、E2% 显著低于对照。禾谷缢管蚜的 np%、C% 和 E_2% 存在极显著差异（$F>F_{0.01}$）。在含单宁酸浓度>0. 04%，np% 显著高于对照，C% 和 E2% 显著低于对照。

表 24. 1 不同浓度单宁酸对麦蚜取食波形历期的影响

麦长管蚜					
单宁酸（%）	np%±SD	C%±SD	E1%±SD	E2%±SD	E1/（E_1+E_2）%
0. 005	53. 80±0. 00bcBC	46. 20±0 . 00aAB	0. 00±0. 00	0. 00±0. 00 b B	0. 00
0. 030	79. 60±9. 98abAB	17. 46±11. 40 bBC	2. 59±5. 75	0. 33±0. 40 bAB	88. 70
0. 040	90. 45±5. 04a A	9. 24±5. 40 bC	0. 28±0. 57	0. 03±0. 07 bB	90. 32
0. 050	87. 59±9. 85a AB	12. 18±9. 84bC	0. 16±0. 31	0. 08±0. 21bB	40. 00
0. 060	97. 13±0. 66aA	2. 87±0. 66 bC	0. 00±0. 00	0. 00±0. 00 bB	0. 00

续表

单宁酸（%）	np%±SD	C%±SD	E1%±SD	E2%±SD	E1/（E1+E2)%
麦长管蚜					
CK	23.15±15.16cC	51.62±6.80aA	3.80±4.45	21.43±5.13a A	15.06
F 显著性	13.24＊＊＊	13.65＊＊＊	0.99 ns	5.48＊＊	
禾谷缢管蚜					
0.030	79.44±10.29 bB	14.97±7.01aA	0.18±0.29	4.66±5.85bB	3.72
0.040	97.37±2.17aA	2.45±2.32bB	0.04±0.08	0.10±0.18bB	28.57
0.050	89.65±12.3abAB	9.19±10.95abAB	0.63±0.67	0.54±0.68bB	58.85
0.060	97.41±1.19aA	2.49±1.29bB	0.06±0.12	0.04±0.08bB	60.00
CK	69.82±11.96bB	12.04±5.40aAB	2.59±3.60	15.55±8.04aA	14.22
F 显著性	6.83＊＊	6.07＊＊	2.76 ns	4.05＊＊	

注：表中波形参数值为平均值±SD，其中 *F* 值及显著性检验：$P<0.001$ 标＊＊＊，$P<0.01$ 标＊＊，$P<0.05$ 标＊，$P>0.05$ 标 ns。np%、C%、E1%、E2%分别表示非刺探波 np、C、E1、E2 波形历时占总历时的百分比。

（2）单宁酸对蚜虫取食波形次数影响：研究结果（图 24.2）表明，在含不同浓度单宁酸的人工饲料上，麦长管蚜在单宁酸浓度>0.03%，E1、E2 波次数显著低于对照；禾谷缢管蚜在单宁酸浓度>0.03%，E1、E2 波次数显著低于对照。上述结果表明，单宁酸浓度>0.03%对两种蚜虫具有明显拒刺探和拒食作用。

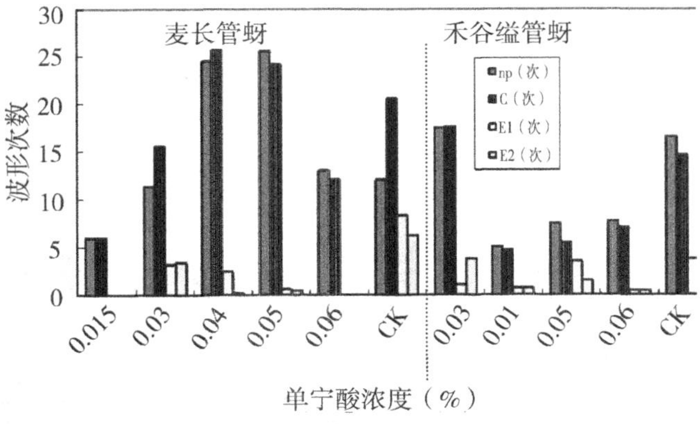

图 24.2 不同浓度的单宁酸对麦蚜取食波形次数的影响

注：np 次、C 次、E1 次、E2 次为相应波形出现的次数。

3. 没食子酸对麦蚜取食行为的影响

（1）没食子酸对蚜虫取食波形历期比例的影响：结果表明（表 24.2）：在不同浓度没食子酸的人工饲料上，麦长管蚜的 np%和 C%存在极显著差异（$F>F_{0.01}$），E1%和 E2%存在显著差异（$F>F_{0.05}$），在 0.05%浓度 np%显著高于对照，C%、E2%显著低于对照。

禾谷缢管蚜的E2%存在差异显著（$F>F_{0.05}$），在0.03%浓度E2%显著低于对照。

表 24.2　不同浓度的没食子酸对麦蚜取食波形历期的影响

没食子酸（%）	np%±SD	C%±SD	E1%±SD	E2%±SD	E1/（E1+E2）%
麦长管蚜					
0.03	81.01±12.36a AB	17.82±12.30 bB	0.86±1.05b	0.32±0.42b	72.88
0.04	51.91±20.76bcBC	31.76±22.82 bAB	3.40±3.26 ab	12.96±21.80a	20.78
0.05	84.43±17.25a A	14.14±15.99 bB	1.09±1.29 b	0.34±0.33b	76.22
0.06	64.62±21.42ab BC	29.41±20.21 bAB	4.79±4.86a	1.63±2.29 b	74.61
CK	23.15±15.16c C	51.62±6.80a A	3.80±4.45 ab	21.43±5.13a	15.06
F 显著性	6.88＊＊	4.96＊＊	2.59 ns	3.04＊	
禾谷缢管蚜					
0.03	92.93±3.43a A	5.15±2.31cC	1.05±0.94	0.87±0.79 b	54.69
0.04	58.71±12.97cC	36.02±10.24 aA	1.90±1.42	3.38±3.78 b	35.98
0.05	80.44±13.37bAB	17.44±12.51 bB	1.02±0.92	1.10±0.94 b	48.11
0.06	75.12±9.84 bBC	18.33±7.15 bB	1.97±1.18	2.01±2.57 b	49.50
CK	69.82±11.96 bBC	12.04±5.40 bcBC	2.59±3.6	15.55±8.04 a	14.22
F 显著性	8.71＊＊＊	12.49＊＊＊	0.82 ns	3.41＊	

注：同表24.1。

（2）没食子酸对蚜虫取食波形次数影响：由图24.3可知，在含不同浓度没食子酸的人工饲料上，没食子酸0.05%浓度对麦长管蚜E2波次数显著低于对照；0.03%浓度对禾谷缢管蚜E2波次数显著低于对照。表明没食子酸0.05%浓度麦长管蚜具有明显拒刺探和拒食作用。

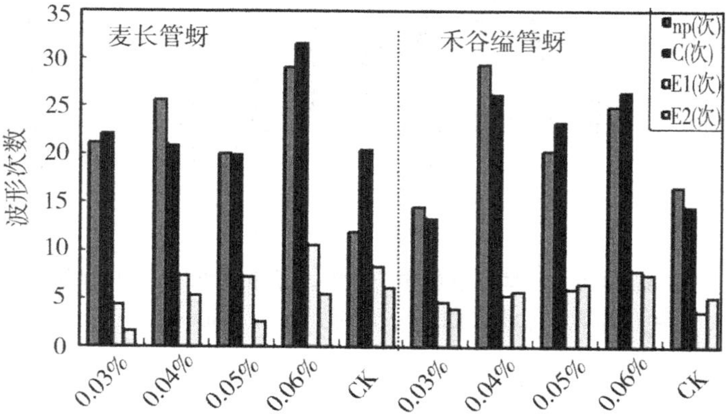

图 24.3　不同浓度的没食子酸对麦蚜取食波形出现次数的影响

4. 几种次生物质混合对麦长管蚜取食行为的影响　从单宁酸、总酚、丁布、门布正交旋转组合试验中，选取对麦长管蚜存活率抑制作用最强的两个组合（17 号、4 号）进行蚜虫刺探电位研究，结果见图 24.4，17 号组合蚜虫的吸食历期 E2%为 2.38%；4 号组合蚜虫吸食历期 E2%为 7.14%；对照（无次生物质的全纯饲料）蚜虫吸食历期比例为 21.43%。两个处理组合 E2 波形比例显著低于全纯饲料的对照。由此可知，次生物质组合对麦长管蚜具有很强的拒食作用。

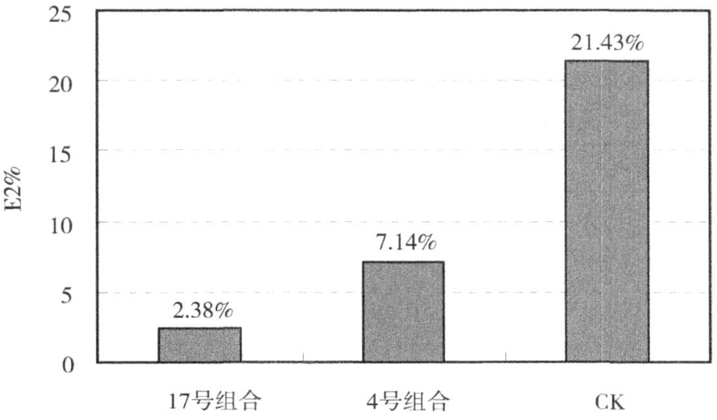

图 24.4　几种次生物质组合对麦长管蚜取食作用的影响

CK 表示无次生物质的纯人工饲料

5. 植物凝集素 ConA 对麦长管蚜取食波形参数影响比较　为了明确伴刀豆凝集素 ConA 对麦长管蚜取食行为的影响，在 0.1~2mg/mL 设置 7 个浓度梯度，研究结果如下。

（1）ConA 对蚜虫取食波形历期比例的影响：研究表明（表 24.3），随着 ConA 浓度增加，蚜虫非取食波 np 波持续历期延长，水溶性唾液 E1 波和被动取食 E2 波持续历期缩短。与对照相比，0.25mg/mL、0.5mg/mL、1.5mg/mL 非取食波 np 波缩短（$P<0.05$）；各测试浓度 ConA（除 1mg/mL 以外）E1 波显著缩短（$P<0.05$）；各测试浓度 ConA E2 波均极显著缩短（$P<0.01$）>1mg/mL 没有检测到取食 E2 波；ConA 对蚜虫刺探波 C 波、吸水过程 G 波没有明显影响。

表 24.3　麦长管蚜在不同浓度 ConA 上取食各波形历期比例

浓度(mg/mL)	np%	C%	E1%	E2%	G%
0	13.93±1.69cA	52.77±3.21a	10.44±1.65aA	10.94±1.65aA	11.92±2.24a
0.1	18.42±2.62bcA	68.95±2.78a	2.81±0.23cAB	4.93±0.91bB	5.13±1.79a
0.25	37.30±3.74aA	54.71±3.69a	2.77±0.57cAB	2.13±0.45bcB	2.84±2.84a
0.5	33.11±2.20abA	55.49±3.71a	1.65±0.42cB	1.01±0.34bcB	8.74±2.21a
0.75	18.50±2.23bcA	61.34±1.99a	3.90±0.72bcAB	1.60±0.36bcB	14.65±1.97a
1	28.56±2.56abcA	51.00±3.63a	8.85±1.68abAB	0.00±0.00cB	11.59±2.92a
1.5	32.62±4.32abA	58.88±3.59a	0.83±0.83cB	0.00±0.00cB	7.64±3.19a
2	26.88±2.43abcA	66.79±2.95a	2.54±0.75cAB	0.00±0.00cB	4.37±0.76a

（2）ConA 对蚜虫食波形出现次数影响：ConA 使非取食波 np 波和刺探波 C 波出现的次数增多，减少了 E1 波和 E2 波出现的次数（表 24.4）。与对照相比，ConA 0.25mg/mL、0.5mg/mL、1.5mg/mL、2mg/mL np 波次数显著增多（$P<0.05$）；0.25mg/mL、1.5mg/mL、2mg/mL 使 C 波次数显著增多（$P<0.05$）；0.5mg/mL、1.5mg/mL、2mg/mL 使 E1 次数显著减少（$P<0.05$）；ConA >0.5mg/mL 使 E2 次数显著减少（$P<0.05$）；0.25mg/mL、1.5mg/mL 处理使 G 波次数显著减少（$P<0.05$）。综上所述，ConA 浓度 1.5mg/mL 对各取食波形出现次数均有显著影响，即增加非取食波与刺探波（np、C）频次，减少取食相关波（E1、E2、G）频次。

表 24.4 麦长管蚜在不同浓度 ConA 上取食各波形出现的次数

浓度（mg/mL）	np 波频次	C 波频次	E1 波频次	E2 波频次	G 波频次
0	12.1±1.1eC	18.9±1.0cB	6.4±0.4aA	3.0±0.3aA	1.9±0.3aA
0.1	21.4±1.7cdeBC	26.0±1.4bcAB	4.9±0.5abcABC	2.8±0.3aA	0.9±0.2abA
0.25	31.5±1.6bcdABC	34.9±1.5abAB	5.3±1.0abAB	2.3±0.4aAB	0.4±0.4bA
0.5	29.1±2.3bcdABC	30.0±2.2abcAB	1.8±0.4bcdABC	0.5±0.1bC	0.8±0.2abA
0.75	16.8±2.0deC	19.3±1.8cB	4.3±0.8abcABC	1.0±0.2bBC	1.0±0.1abA
1	32.0±3.9bcABC	31.0±4.1abcAB	3.0±0.6abcdABC	0.0±0.0bC	0.9±0.2abA
1.5	46.5±2.4aA	43.9±2.8aA	0.1±0.1dC	0.0±0.0bC	0.3±0.1bA
2	40.8±3.2abAB	39.5±2.8abA	1.4±0.5cdBC	0.0±0.0bC	2.0±0.4aA

（3）麦长管蚜在不同浓度 ConA 上 E 波细节比较（表 24.5）：E2 波持续时间>10min［即 E2（10）］的事件作为蚜虫成功取食主要参数。与对照相比，不同浓度 ConA 处理极显著降低蚜虫 E2（10）历期比例（$P<0.01$）和次数（$P<0.01$），且 ConA 浓度 >1.0mg/mL 未检测 E2（10），表明 ConA 抑制蚜虫的成功取食。ConA 处理还极显著减小 E 波（即 E1+E2）历期比例（$P<0.01$），而在 1.5mg/mL E 波历期比例最低。ConA 浓度>1.0mg/mL 时，未检测 E2 波，E1 波历期在 E 波的比例［即 E1/（E1+E2）］显著高于对照（$P=0.05$）。

表 24.5 麦长管蚜在不同浓度 ConA 上取食波形首次出现的时间（min）

浓度（mg/mL）	np（min）	C（min）	E1（min）	E2（min）	G（min）
0	0.46±0.18bA	8.30±3.84a	152.49±16.93bB	196.35±15.56a	263.42±26.251a
0.1	1.12±0.46bA	29.14±12.14a	130.48±16.21bB	226.19±26.93a	360.86±36.56a
0.25	0.00±0.00bA	1.90±0.35a	134.30±20.75bB	171.29±23.22a	285.76±0.00a
0.5	0.00±0.00bA	22.84±5.85a	192.96±26.10bB	177.56±14.87a	318.79±29.45a
0.75	9.32±3.71aA	4.50±1.21a	159.41±36.64bB	393.59±27.15a	296.15±22.67a
1	0.37±0.19bA	15.85±3.62a	92.05±734.00bB		250.70±15.01a
1.5	0.36±0.12bA	2.46±0.44a	448.12±0.00aA		297.90±33.23a
2	0.46±0.15bA	2.23±0.73a	246.44±42.29bAB		363.08±19.43a

通常 E2 波跟随 E1 波而出现，不同浓度 ConA 干扰了麦长管蚜 E2 波跟随 E1 波的事件，结果（表 24.6）表明，随着 ConA 浓度增加，E2 波跟随的 E1 波的持续时间显著缩短（$P<0.05$）；在 ConA 浓度 0.5~1mg/mL，该事件出现次数显著减少（$P<0.05$），当浓度>1.0mg/mL 完全抑制了 E2 波的出现，即 E2 波跟随 E1 波的持续时间和次数均为 0。出现 E 波的蚜虫个体比例反映了 E 波取食阶段的难易程度，结果显示，ConA 1.5mg/mL，出现 E 波蚜虫个体比例显著少于对照（$P<0.05$）（图 24.5）。ConA 处理使出现被动取食 E2 波的蚜虫个体比例减少（图 24.6），1mg/mL 以上的 ConA 完全抑制了 E2 波的出现（$P<0.05$）。

表 24.6　不同浓度 ConA 对 E 波的影响

浓度（mg/mL）	E2（10）%	E2（10）次数	E%	E1/（E1+E2）%
0	9.21±1.67aA	1.3 ±0.6aA	21.38±2.05aA	47.85±6.07bB
0.1	3.73±0.98bB	0.6 ±0.2abAB	7.74±1.03bcB	47.27±3.96bB
0.25	0.72±0.24bB	0.25 ±0.16bAB	4.89±0.97bcB	61.91±4.43bAB
0.5	0.68±0.68bB	0.1 ±0.1bB	2.66±0.63bcB	74.76±6.84abAB
0.75	1.25±0.34bB	0.5 ±0.3abAB	5.50±1.06bcB	74.61±3.35abAB
1	0.00±0.00bB	0.0 ±0.0bB	8.85±1.68bB	100.00±0.00aA
1.5	0.00±0.00bB	0.0 ±0.0bB	0.83±0.83cB	100.00±0.00aA
2	0.00±0.00bB	0.0 ±0.0bB	2.54±0.75bcB	100.00±0.00aA

表 24.7　麦长管蚜在不同浓度 ConA 上取食 E2 波跟随 E1 波的事件

浓度（mg/mL）	无 E2 跟随的 E1 的时间（min）	无 E2 跟随的 E1 的次数	有 E2 跟随的 E1 的时间（min）	有 E2 跟随的 E1 的次数
0	27.53±3.78abAB	3.63 ±0.75aA	22.57±5.69aA	3.00 ±0.60aA
0.1	6.59 ±0.92bB	2.13 ±0.69abcA	6.88 ±0.90bAB	2.75 ±0.53aA
0.25	9.35 ±2.37bB	2.88 ±1.42abcA	3.94±0.60bB	2.25 ±0.70aAB
0.5	5.4 ±1.540bB	1.25 ±0.62abcA	2.34±0.73bB	0.50 ±0.27bC
0.75	12.55±2.71bB	3.38 ±1.28abA	5.66 ± 1.21bB	1.00 ±0.46bBC
1	42.50±8.06aA	3.00 ±1.20abcA	0.00 ±0.00bB	0.00 ±0.00bC
1.5	3.98 ±3.98bB	0.13 ±0.13cA	0.00 ±0.00bB	0.00 ±0.00bC
2	12.17±3.594bB	0.50 ±0.19bcA	0.00 ±0.00bB	0.00 ±0.00bC

（四）讨论

应用 EPG 技术可测定麦蚜在人工饲料上取食行为、建立标准检测和统计分析方法，由于人工饲料组分已知，为外源次生物质、毒蛋白等抗蚜评价提供均一的介质，避免因不同批次小麦植株等因营养成分不同而造成对抗性程度变异的误差。从该研究看出，

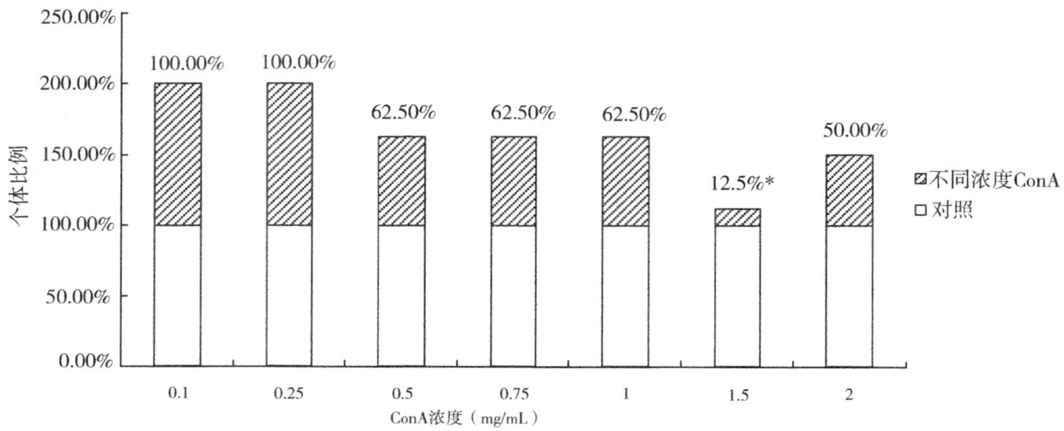

图 24.5　出现 E 波的个体比例

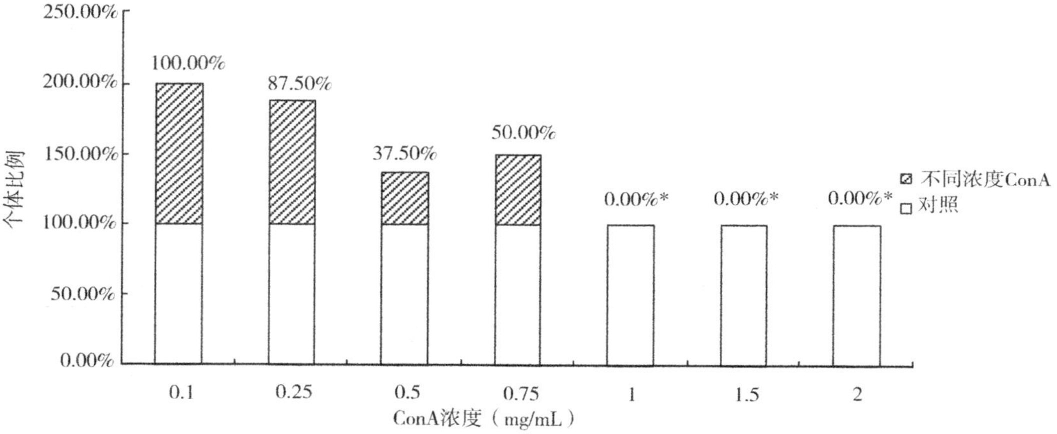

图 24.6　出现 E2 波的个体比例

次生物质单宁酸浓度>0.03%时可使麦长管蚜及禾谷缢管蚜的非刺探波形比例 np%显著增加，路径波形 C%、唾液分泌波形 E1%和吸食波形 E2%比例均显著低于对照；并且 E2%比例在 0~0.33%（麦长管蚜），0.04%~4.66%（禾谷缢管蚜），而非刺探波形占 53.8%~97.13%（麦长管蚜）和 79.44%~97.41%（禾谷缢管蚜），表明单宁酸对麦蚜具有极强的拒刺探和拒食作用。同样，总酚（没食子酸）对两种麦蚜也具有拒刺探和拒食作用，但抑制强度比单宁酸弱。本文利用人工饲料与 EPG 技术结合，对多种次生物质混合对蚜虫取食行为的影响进行初步研究，这是模拟植物中多种次生物质并存环境对蚜虫取食复合作用做了有益探索。

伴刀豆凝集素 ConA 对蚜虫有毒性，但不同蚜虫敏感性不同。Rahbé 等（1995）报道 ConA 对豌豆蚜、桃蚜、羽扇豆蚜、棉蚜、马铃薯长管蚜、茄无网蚜 6 种蚜虫有毒性，其中豌豆蚜和羽扇豆蚜对 ConA 最敏感。半致死浓度 LC_{50}豌豆蚜 310μg/mL、羽扇豆蚜 330μg/mL、桃蚜 1 200μg/mL，其余三种>1 500μg/mL；在 ConA >750μg/mL 的人工饲料上，豌豆蚜和羽扇豆蚜发育至成虫期前全部死亡。人工饲料中含有 0.01%~0.1%（*W/V*）的 ConA 能减少桃蚜的累积产仔数 35%~65%，延长发育历期，个体大小减小了 30%；转 ConA 的马铃薯上桃蚜若虫数量减少了 15%~45%（Gatehouse et al.，1999）。邓青（2012）通过人工饲料添加 ConA 对麦长管蚜的 LC_{50}为 1.45mg/mL。高浓度 ConA 对麦长管蚜有抑制发育及致死效应。在饲喂的最高浓度 2mg/mL 时，ConA 在饲喂第 5 天致蚜虫全部死亡。

植物凝集素 ConA（Sauvion 等，2003）、PHA（Sprawka et al.，2010）干扰其正常生理代谢和细胞功能，能影响豌豆蚜取食行为，非取食波 np 波和摄食阶段 E2 波持续时间缩短。本研究结果表明，ConA 能使麦长管蚜的非取食波 np 波历期、被动取食阶段 E2 波历期缩短，当 ConA 浓度>1mg/mL 时没有 E2 波出现，表明高浓度的 ConA 能完全抑制蚜虫的被动取食 E2 波的出现。因此，在植物中高表达凝集素蚜虫防治有重要意义。

本实验中采用 3~4 龄的若蚜为研究对象研究次生物质的影响，用无翅成蚜研究凝集素的影响。麦蚜的成蚜与若蚜之间刺探行为是否相同尚无报道。因此，今后需对麦蚜成蚜、若蚜之间取食行为差异进一步探讨，以及分泌蜜露、产仔时的波形进行记录研究，将有助于进一步完善抗性品种的室内刺探电位鉴定标准。

参考文献

[1] 陈巨莲，倪汉祥，丁红建，等．麦长管蚜全纯人工饲料的研究［J］．中国农业科学，2000，33（3）：54-59.

[2] 陈巨莲．小麦抗蚜生化机制和 G 蛋白在离体昆虫细胞系中表达、纯化及活性测定［D］．北京：中国农业科学院，2001.

[3] 邓青．转基因抗蚜小麦抗麦长管蚜效果的评价技术研究．中国农业科学院硕士学位论文，2012.

[4] 李巧丝，武予清，李素娟，等．麦蚜危害对优质面包小麦品质的影响．植物保护，2003，29（1）：43-44.

[5] 杨益众，印毅，王红，等．蜂蜜模拟麦蚜蜜露危害对小麦产量和主要营养品质的影响．中国农学通报，2005，21（1）：268-271.

[6] 杨益众，戴志一．麦蚜的阶段性为害对小麦产量和品质影响的研究．昆虫知识，1995，32（1）：10-13.

[7] 袁浩．麦长管蚜危害对优质小麦产量和面粉品质特征的影响．山东农业大学硕士研究生论文，2013.

[8] 张向才，周广和，史明，等．麦蚜远距离迁飞和传毒规律的研究．植物保护学报，1985，12（1）：9-15.

[9] HILDER V A，POWELL K S，GATEHOUSE A M R，et al. Expression of snowdrop lectin in transgenic tobacco plants results in added protection against aphids. *Transgenic Research*，1995，4（1）：18-25.

[10] GATEHOUSE A M R，DAVISON G M，STEWART J N. Concanavalin A inhibits development of tomato moth（*Lacanobia oleracea*）and peach-potato aphid（*Myzus persicae*）when expressed in transgenic po-

tato plants. *Molecular Breeding*, 1999, 5 (2): 153-165.

[11] DUTTA I, MAJUMDER P, SAHA P, et al. Constitutive and phloem specific expression of *Allium sativum* leaf agglutinin (ASAL) to engineer aphid (*Lipaphis erysimi*) resistance in transgenic Indian mustard (*Brassica juncea*). *Plant Science*, 2005, 169 (6): 996-1007.

[12] LESZCZYNSKI B, TJALLINGII W F, DIXON A F G, et al. Effect of methoxyphenols on grain aphid feeding behaviour. *Ent. Exp. Appl.*, 1995, 76: 157-162.

[13] LESZCZYNSKI B., URBANSKA A., ROZBICKA B., et al., Application of EPG in determination of cereal resistance to aphid. *Progress in Plant Protection*, 1997, 37 (2): 40-43.

[14] RAHBÉ Y, SAUVION N, FEBVAY G, et al. Toxicity of lectins and processing of ingested proteins in the pea aphid *Acyrthosiphon pisum*. *Entomologia Experimentalis et Applicata*, 1995, 76 (2): 143-155.

[15] SAUVION N, CHARLES H, FEBVAY G, et al. Effects of jackbean lectin (ConA) on the feeding behaviour and kinetics of intoxication of the pea aphid, *Acyrthosiphon pisum*. *Entomologia Experimentalis et Applicata*, 2003, 110 (1): 31-44.

[16] SPRAWKA I & GOLAWSKA S. Effect of the lectin PHA on the feeding behavior of the grain aphid. *Journal of Pest Science*, 2010, 83 (2): 149-155.

(陈巨莲　邓　青　范　佳　孙京瑞)

第二十五章　EPG 在灰飞虱传播水稻条纹病毒（RSV）研究中的应用

（一）研究背景

灰飞虱 *Laodelphax striatellus*（半翅目飞虱科）是亚洲地区水稻上的重要害虫。它不仅刺吸水稻汁液造成危害，更重要的是传播水稻条纹叶枯病毒（rice stripe virus，RSV）造成水稻条纹叶枯病的流行。2004 年江苏省仅灰飞虱传播的水稻条纹叶枯病造成的经济损失就高达 100 亿元（张景飞等，2005）。RSV 是纤细病毒属（*Teuivirus*）的代表种（Ramirez & Haenni，1994），由灰飞虱以持久、循回增殖型方式传播（Toriyama，1986）。灰飞虱取食获毒后，RSV 在介体内增殖、复制，可以经卵传给子代（Toriyama，1986）。Deng 等（2013）通过研究对 RSV 经卵传递的途径提出假设，认为 RSV 复制后最初聚集在卵泡细胞，而后随着营养扩散穿过卵壳在卵母细胞扩散。灰飞虱取食水稻植株时的获毒、传毒环节缺乏研究。

刺吸电位技术（electrical penetration graph，EPG）是研究植食性刺吸式口器昆虫在寄主植物上刺探和取食行为的电生理技术（闫凤鸣，2011）。迄今，EPG 技术已被用于昆虫对寄主植物的选择性（Gabrys et al.，1997；Gabrys & Pawluk，1999）、植物抗虫性机制（Miao & Han，2007；Seo et al.，2010）、拒食因子（Zapata et al.，2010）和农药（何月平等，2010；Polston & Sherwood，2003）对昆虫取食行为的影响等研究领域。EPG 研究的结果表明，蚜虫对于持久性病毒大麦黄矮病毒（barley yellow dwarf virus，BYDV）（Prado & Tjallingii，1994）、粉虱对于持久性病毒番茄黄化曲叶病毒（tomato yellow leaf curl virus，TYLCV）（Jiang et al.，2000）以及半持久性病毒莴苣萎黄病毒（lettuce chlorosis virus，LCV）（Johnson et al.，2002）的传播都是在介体持续吸食前唾液分泌（E1）阶段进行。稻飞虱传递病毒的取食行为细节尚缺乏研究。最近，我们利用刺吸电位技术（EPG）研究了灰飞虱 EPG 波形与其取食行为之间的对应关系（荆裴等，2013）。在此基础上，研究了灰飞虱获得 RSV 以及传递 RSV 的取食行为环节，并进一步研究了抗性品种对灰飞虱取食行为及传毒率的影响（Jing et al.，2015），为通过调控介体昆虫的取食行为从而控制水稻病毒的危害提供依据。

（二）材料与方法

1. 植物和昆虫　供试植物选用武育粳 3 号水稻健康植株（3 片真叶期）及发病植株、盐稻 8 号水稻健康植株（3 片真叶期）及发病植株、田间采集的发病植株（用于灰飞虱获毒的取食行为环节判定试验)。

灰飞虱采自扬州大学实验田，在温度25℃±1℃，65%±5% RH，光周期12 L：12 D，光照强度3 成品000 ~ 4 000 Lux的温室内以感虫水稻品种武育粳3号饲养。参照刘海建等（2007）方法，在实验室内进行无毒虫及带毒虫筛选。将雌雄灰飞虱单对饲养于盆栽（直径5cm，高7cm）（编号）的武育粳3号（感虫）水稻上。等雌虫产卵一段时间之后收集雌虫，并检测雌虫是否带毒。带毒雌虫后代即为带毒灰飞虱群体，无毒雌虫后代即为无毒灰飞虱群体。

2. EPG记录 将灰飞虱饲水饥饿1h后，迅速用可溶导电胶将灰飞虱的中胸背板与一根直径为20μm，长2~3cm的金丝粘合，再通过铜丝连接于8通道昆虫刺吸信号电子记录仪上（型号：CR-8 DC-EPG Ⅰ，购自武汉普赛斯电子技术有限公司）。点击开始采集按钮后，将试虫放置在供试植株的茎秆上进行取食。记录过程在温室内的屏蔽罩中进行。

3. 灰飞虱取食的EPG波形认定 利用EPG技术观察、记录不同虫龄和性别的灰飞虱在取食过程中产生的EPG波形。参照褐飞虱 *Nilaparvata lugens* 取食行为与EPG波形的对应关系（Seo et al.，2009)，对灰飞虱的取食波形进行类比、分析及认定。

4. 无毒灰飞虱获毒的取食行为环节判定 主要试验方法采用即时中断技术（张鹏飞等，2001)。试验开始记录后，将粘有金丝的灰飞虱放在带毒植株的叶片上进行取食。实时观察回路中电信号的变化，当出现某一特定波形（N1波，N2波，N3波或N4波）结束时，迅速中断灰飞虱的取食。将中断取食后的灰飞虱置于另一株健康水稻上，继续EPG记录5min，检验灰飞虱的口针是否受损伤。将口针未受损伤的灰飞虱置于健康植株上饲养10~15d以确保病毒循回增殖。用Dot-ELISA法检测供试灰飞虱的带毒情况（ELISA试剂盒购自江苏省农科院)，统计各个处理灰飞虱的获毒率。获毒率=获毒虫数/总虫数。单头灰飞虱为一个重复，每个处理至少设置30个重复。灰飞虱带毒情况检测采用Dot-ELISA法，参照周益军等（2004）。

5. 带毒灰飞虱传毒的取食行为环节判定 主要试验方法是采用即时中断技术（张鹏飞等，2001)。将带毒灰飞虱放在健康植株上进行取食。实时观察回路中电信号的变化，当出现某一特定波形结束时，迅速中断灰飞虱的取食。供试灰飞虱再次检测带毒率，阳性结果视为有效虫。将带毒灰飞虱取食处理后的单株水稻苗编号并继续培养10~15d以确保病毒循回增殖（28~30℃，12L：12D)。用液氮采集各处理水稻叶片，采用RT-PCR方法（嵇朝球等，2005）检测供试水稻植株的带毒情况（总RNA提取试剂盒购自天根生化科技有限公司，引物购自上海生物工程技术有限公司)，统计各个处理灰飞虱的传毒率。传毒率=带毒苗数/总苗数。每个处理至少设置30个重复。

6. 无毒灰飞虱在不同抗性水稻品种上的取食行为和获毒率研究 利用EPG分别记录无毒灰飞虱在发病盐稻8号（抗病抗虫）及武育粳3号（感病感虫）水稻植株上连续12h的取食情况，之后将灰飞虱置于健康植株上饲养10~15d以确保病毒循回增殖。用Dot-ELISA法测定灰飞虱的获毒情况并分析水稻品种的抗感性对灰飞虱的取食行为以及获毒率的影响。单头灰飞虱为一个重复，每个处理至少25个重复。

7. 带毒灰飞虱在不同抗性水稻品种上的取食行为和传毒率研究 利用EPG分别记

录带毒灰飞虱在健康盐稻8号（抗病抗虫）及武育粳3号（感病感虫）水稻幼苗上连续12 h的取食情况。供试灰飞虱再次检测带毒率，阳性结果视为有效虫。将处理后的单株水稻苗编号并继续培养10~15d以确保病毒循回增殖（28~30℃，12L：12D）。测定水稻植株的带毒情况（同3.6）并分析水稻品种的抗感性对灰飞虱的取食行为以及传毒率的影响。每个处理设置至少30个重复。

8. 数据处理　用DPS统计软件对不同虫龄、性别灰飞虱取食行为各个阶段的持续时间及灰飞虱在不同水稻抗性品种取食时的各个环节的持续时间分别进行单因素试验统计分析。处理结果均以“平均值±标准误”表示。

本试验中关于灰飞虱传播条纹叶枯病行为机制及不同抗性水稻品种对灰飞虱获毒效率和传毒效率影响中的获毒率、传毒率数据均用DPS统计软件进行χ^2独立性检验，从而分析其是否有显著差异，进而判定灰飞虱获毒和传毒的取食行为环节及不同抗性品种对灰飞虱的获毒效率和传毒效率的影响。

（三）结果

1. 灰飞虱取食的EPG波形认定　参照Seo等（2009）对褐飞虱取食过程中EPG波形的命名方法，依据波谱的形状和频率，将灰飞虱在取食过程中产生的EPG波形分为7种类型：即NP、N1、N2-a、N2-b、N3、N4和N5（图25.1）。它们分别代表灰飞虱口针在水稻表面和组织内部刺探、移动、分泌唾液和取食的典型波形。

其中，NP波表示灰飞虱的口针尚未产生刺探行为，此时灰飞虱在水稻表面停留或爬动，其产生的电势水平几近为零，EPG波形图谱近似直线（图25.1B）。N1波表示口针的刺探行为，当灰飞虱口针刺入植物组织内部时，其电势水平会根据植物本身的电势水平产生急剧的升高或降低，但其刺探过程通常只维持数秒（图25.1C）。刺探之后，灰飞虱的口针开始向维管束移动并伴随有唾液的分泌，在此阶段会形成一段无规律且振幅较小的波形，我们将此波形定义为N2-a波（图25.1C）。在N2-a波之后，灰飞虱开始持续分泌唾液，产生一段比较规律的波形即N2-b波（图25.1D）。通常N2波之后会紧跟着N3波或N5波，N3波是口针在临近韧皮部细胞外面移动的波形，此波形通常每3~7 s重复一次，并且与其他波形相比波形的电势水平变化幅度较大（图25.1E）。N4波与韧皮部取食汁液有关，此波形在N3波之后产生，是一段频率较高（3~8 Hz）、振幅较小、波峰向上的有规律重复的波形，通常N4波能持续几分钟或几小时（图25.1F）。N5波表示吸食木质部导管汁液，此波形的波峰向下并且是有规律重复的，其频率可高达18 Hz（图25.1G）。

综合以上EPG波形图谱信息，确立了灰飞虱EPG波形与其刺探、取食行为的对应关系（表25.1），初步明确了灰飞虱取食波形的生物学意义。

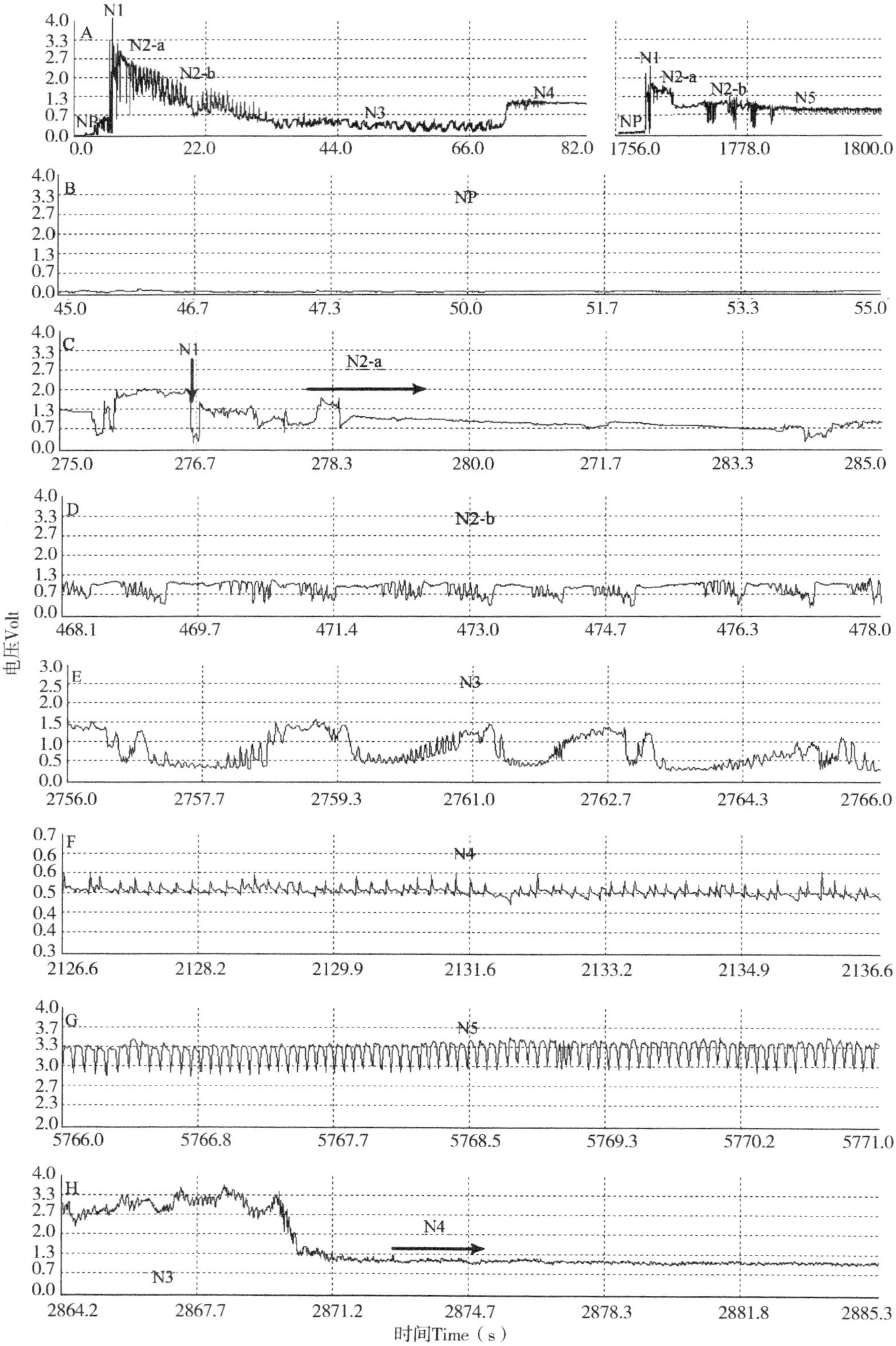

图 25.1　灰飞虱 EPG 波形图谱（荆裴等，2013）

表 25.1　灰飞虱口针刺探行为与对应的 EPG 波形（荆裴等，2013）

EPG 波形		刺探行为	频率（Hz）
NP		口针未刺探 non-penetration	-
N1		口针刺探进入水稻表皮 the stylet penetrating into rice epidermis	-
N2	N2-a	分泌唾液及口针移动 salivation and movement of stylet	-
	N2-b	持续分泌唾液 sustained salivation	0.8～1
N3		口针在临近韧皮部的细胞外移动 extracellular movement of stylet near the phloem region	0.14～0.3
N4		吸食韧皮部汁液 sap ingestion in phloem	3～8
N5		吸食木质部水分 water ingestion in xylem	15～18

2. 无毒灰飞虱获毒的取食行为环节判定　所有灰飞虱在口针刺探（N1 波）阶段结束后中断取食，其获毒率为 0，表明口针刺探阶段不能获毒。从唾液分泌阶段（N2 波）到韧皮部取食阶段（N4 波）结束后中断取食，其获毒率介于 13.9%～42.9%，表明灰飞虱在唾液分泌阶段及其随后的取食环节阶段均可从带毒植株体内获取病毒并开始增殖。在 N4 波结束中断取食，灰飞虱的获毒率较 N2 波中断时的高出 2.1 倍。与 N2 结束后中断取食的获毒率相比，在 N3、N4 波结束后中断取食的获毒率无显著差异（表 25.2）。

表 25.2　不同取食阶段结束后中断无毒灰飞虱取食的获毒率（Jing et al.，2015）

EPG 波形	有效试虫（头）	获毒灰飞虱虫数（头）	获毒率（%）	χ^2 检测		
				处理	χ^2值	*P* 值
N1	30	0	0	—	—	—
N2	36	5	13.9	N2 vs N4	6.00	0.0143
N3	30	9	30.0	N2 vs N3	1.67	0.1964
N4	35	15	42.9	N3 vs N4	0.66	0.4162

注：自由度为 1。N1 表示口针刺探进入水稻表皮；N2 表示唾液分泌；N3 表示口针在临近韧皮部的细胞外移动；N4 表示吸食韧皮部汁液。

3. 带毒灰飞虱传毒的取食行为环节判定　在口针刺探阶段结束中断取食的灰飞虱传毒率为 0，显著低于在唾液分泌（N2）及 N3、N4 阶段中断取食后灰飞虱的传毒率（N2：33.33%，N3：46.67%，N4：50.00%）。由此得知：灰飞虱在口针刺探阶段不能传递病毒，其传毒的取食行为主要发生在唾液分泌阶段、口针在韧皮部细胞外面的探索阶段及韧皮部持续取食阶段。在 N2 波阶段与 N3 波阶段、N2 波阶段与 N4 波阶段、N3 波阶段与 N4 波阶段中断的灰飞虱传毒率都没有显著性差异（表 25.3）。

表 25.3 不同取食阶段中断带毒灰飞虱取食后的传毒率（Jing et al.，2015）

EPG 波形	有效试虫（头）	感染病毒植株数（株）	传毒率（%）	χ^2检测		
				处理	χ^2值	P 值
N1	30	0	0	—	—	—
N2	30	10	33.3	N2 vs N4	1.10	0.2949
N3	30	14	46.6	N2 vs N3	0.63	0.4292
N4	30	15	50.0	N3 vs N4	0	0.9999

注：自由度为 1。N1 表示口针刺探进入水稻表皮；N2 表示唾液分泌；N3 表示口针在临近韧皮部的细胞外移动；N4 表示吸食韧皮部汁液。

4. 无毒灰飞虱在不同抗性水稻品种上的取食行为和获毒率 在取食的 12h 内，无毒灰飞虱在盐稻 8 号上的刺探次数（39.0 ± 3.6）与武育粳 3 号上（40.2 ± 3.6）无显著差异（F = 0.06，df = 1，49，P = 0.813 9）。盐稻 8 号上灰飞虱非刺探时间比武育粳 3 号的延长 45.5%（F = 9.76，df = 1，49，P = 0.003 0）。盐稻 8 号上灰飞虱的 N2 波、N3 波持续的时间与武育粳 3 号上无显著差异（N2：F = 2.51，df = 1，49，P = 0.119 6；N3：F = 2.36，df = 1，49，P = 0.131 1）。而盐稻 8 号上 N4 波的持续时间显著短于武育粳 3 号（F = 4.63，df = 1，49，P = 0.036 4）。两个品种上灰飞虱 N5 持续的时间极短且无显著差异（F = 0.01，df = 1，49，P = 0.918 3）（图 25.2）。这表明，抗性品种盐稻 8 号植株的表皮细胞、维管束、韧皮部细胞外围及木质部内存在抗性因子对灰飞虱口针在其之内的行为产生阻碍。

无毒灰飞虱在带毒盐稻 8 号上取食 12h 的获毒率为 12.0%，仅为感性品种武育粳 3 号上获毒率（42.3%）的 28%（表 25.4）。结果表明抗性品种抑制灰飞虱获毒。

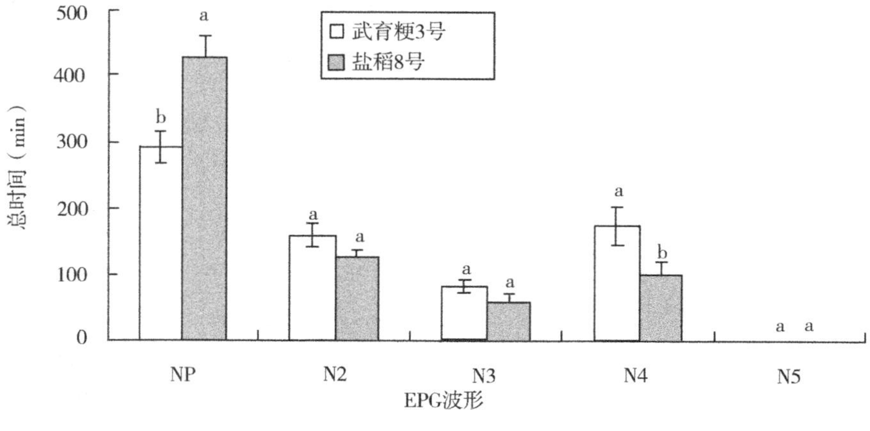

图 25.2 无毒灰飞虱在染毒的不同抗性品种上取食 12h 的行为参数（Jing et al.，2015）

注：NP 表示无刺探；N2 表示唾液分泌；N3 表示口针在临近韧皮部的细胞外移动；N4 表示吸食韧皮部汁液；N5 表示吸食木质部汁液。不同字母表示处理间有显著差异（P<0.05）。

表 25.4 无毒灰飞虱在带毒不同抗性水稻植株上取食12h后的获毒率（Jing et al.，2015）

水稻品种	有效试虫（头）	获毒灰飞虱（头）	获毒率（%）	χ^2 检测
武育粳3号	26	11	42.3	χ^2=4.46，df=1，
盐稻8号	25	3	12.0	*P* =0.034 8

5. 带毒灰飞虱在不同抗性水稻品种上的取食行为和传毒率 带毒灰飞虱在健康盐稻8号上的刺探次数（54.3±9.3）与武育粳3号上接近［（36.4±4.2）min］（*F* = 3.03，df=1，58，*P* = 0.087 0）。盐稻8号上的非刺探波形时间达（429.46 ± 24.96）min，比武育粳3号上［（321.34 ± 29.91）min］的长0.34倍（*F* = 7.70，df = 1，58，*P* = 0.007 4）。盐稻8号上N4波较武育粳3号上的短0.55倍（*F* = 10.71，df = 1，58，*P* = 0.001 8）。盐稻8号与武育粳3号上N2、N3及N5波形持续时间无显著差异（N2：*F* = 0.16，df = 1，58，*P* = 0.694 4；N3：*F* = 0.68，df = 1，58，*P* = 0.413 7；N5：*F* = 0.03，df=1，58，*P* = 0.955 3）（图25.3）。

带毒灰飞虱在健康盐稻8号上取食12h后可传毒，传毒率为63.3%，较在武育粳3号上的传毒率（100.0%）降低了26.7%（表25.5）。结果表明盐稻8号不适宜灰飞虱传毒。

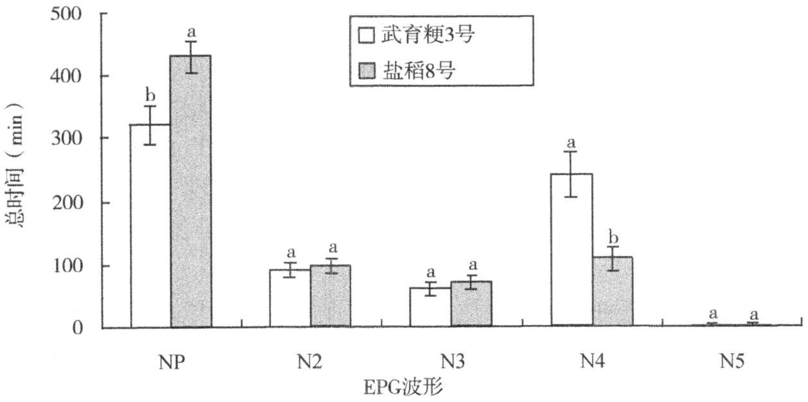

图 25.3 带毒灰飞虱在健康的不同抗性品种上取食12h的行为参数（Jing et al.，2015）

注：NP表示无刺探；N2表示唾液分泌；N3表示口针在临近韧皮部的细胞外移动；N4表示吸食韧皮部汁液；N5表示吸食木质部汁液。不同字母表示处理间有显著差异（*P*<0.05）。

表 25.5 带毒灰飞虱在健康的不同抗性水稻植株上取食12h的传毒率（Jing et al.，2015）

水稻品种	有效试虫（头）	感染植株	传毒率（%）	χ^2检测
武育粳3号	30	30	100.0	χ^2= 11.12，df =1，
盐稻8号	30	19	63.3	*P* = 0.000 8

（四）讨论

1. 与灰飞虱传毒相关的取食行为环节 我们的研究发现灰飞虱在取食过程中产生7种典型的EPG波形，即NP、N1、N2-a、N2-b、N3、N4和N5，与褐飞虱的基本相同（Kimmins，1989；Seo et al.，2009），所不同的是，其在产生代表口针移动及分泌唾液的波之后会经历一段规律重复的波形，然后再进入N3波或N5波阶段，我们将这段规律重复的波形定义为N2-b波，表示灰飞虱开始持续分泌唾液。

灰飞虱在取食过程中产生的7种典型的EPG波形，我们重点关注了N1、N2、N3、N4等4种波形与灰飞虱获毒、传毒的关联性。NP为非刺探波形，不可能与获毒、传毒有关。代表木质部吸食N5出现的时间极短，因此亦未加以考虑。Sogawa（1982）认为褐飞虱分泌2种唾液，一种为可凝结的唾液，形成口针鞘；另一种为水样唾液，含有消化酶。Pirone & Perry（2002）研究发现蚜虫口针进入细胞内时水样唾液分泌与传毒有关。我们推测灰飞虱的传毒很可能发生在N2b水样唾液分泌阶段。因此，实验中N2波的中断是指N2b结束时中断取食，研究唾液分泌与传毒的关系。

一般认为，由同翅目昆虫传播的位于韧皮部病毒的获得发生在韧皮部吸食阶段（Prado & Tjallingii，1994；Limburg et al.，1997）。我们的研究发现，灰飞虱获得RSV确实可以发生在韧皮部吸食（N4）阶段。同时，我们发现灰飞虱唾液分泌阶段（N2）及随后的口针在细胞间移动（N3）亦能获毒。Liang等（2005）包含有RSV的无定形半透明内含体（amorphous semi-electron-opaque inclusion bodies，ASOs）主要分布在叶肉细胞中，同时，表皮细胞、韧皮部组织及维管束的薄壁细胞中亦有分布。在唾液分泌或口针细胞间移动时，灰飞虱的口针已刺破表皮细胞，所以此阶段灰飞虱能获毒就不难理解了。同时，我们发现韧皮部吸食中断取食的获毒率明显高于唾液分泌后中断取食的获毒率（表25.2），推测灰飞虱在大量吸食韧皮部汁液时获得了更多的病毒内含体。

我们的研究明确了唾液分泌阶段、口针在细胞间移动以及韧皮部吸食结束后中断取食的灰飞虱均能传毒。对灰飞虱而言，口针在细胞间移动、韧皮部吸食之前必然有唾液分泌环节（Sogawa，1982；Seo et al.，2009），意即唾液分泌是包含于口针在细胞间移动、韧皮部吸食过程的。因此，唾液分泌对灰飞虱传递RSV至关重要。这与先前的报道认为持久性病毒及半持久性病毒的传递主要发生在唾液分泌阶段一致（Jiang et al.，2000；Johnson et al.，2002；Prado & Tjallingii，1994）。

2. 抗性品种对灰飞虱取食行为及随后的传毒效率的影响 品种抗性对无毒、带毒灰飞虱的取食行为均有显著影响。在抗性品种盐稻8号上，灰飞虱的非刺探时间显著长于感性品种武育粳3号。灰飞虱在2个水稻品种上均能实现韧皮部持续吸食，但在抗性品种盐稻8号上吸食的时间显著短于感性对照武育粳3号。Khan & Saxena（1984）报道，白背飞虱 *Sogatella furcifera* 在感性品种TN1上刺探时间短、吸食时间长，而在抗性品种IR 2035-117-3上反复刺探、吸食时间短。进一步的研究发现，水稻茎秆蒸馏提取物影响了飞虱的取食行为（Khan & Saxena，1986）。抗性水稻品种IR46和IR62能阻止褐飞虱持续吸食（Kimmins，1989）。我们推测抗性品种存在1个或多个因子阻碍灰飞虱持续吸食。盐稻8号抑制灰飞虱取食的机制有待进一步研究。

无毒灰飞虱在盐稻8号上取食12h后的获毒率显著低于武育粳3号。我们的研究结果显示无毒灰飞虱在盐稻8号上持续吸食时间明显较短，而持续吸食环节与获毒相关。我们推测盐稻8号上较低的获毒率源于持续吸食时间较短。这一推测尚待进一步的研究证实。

一些研究表明，在接种了相同量的病毒后，抗性品种的病毒感染率显著低于感性品种（孙黛珍等，2006；Ise et al.，2002）。杨金广等（2008）的研究发现RSV在抗性品种KT95-418内的复制慢于感性品种武育粳3号。显然，上述品种的抗性主要是对RSV的抗性或耐受性。本研究中，灰飞虱在抗性品种盐稻8号取食12h后的传毒率显著低于武育粳3号，且抗性品种上与传毒相关的环节持续吸食时间较短。因此，盐稻8号上的传毒率较低可能与介体昆虫持续吸食时间短有关。我们下一步的研究将进一步明确抗性品种上取食行为与传毒效率之间的关系。

参考文献

[1] DEBG J H, LI S, HONG J, et al. Investigation on subcellular localization of rice stripe virus in its vector small brown planthopper by electron microscopy [J]. *Virology Journal*, 2013, 10: 310.

[2] GABRYS B, PAWLUK M. Acceptability of different species of Brassicaceae as hosts for the cabbage aphid [J]. *Entomologia Experimentalis et Applicata*, 1999, 91: 105-109.

[3] GABRYS B, TJALLINGII W F, VAN BEEK T A. Analysis of EPG recorded probing by cabbage aphid on host plant parts with different glucosinolate contents [J]. *Journal of Chemical Ecology*, 1997, 23 (7): 1 661-1 673.

[4] ISE K, ISHIKAWA K, LI C Y et al. Inheritance of resistance to rice stripe virus in rice line 'BL 1' [J]. *Euphytica*, 2002, 127: 185-191.

[5] JIANG Y X, DE BIAS C, BARRIOS L et al. Correlation between whitefly (Homoptera: Aleyrodidae) feeding behavior and transmission of tomato yellow leaf curl virus [J]. *Annals of the Entomological Society of America*, 2000, 93: 573-579.

[6] JING P, BAI S F, LIU F. Effects of rice resistance on the feeding behavior and subsequent virus transmission efficiency of *Laodelphax striatellus* [J]. *Arthropod-Plant Interactions*, 2015, 9: 97-105.

[7] JOHNSON D D, WAIKER G P, CREAMER R. Stylet penetration behavior resulting in inoculation of a semipersistently transmitted closterovirus by the whitefly *Bemisia argentifolii* [J]. *Entomologia Experimentalis et Applicata*, 2002, 102: 115-123.

[8] KHAN Z R, SAXENA R C. Effect of steam distillate extracts of resistant and susceptible rice cultivars on behavior of *Sogatella furcifera* (Homoptera: Delphacidae) [J]. *Journal of Economic Entomology*, 1986, 79: 928-935.

[9] KHAN Z R, SAXENA R C. Electronically recorded waveforms associated with the feeding behavior of *Sogatella furcifera* (Homoptera: Delphacidae) on susceptible and resistant rice varieties [J]. *Journal of Economic Entomology*, 1984, 77: 1479-1482.

[10] KIMMINS F M. Electrical penetration graphs from *Nilaparvata lugens* on resistant and susceptinle rice varieties [J]. *Entomologia Experimentalis et Applicata*, 1989, 50: 69-79.
[11] LIANG D, QU Z, MA X, et al. Detection and localization of rice stripe virus gene products in vivo [J]. *Virus Genes*, 2005, 31: 211-221.
[12] LIMBURG D D, MAUK P A, GODFRUY L D. Characteristics of Beet yellows closterovirus transmission to sugar beets by *Aphis fabae* [J]. *Phytopathology*, 1997, 87: 766-771.
[13] MIAO J, HAN B Y. Probing behavior of the tea green leafhopper on different tea plant cultiars [J]. *Acta Ecologica Sinica*, 2007, 27 (10): 3 973-3 982.
[14] PIRONE T P, PERRY K L. Aphids: non - persistent transmission [J]. *Advances in Botanical Research*, 2002, 36: 1-19.
[15] POLSTON J E, SHERWOOD T. Pymetrozine interferes with transmission of tomaoto yellow leaf curl virus by whitefly Bemisia tabaci [J]. *Phytoparasitica*, 2003, 31 (5): 490-498.
[16] PRADO E, TJALLINGII W F. Aphid activities during sieve element punctures [J]. *Entomologia Experimentalis et Applicata*, 1994, 72: 157-165.
[17] RAMIREZ B C, HAENNI A L. Molecular biology of *Tenuiviruses*: a remarkable group of plant viruses [J]. *Journal of General Virology*, 1994, 75: 467-475.
[18] SEO B Y, KWON Y H, JUNG J K, et al. Electrical penetration graphic waveforms in relation to the actual positions of the stylet tips of *Nilaparvata lugens* in rice tissue [J]. *Journal of Asia-Pacific Entomology*, 2009, 12: 89-95.
[19] SOGAWA K. The rice planthopper: feeding physiological and host plant interactions [J]. *Annal Review of Entomology*, 1982, 27: 49-73.
[20] TORIYANA S. Rice stripe virus: prototype of a new group of viruses that replicate in plants and insects [J]. *Microbiology Science*, 1986, 3 (11): 347-351.
[21] ZAPATA N, VARGAS M, MEDINA P, et al. The activity of a selected extract of *Drimys winteri* bark and polygodial on settling and probing behavior of the lettuce aphid *Nasonovia ribisnigri* [J]. *Phytoparasitica*, 2010, 38: 191-199.
[22] 何月平，陈利，陈建明，等．吡蚜酮对水稻褐飞虱取食行为的影响 [J]．中国水稻科学，2010，24 (6)：635-640.
[23] 嵇朝球，钟环，钟佩英，等．水稻条纹叶枯病病毒 RT-PCR 快速检测研究 [J]．上海交通大学学报：农业科学版，2005，23 (2)：188-191.
[24] 荆裴，白素芬，刘芳．灰飞虱取食行为刺吸电位波形的初步研究 [J]．应用昆虫学报，2013，50 (3)：758-763.
[25] 刘海建，程兆榜，王跃，等．灰飞虱传递水稻条纹病毒研究初报 [J]．江苏农业学报，2007，23 (5)：492-494.
[26] 孙黛珍，江玲，张迎信，等．8 个水稻品种的条纹叶枯病抗性特征 [J]．中国水稻科学，2006，20 (2)：219-222.
[27] 闫凤鸣．化学生态学 [M]．北京：科学出版社，2003.
[28] 杨金广，方振兴，张孟倩，等．应用 Real-Time RT-PCR 鉴定 2 个水稻品种（品系）对水稻条纹病毒的抗性差异 [J]．华南农业大学学报，2008，29 (3)：25-28.
[29] 张景飞，龚林根，瞿燕，等．2004 年常熟市 5、6 代灰飞虱严重危害水稻 [J]．中国植保导刊，2005，25 (4)：39.
[30] 张鹏飞，陈建群，张闲，等．棉蚜获得黄瓜花叶病毒的行为与取食过程的关系 [J]．昆虫学报，

2001, 44（4）：395-401.
[31] 周益军，刘海建，王贵珍，等. 灰飞虱携带的水稻条纹病毒免疫检测［J］. 江苏农业科学，2004，1：50-51.

（刘　芳）

中文索引

九画

十画

十一画

十二画

十三画

十四画

十五画

十六画

英文索引

R

S

T

W

后　　记

本书出版之际，恰逢我利用 EPG 开展研究工作 20 年！这里回顾一下我利用 EPG 开展研究的经历，一方面可以梳理我国 EPG 应用的历史和现况，更重要的是，对给予我帮助的专家、同事和参与研究的学生表示感谢！

对于我来说，EPG 技术的入门和提高，特别要感谢荷兰瓦赫宁根大学（Wageningen University，the Netherlands）的 W. F. Tjallingii 教授和美国密苏里大学哥伦比亚分校的 Elaine A. Backus 教授（现在美国农业部工作）的指导和资助。我于 1995 年到荷兰瓦赫宁根农业大学 W. Freddy. Tjallingii 实验室做博士后的时候开始接触 DC-EPG，当时开展的工作是利用 EPG 记录甘蓝蚜 *Brevicoryne brassicae* 在芥菜 *Sinapis alba* 上的取食行为，同时利用 HPLC 进行芥子油苷的化学分析，试图阐明甘蓝蚜偏好在花柄上取食的机制。在荷兰一年的时间里，同在一个实验室的雷宏博士给了我很多 EPG 方面的指导和帮助；1996 年到法国里昂参加 EPG 培训班，认识了当时在 INRA（法国农科院）读博士的来自南京大学的陈建群，我俩从此建立了友谊，他当时利用 EPG 进行昆虫传毒机制的研究，回国后继续进行相关工作，并研发了国内第一款 EPG 应用软件。1997 年 3~7 月，我受邀到希腊色萨利大学 John Tsitsipis 教授的实验室指导 EPG 研究，帮助其建立了 EPG 实验室。2002~2004 年我又受邀到美国密苏里大学哥伦比亚分校（University of Missouri Columbia）Elaine A. Backus 教授的实验室进行合作研究，利用 AC-EPG 结合实时图像记录技术研究玻璃叶蝉（glassy-winged sharpshooter）*Homalodisca coagulata* 传播葡萄细菌病（Pierce's disease）木质部难养菌的机制。

我于 1998 年开始在北京大学利用 EPG 开展植物与昆虫关系的研究，研究内容主要是以烟粉虱和棉蚜为材料研究转基因棉的非靶标效应、化学生态学、植物与昆虫的关系、植物抗性机制等，薛堃、殷海娣的博士学位论文和王晓颖的硕士论文都涉及 EPG 技术的应用。北京大学生命科学学院的许崇任教授、吴才宏教授、李绍文教授、饶广远教授和王戎疆博士对我的研究给予了大力支持。2007 年以后，我调到河南农业大学工作，集中进行烟粉虱与植物相互关系的研究工作，我的研究生张大山、姬琨、李晓敏、梁相志、李静静、卢少华、刘明杨、王雪丽等先后利用 EPG 开展了植物与昆虫的关系、转基因作物安全性评价、传毒机制、植物次生物质生物活性、人工饲料结合 EPG 等研究内容，汤清波教授、郭线茹教授的工作和他们的部分研究生的论文内容也涉及 EPG 技术。和我进行科研合作的中国农业科学院植物保护研究所的陈巨莲研究员，在利用 EPG 结合人工饲料开展工作方面给予了我们很多启发。

对我的研究工作大力支持并在研究中相互探讨的同事有：浙江大学刘树生教授、

刘银泉博士，武汉大学何光存教授、杜波博士，中国科学院动物研究所戈峰研究员、王琛柱研究员，中国农业科学院植物保护研究所张永军研究员、陈巨莲研究员，西北农林科技大学张雅林教授、花保祯教授、刘同先教授，北京市农林科学院罗晨研究员，中国农业大学张青文教授、张龙教授、高希武教授、彩万志教授、赵章武教授，华南农业大学张茂新教授、岑伊静教授，浙江农科院的吕要斌研究员，贵州大学郅军锐教授，华中农业大学王满囷教授、杨凤连博士，扬州大学杜予州教授、新疆农业科学院的王素梅研究员等。这些同事和专家的鼓励是我能够坚持进行 EPG 研究的动力，在此向他们表示衷心的感谢！

由于我利用 EPG 开展相关研究的经历，这些年来国内许多正在使用 EPG 开展科研工作的年轻学者和研究生遇到问题后，不断和我联系，询问解决办法。联系较多的学生包括中国农业科学院的刘佰明（目前在天津农业科学院工作）、湖南农业大学的孙凯、贵州大学的温娟、浙江大学的何文波、华中农业大学的孙泽等。这些同学所提出的问题，多是 EPG 原理、操作等的常见问题，于是我就考虑以举办培训班的形式，当面指导原理和操作，因此我就和荷兰的 W. F. Tjallingii 教授于 2007 年在郑州、2010 年在杭州举办了第一和第二届“全国 EPG 技术培训班和研讨会”，2012 年我联合国内有经验的年轻博士如中国科学院动物研究所的罗岚工程师、浙江省农业科学院的黄芳博士等，在郑州举办了第三届 EPG 培训班和研讨会。加上在 2009 年的全国首届化学生态学培训班和研讨会中 EPG 技术的培训，4 次 EPG 培训班为全国培养了一批 EPG 技术力量，一定程度上推动了 EPG 技术在国内的应用。本书的许多作者都是 EPG 培训班的学员。他们的一些经验和体会，在本书中都有所体现，期望对利用 EPG 从事研究的同行和研究生有所启发和帮助。

尽管 EPG 技术在我国得到了广泛应用并在很多方面发挥了巨大作用，但我们在应用中深切感到，这项技术在我国的应用还存在很多值得改进的地方。首先，由于我国应用的 EPG 仪器均是荷兰生产的直流型的，设备和消耗品（金丝、银胶等）的购买均需要进口，很费时间，维修也很不方便；第二，对于从没有进行过 EPG 记录的昆虫种类，需要借助其他技术进行波形的生物学意义的阐释；第三，DC-EPG 的记录需要在法拉第金属笼内进行以屏蔽噪声，为实验造成很多麻烦，特别是有的记录波形由于噪声干扰难以识别；第四，记录波形的分析需要人工进行，很费时间和精力，很多学生对分析波形感到很头疼但又无可奈何。因此，对于这项发明于 20 世纪中期的技术进行改进甚至全面更新，就成为大家的共同认识。

鉴于此，我们做出了一些努力。首先，银胶配方试验成功了，申请了国家发明专利（专利号 ZL201010610889. 7），提供给国内使用，节约了时间和成本；其次，我们和河南农业大学理学院的潘建斌教授、吴莉莉博士和邢玉清博士合作，对 EPG 进行了改进，使得其在波形精细度和抗噪声方面得到了很大改善。此外，我们利用镀银碳纤维替代金丝进行了测试，取得了初步成功。目前，我们正在研发适合目前 EPG 仪器的波形自动识别软件，同时考虑设计全新的昆虫刺吸行为可视化记录系统。希望我们的努力能够为刺吸式昆虫取食行为、昆虫与植物的关系、植物抗性机制、昆虫传毒机制、刺吸式害虫及其所传播的动植物病毒病的防控等领域提供有力的技术支撑，也希望与更多的同行和年轻人进行合作和交流。

我多年在 EPG 方面的科研工作，得益于北京大学生命科学学院、河南农业大学所提供的条件，得到了下列项目的支持和资助：荷兰皇家科学基金（Dutch Royal Academic Grant）、美国加利福尼亚州葡萄皮尔氏病害研究项目（California Pierce's Disease Research Program）、北京大学青年基金项目、教育部归国人员科研启动基金项目、科技部中瑞合作项目、国家 863 子项目、国家 973 项目（G2000046803，001CB109007）、国家自然科学基金项目（39970153，30571219，31471776）、国家转基因新品种培育重大专项（2009ZX08012-007B）、国家公益性行业（农业）科研专项（201203036，201303030）、河南省重大公益性项目（081100911300）等，在此一并表示感谢！

闫凤鸣
2016 年 1 月